PROGRESS

IN

HETEROCYCLIC CHEMISTRY

Volume 4

Related Pergamon Titles of Interest

Books

Organic Chemistry Series

CARRUTHERS: Cycloaddition Reactions in Organic Synthesis
DAVIES: Organotransition Metal Chemistry: Applications to Organic Synthesis
DEROME: Modern NMR Techniques for Chemistry Research
DESLONGCHAMPS: Stereoelectronic Effects in Organic Chemistry
GIESE: Radicals in Organic Synthesis: Formation of Carbon–Carbon Bonds
PAULMIER: Selenium Reagents and Intermediates in Organic Synthesis
PERLMUTTER: Conjugate Addition Reactions in Organic Synthesis
WILLIAMS: Synthesis of Optically Active α-Amino Acids
WONG & WHITESIDES: Enzymes in Synthetic Organic Chemistry

Major Works of Reference

ALLEN & BEVINGTON: Comprehensive Polymer Science
BARTON & OLLIS: Comprehensive Organic Chemistry
HANSCH *et al*: Comprehensive Medicinal Chemistry
KATRITZKY & REES: Comprehensive Heterocyclic Chemistry
TROST & FLEMING: Comprehensive Organic Synthesis
WILKINSON *et al*: Comprehensive Organometallic Chemistry
WILKINSON *et al*: Comprehensive Coordination Chemistry

Also of Interest

KATRITZKY: Handbook of Heterocyclic Chemistry
NICKON & SILVERSMITH: Organic Chemistry—The Name Game
PERRIN & ARMAREGO: Purification of Laboratory Chemicals, 3rd edition
RIGAUDY & KLESNEY: Nomenclature of Organic Chemistry ('The Blue Book')

Journals

TETRAHEDRON (rapid publication primary research journal for organic chemistry)
TETRAHEDRON: ASYMMETRY (international journal for rapid publication on all aspects of asymmetry in organic, inorganic, organometallic, physical chemistry and bio-organic chemistry)
TETRAHEDRON LETTERS (rapid publication preliminary communications journal for organic chemistry)
BIO-ORGANIC & MEDICINAL CHEMISTRY LETTERS (new journal for rapid dissemination of preliminary communications on all aspects of bio-organic chemistry, medicinal chemistry & related disciplines)

Full details of all Pergamon publications, and a free specimen copy of any Pergamon journal, are available on request from any Pergamon office

PROGRESS

IN

HETEROCYCLIC CHEMISTRY

Volume 4

A critical review of the 1991 literature preceded by two chapters on current heterocyclic topics

Editors

H. SUSCHITZKY

Department of Chemistry and Applied Chemistry, University of Salford, UK

and

E. F. V. SCRIVEN

Reilly Industries Inc., Indianapolis, Indiana, USA

PERGAMON PRESS

OXFORD · NEW YORK · SEOUL · TOKYO

U.K.	Pergamon Press Ltd, Headington Hill Hall, Oxford OX3 0BW, England
U.S.A.	Pergamon Press Inc., 660 White Plains Road, Tarrytown, New York 10591-5153, USA
KOREA	Pergamon Press Korea, KPO Box 315, Seoul 110-603, Korea
JAPAN	Pergamon Press Japan, Tsunashima Building Annex, 3–20–12 Yushima, Bunkyo-ku, Tokyo 113, Japan

First edition 1992

Library of Congress Cataloging in Publication Data

Progress in heterocyclic chemistry: a critical review of the 1991 literature preceded by a chapter on current heterocyclic topics/editors, H. Suschitzky and E.F.V. Scriven.—1st ed.
p. cm.
Includes index.
1. Heterocyclic compounds.
I. Suschitzky, H. II. Scriven, Eric F.V.
QD399.P76 1992 547′.59—dc20 89-8531

British Library Cataloguing in Publication Data

Progress in heterocyclic chemistry.
Vol. 4
I. Suschitzky, H. II. Sciven, E.F.V.
547′.59

ISBN 0-08-0406122 (Hardcover)
ISBN 0-08-0406130 (Flexicover)

Printed in Great Britain by BPCC Wheatons Ltd, Exeter

Contents

Foreword

Progress in Heterocyclic Chemistry (PHC), Volume 4 describes new and significant advances in the field of heterocyclic chemistry mainly based on the 1991 literature. It also presents the first review on the widely studied SO_2-extrusion from heterocyclic compounds, written by R.A. Aitken, Sir John Cadogan and I. Gosney, which is of considerable mechanistic and preparative interest. We much regret that the promised article on 2-isoxazolines by Sh. Shatzmiller was not submitted.

The chapters are arranged according to ring-size as in the previous volumes and backed up by references, a subject index and clear and numerous diagrams. These should render the search for new developments a pleasant task. The reference system in the text is that used in *Comprehensive Heterocyclic Chemistry* (Pergamon, 1984). This system is becoming widely accepted because of its logical simplicity.

The treatment and emphasis of the new chemistry cannot be expected to be uniform in a multi-author work. For instance, some reporters decided to emphasize advances in synthetic methods of a particular heterocycle rather than to treat the subject comprehensively. This approach is justifiable, since most readers have access to automatic searching and may prefer to be alerted to new developments in a specific area.

As editors we would very much welcome comments, criticism and suggestions from our readers with regard to possible topics and authors of review articles.

We are grateful to our authors for submitting well-presented and concise manuscripts, and to Dr. Helen McPherson of Pergamon Press for much editorial help and advice which made our own task a pleasure.

We are confident that the International Society of Heterocyclic Chemistry by commissioning this annual publication is fulfilling the most important function of a "Learned Society", namely to assist scientists in their search for advances in their chosen field.

H. SUSCHITZKY
E. F. V. SCRIVEN

Advisory Editorial Board Members

CHAPTER 1

Extrusion of SO_2 from Heterocyclic Compounds, Part 1: Systems other than Five-Membered Rings

R. ALAN AITKEN*
University of St Andrews, UK

IAN GOSNEY
University of Edinburgh, UK

and

Sir JOHN CADOGAN
B.P. Research Centre, Sunbury on Thames, UK

1.1 INTRODUCTION

Together with N_2, CO_2, and CO, sulphur dioxide is among the commonest small neutral fragments to be eliminated in thermal extrusion reactions. Extrusion of SO_2 from cyclic compounds has been much studied and the enormous variety of reactions discovered includes many of considerable preparative importance as well as mechanistic interest. In this review we have aimed to survey comprehensively, for the first time, extrusion of SO_2 from all types of heterocyclic compounds. While the reactions are mainly thermal, photochemical and, occasionally, chemically-induced extrusion are also considered. For reasons of space, the review has been divided into two parts: the present chapter covers extrusion from 3-, 4-, 6-, 7-, and 8-membered and larger rings together with bi- and polycyclic systems. "Part 2: Extrusion from Five-Membered Rings" will appear in *Progress in Heterocyclic Chemistry*, Volume 5.

1.2 THREE-MEMBERED RINGS

1.2.1 Thiirane Dioxides

The first observation of SO_2 extension from a three-membered ring was made by Staudinger and Pfenninger in 1916 [16CB1941]. They found that thiirane dioxide (**1**), prepared from diphenyldiazomethane and SO_2, lost SO_2 on heating above its melting point to give tetraphenylethylene. Less highly substituted thiirane dioxides are more unstable and decompose slowly at room temperature. Thermolysis of thiirane dioxides as a synthetic route to alkenes has been reviewed by Fischer [70S393] and gives good yields of unsymmetrically substituted alkenes (**2**) by the route shown [65AG(E)70]. Recent examples include the reaction of trimethylsilyldiazomethane with sulphenes to give vinyl silanes [83CPB2957].

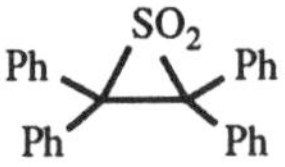

(**1**)  (**3**) 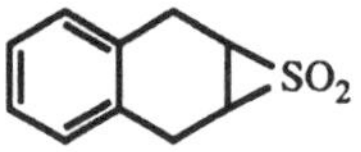 (**4**)

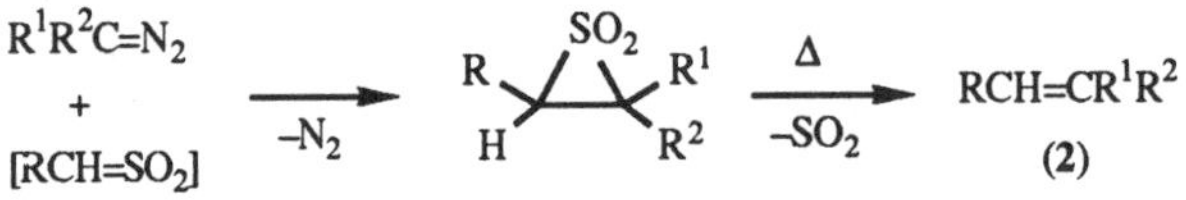

$R^1R^2C{=}N_2$ + $[RCH{=}SO_2]$ $\xrightarrow{-N_2}$ thiirane dioxide $\xrightarrow[-SO_2]{\Delta}$ $RCH{=}CR^1R^2$ (**2**)

Treatment of primary sulphonyl chlorides, RCH_2SO_2Cl, with Et_3N at –40°C gives the isolable thiirane dioxides (**3**) among other products and these slowly lose SO_2 to afford RCH=CHR on warming to room temperature [90CB1989]. The thiirane dioxide (**4**) was thought to be formed by an unusual cycloaddition reaction between benzyne and divinyl sulphone but it decomposed under the reaction conditions to give a mixture of 1,4-dihydronaphthalene and naphthalene [76TL4199]. Although the *trans*-distyrylthiirane dioxide (**5**)

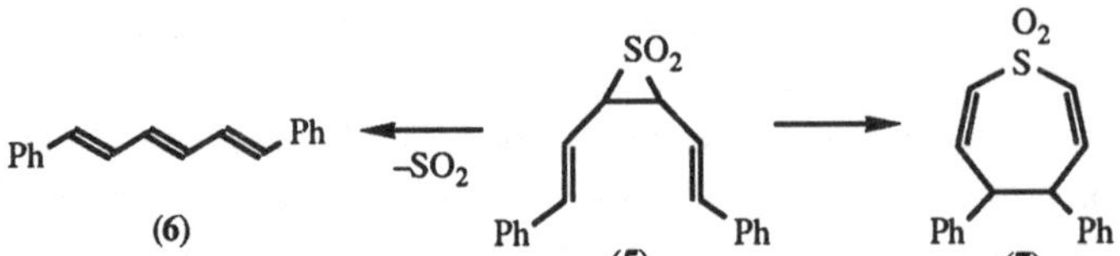

(**6**) $\xleftarrow{-SO_2}$ (**5**) $\longrightarrow$ (**7**)

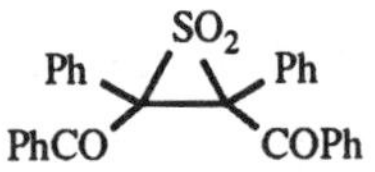

(**8**) 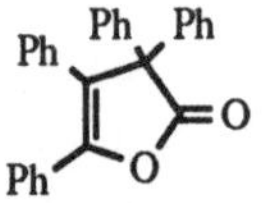 (**9**) 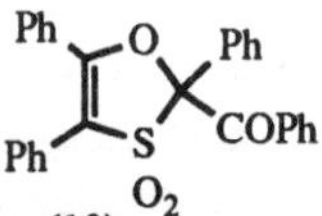 (**10**)  (**11**)

extrudes SO_2 at room temperature to form triene (**6**) in the normal way, its *cis*-isomer instead undergoes a Cope Rearrangement to form the stable dihydrothiepine dioxide (**7**) in 32% yield [84TL4553].

The anomalous thermal decomposition of the compound originally assigned the thiirane dioxide structure (**8**), to give diphenylacetylene, benzil and compound (**9**) [69JA2097], was subsequently explained by the finding that the starting compound actually had structure (**10**), although (**8**) is still postulated as an intermediate in the formation of (**9**) [74JOC2722]. Hydride ion induced SO_2 extrusion has been observed for *cis*-diphenylthiirane dioxide (**11**). While treatment with NaH or LiH gave a quantitative yield of *Z*-stilbene, use of $NaBH_4$ generally also produced some dibenzyl sulphone [66TL3929].

It is widely accepted that thiirane dioxides are produced during the Ramberg-Bäcklund reaction of α-halosulphones (**12**). Paquette has discussed the evidence for this in his classic review of the scope and mechanism of the reaction [77OR1]. Under the strongly basic reaction conditions however, they cannot usually be isolated and lose SO_2 to give the alkene (**13**). A recent report

(**12**) X = Cl, Br, I (**13**)

(**14**) (**15**) (**16**) (**17**)

indicates that the extrusion step is likely to occur from an intermediate such as (**14**) [85CJC1]. The reaction has been widely used and is particularly useful in preparing strained alkenes. For example, the bromosulphone (**15**) reacts with aqueous potassium hydroxide to give an 80% yield of the bicyclooctene (**17**), *via* thiirane dioxide (**16**) [69JOC1233].

The evidence for thiirane dioxide involvement has been further strengthened recently by the fortuitous isolation of the intermediate in one case. Thus when compound (**18**) was treated with 1.5 eq. of $KOBu^t$ at –78°C, (**19**) was obtained as a white solid, stable for a period of months at –20°C. Conversion to (**20**) could be achieved either by heating to 100°C or by treatment with additional $KOBu^t$ at –20°C [89TL3267].

The reactions of several nitrogen-containing compounds with ketene in liquid SO_2 are believed to involve 2-oxothiirane dioxide (**21**) as an intermediate.

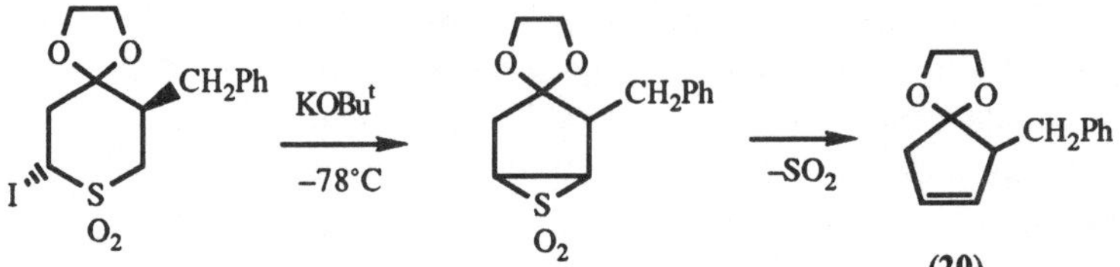

This was isolated at –78°C as a white solid but decomposed with evolution of SO_2 to give a tar on warming to room temperature [73JOC2652].

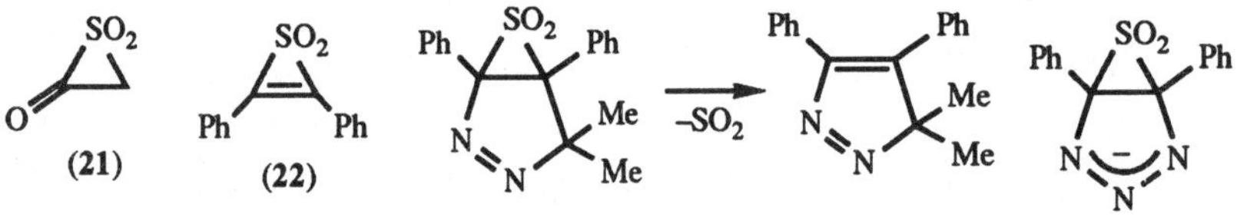

Azolothiirane dioxides are intermediates in the reaction of diphenylthiirene dioxide (**22**) with a number of 1,3-dipoles. Thus, treatment of (**22**) with diazoalkanes provides a useful synthesis of pyrazoles by loss of SO_2 from intermediates such as (**23**) [71JA476][80CB1632]. Reaction of (**22**) with lithium azide afforded a mixture of products including 4,5-diphenyl-1,2,3-triazole in 8% yield, presumably formed by loss of SO_2 from the anionic intermediate (**24**) [80JOC2604]. The reaction of both (**22**) and its dimethyl analogue with diphenylnitrile imine has been reported to give good yields of tetrasubstituted pyrazoles by SO_2 extrusion from intermediates (**25**) [84JOC1300]. With nitrile ylides the yields of the corresponding pyrroles (**26**) were lower, and in the case of R = Ph, compound (**27**) was also isolated due to the competing ring-opening reaction of (**28**) [84JOC1300].

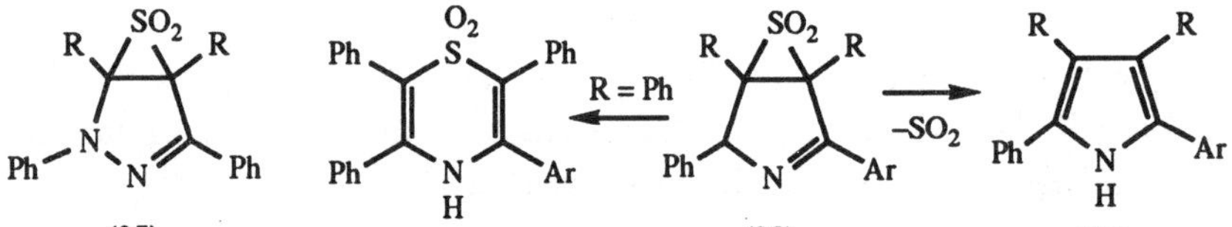

1.2.2 Thiirene Dioxides

The first thiirene dioxide (**22**) was obtained by treating α,α′-dibromodibenzylsulphone with triethylamine [65JA5804] and the unsymmetrical α,α-dichlorodibenzylsulphones can also be used as precursors [71CC22].

The thiirene dioxides are generally much more stable than their saturated analogues but they do undergo clean decomposition on heating, to give an acetylene and SO_2. Kinetic evidence has been obtained that this reaction is stepwise rather than concerted [77CC713]. The thiirene dioxides also lose SO_2 readily under mass spectrometric conditions so that only the peak for the resulting acetylene is normally seen [74JOC3777][75JHC21]. A particularly interesting example is the formation of acenaphthylyne (**31**) by the Ramberg-Bäcklund type reaction of dibromosulphone (**29**) which presumably involves the strained thiirene dioxide (**30**) as an intermediate. Under the conditions the

KOBu^t → −SO_2 →

(**29**) (**30**) (**31**)

acenaphthylyne trimerised to give the benzenoid hydrocarbon in 5% yield [83JOC60]. Because of the limited number of methods for their preparation little synthetic use has been made of thiirene dioxides.

1.2.3 Three-Membered Rings Containing Two or Three Heteroatoms

A number of hetero-thiirane and -thiirene dioxides are known, both as intermediates and stable compounds and these have been reviewed by Quast [80H1677].

Reaction of the substituted α-halosulphonamide (**32**) with base resulted in a Ramberg-Bäcklund type reaction to give the imine PhCH=NBu^t *via* a thiaziridine dioxide (**33**), and the dihalosulphonamide (**34**) similarly gave benzonitrile *via* a thiazirine dioxide (**35**) [74JOC1817]. Reaction of N-sulphonylamines with diazoalkanes at −78°C gives the thiaziridine dioxides (**36**)

base → base →

(**32**) (**33**) (**34**) (**35**)

$R^1N{=}SO_2$ + $R^2CH{=}N_2$ → → $R^2CH{=}NR^1$

(**36**) (**37**) (**38**)

which lose SO_2 on warming to room temperature as shown [81CB774]. In some cases, such as for (**36**, R^1=R^2=Bu^t) the thiaziridine dioxides have been isolated [74AG(E)742].

The thiadiaziridine dioxides (**37**) have also been postulated as reaction intermediates [65AG(E)433], and when isolated they did extrude SO_2 [73JA634], although there was competing cleavage of the N-N bond in some cases [77CB1780]. Treatment of a sulphene with triphenylphosphine gave a phosphorus ylide which was isolated as the quaternary phosphonium salt, presumably *via* the thiaphosphiridine dioxide (**38**) [72CJC866].

1.3 FOUR-MEMBERED RINGS

1.3.1 Thietane Dioxides

The first report of SO_2 extrusion from a four-membered ring appeared in 1963 when Dodson and Klose prepared the diphenylthietane dioxide (**39**) [63CI(L)450]. This compound loses SO_2 on heating at 250°C for 1 hour to give a mixture of isomeric diphenyl cyclopropanes (**40**). The extrusion is likely to be radical in nature since both *cis*- and *trans*-(**39**) give the same mixture of *cis*- and *trans*-(**40**). Despite this lack of stereospecificity, the use of this extrusion as

Ph SO$_2$ Ph (**39**) —$-SO_2$→ Ph Ph + Ph Ph (**40**)

a synthetic route to substituted cyclopropanes both by thermolysis [71JA676] and photolysis [78S579] has been reported. In the case of 3,3-diethoxy-2-phenylthietane dioxide the cyclopropane formed on pyrolysis was unstable and the products isolated were ethanol together with a low yield of ethyl cinnamate [63JA3236].

Several thietane dioxides have been pyrolysed in an attempt to form sterially hindered diradicals. Thus Bushby used (**41**) as a source of the substituted "2-methylene-trimethylene" diradical which isomerised to (**42**) [75JCS(P1)2513]. The stepwise loss of N_2 and SO_2 has been observed for (**43**) which at 400°C loses only N_2 under flash vacuum pyrolysis conditions to give (**44**). At 550°C, SO_2 is also lost to give (**45**) which isomerises to (**46**) at higher temperatures [84JCS(P1)2457]. Pyrolysis of (**47**) at 650°C gives a mixture of (**44**) and (**46**) [84JCS(P1)2457]. Vacuum pyrolysis of (**48**) at 110-160°C results in extrusion of SO_2 and N_2 to give a 55% yield of tetramethylallene [67JA2781].

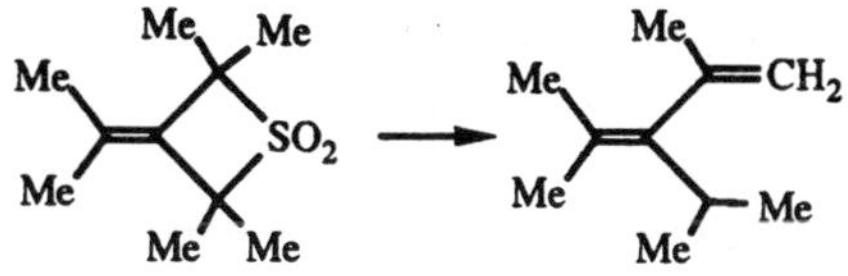

(41) (42)

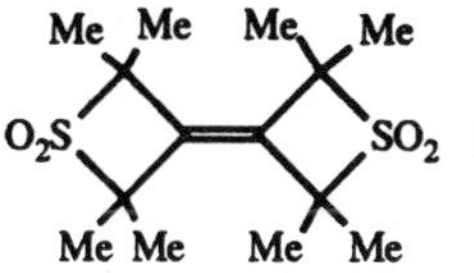

(47)

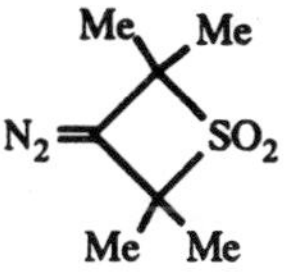

(48)

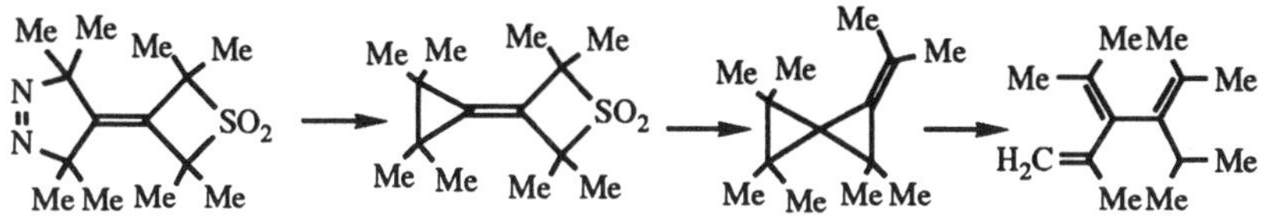

(43) (44) (45) (46)

Irradiation of 3-oxothietane dioxide (**49**) does not lead to extrusion of SO_2. Instead a cycloreversion takes place to give ketene and sulphene which then react further [74JA2964]. In contrast, flash vacuum pyrolysis of (**49**) at 940°C results in clean decomposition to SO_2, CO and ethylene [69CC32][70CJC3704]. In the pyrolysis of 3-hydroxythietane dioxide (**50**) three competing processes are observed: dehydration to thiete dioxide which decomposes as described in Section 1.3.2, cycloreversion reaction to give acetaldehyde and sulphene, and, the major process, loss of SO_2 followed by rearrangement of diradical (**51**) to give acetone and propionaldehyde, isolated in a combined yield of nearly 50% [69CC32] [70CJC3704].

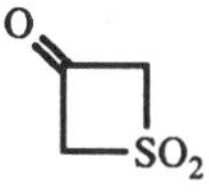

(49)

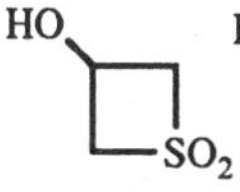

(50)

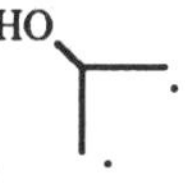

(51)

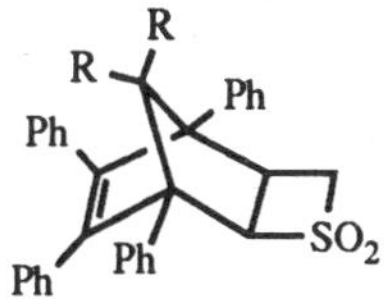

(52) R_2=O
(53) R = H

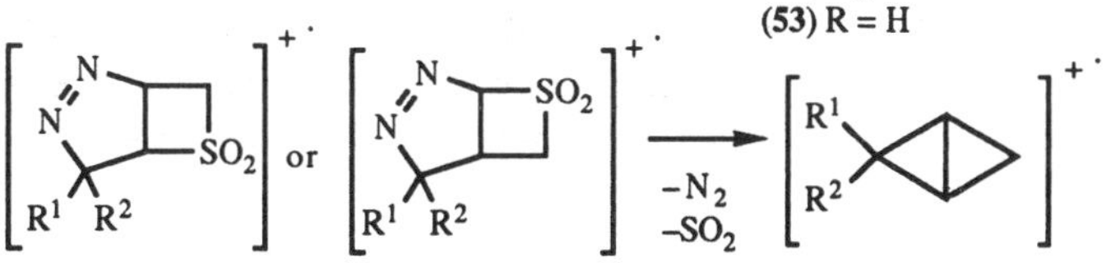

(54) (55)

Reaction of thiete dioxide (**56**) with tetracyclone gave tetraphenylcycloheptatriene as the main product *via* loss of CO and SO_2 from intermediate tricyclic thietane dioxide (**52**). With tetraphenylcyclopentadiene the adduct (**53**) was isolated but on heating it underwent a retro Diels-Alder reaction to give tetraphenylcyclopentadiene and a tar derived from the thiete dioxide [72JOC225]. Under mass spectral conditions the bicyclic thietane dioxides (**54**), formed by addition of a diazoalkane to the double bond of thiete dioxide, lose both N_2 and SO_2 to give the bicyclo[1.1.0]butane molecular ion (**55**) [80JOC3646].

1.3.2 Thiete Dioxides

In contrast to their saturated analogues the thiete dioxides do not usually lose SO_2. Instead they undergo a ring expansion on heating to give the five-membered cyclic sultines (**58**), probably *via* a vinyl sulphene (**57**). This transformation has been reported for the parent compound (**56**) [70CJC3704], as

(**56**) (**57**) (**58**)

well as for the benzo- [76JOC3044], 2,3-naphtho- [69JOC1310], and 1,8-naphtho- [67LA(703)96] fused analogues. On further pyrolysis the parent sultine (**58**) loses sulphur monoxide to give a good yield of acrolein. Loss of SO to form carbonyl compounds has also been observed on photolysis of 2-phenylthiete dioxides [72TL4781] and on thermolysis of 2,4-diphenylthiete dioxide [77JOC3506] and 4,4-dimethylthiete dioxide [69CC32] [70CJC3704].

While flash vacuum pyrolysis of the 1,8-naphthosulphone (**59**) caused mainly ring expansion to the sultine, heating the compound to 240°C did lead to SO_2 loss and dimerisation of the resulting diradicals to produce perylene (**60**) [67LA(703)96].

FVP; 240°C, $-SO_2$

(**59**) (**60**)

1.3.3 1,2-Oxathietane and 1,2-Dithietane Systems

The extrusion of SO_2 from 1,2-oxathietane S-oxides (**62**), isomers of the thiirane dioxides, has been observed. These compounds were postulated as intermediates in the reaction of β-hydroxyalkyl t-butyl sulphoxides (**61**) with sulphuryl chloride to produce alkenes [73JA3420], as well as in other related

SO_2Cl_2; $-SO_2$

(**61**) (**62**) (**63**)

reactions [75TL2179]. The isolable oxathietane oxide (**63**) has been prepared and loses SO_2 at 30°C to give the alkene $Ph_2C{=}CMe_2$ [75CC724].

1,2-dithietane 1,1-dioxides, which may be formed by dimerisation of electron deficient sulphines, readily lose SO_2 and sulphur to give alkenes. Thus either conventional or flash vacuum pyrolysis of (**64**) gives (**66**), presumably *via* the intermediate (**65**) [85CB4553]. Likewise, generation of (**67**) was accompanied by formation of (**69**), and (**68**) could be detected as an intermediate although attempts to isolate it only gave (**69**) [85AG(E)855].

(**64**) R = CF_3
(**67**) R = CO_2Et

(**65**) R = CF_3
(**68**) R = CO_2Et

$-SO_2$
$-S$

(**66**) R = CF_3
(**69**) R = CO_2Et

1.3.4 1,3-Dithietane Systems

Block has reported the synthesis of 1,3-dithietane and its oxidation to sulphones (**70**) and (**71**) whose thermal decomposition was examined by photoelectron spectroscopy [82JA3119]. The monosulphone (**70**) was found to lose SO_2 cleanly at 470°C and 10^{-2}mmHg to give thiirane. In the case of the disulphone (**71**), SO_2 was again lost, at 600°C, to give ethylene and formaldehyde among the products. Several tetrahaloderivatives of both (**70**) and (**71**) have also been prepared and the tetrachloro derivative of compound (**71**) loses SO_2 to form tetrachloroethylene on heating above 120°C [80AG(E)203].

(**70**) (**71**)

(**72**) X = Cl
(**74**) X = F

(**73**) X = Cl
(**75**) X = F

(**76**)

The corresponding monosulphone (**72**) undergoes clean SO_2 extrusion upon flash vacuum pyrolysis at 450°C to afford tetrachlorothiirane (**73**) in 85% yield [85CB1415], while the fluorinated derivative (**74**) gives both (**75**) and the rearranged product (**76**) [87CB1499]. Oxidation of the 1,3-dithietanes obtained by dimerisation of halogenated thiocarbonyl compounds gave monosuolphones such as (**77**), (**78**) and (**79**). On pyrolysis at 550°C and 1 mmHg these all lost SO_2 to give the corresponding thiiranes in good yield [63USP3136781]. Flash vacuum pyrolysis of (**80**) gave a complex mixture of products whose mass spectrometric analysis showed the presence of fragments such as CH_2OS_2, CH_2SO and S_2O [85CB2852].

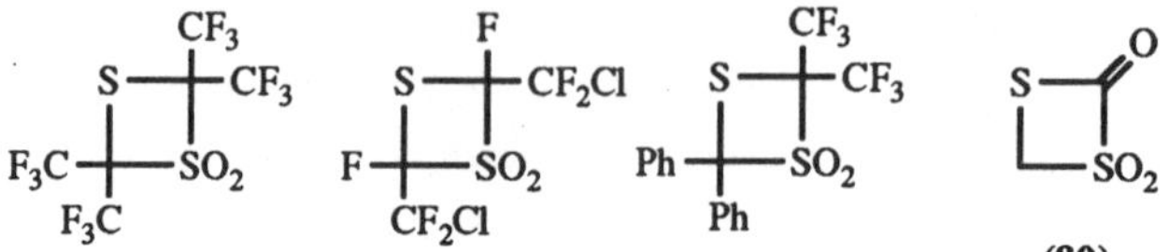

(77) (78) (79) (80)

1.3.5 Dioxathietane and Oxadithietane Systems

The photo-oxygenation reaction of sulphines proceeds by SO_2 extrusion from an intermediate 1,2,3-dioxathietane oxide (**81**) to give good yields of carbonyl compounds in some cases [70TL4683].

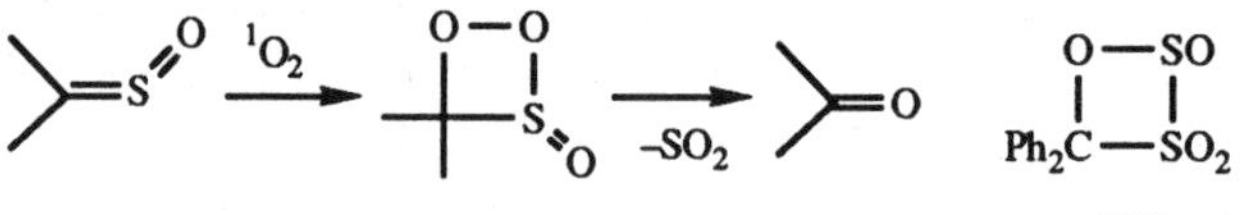

$^1O_2$; $-SO_2$

(81) (82)

The reaction of diphenyldiazomethane with excess SO_2 to produce benzophenone was originally formulated as involving extrusion of SO and SO_2 from 1,2,3-oxadithietane trioxide (**82**) although there is no direct evidence for this [16CB1941].

1.3.6 Four-Membered Rings Containing Nitrogen

The thermal decomposition of thiazetidine dioxides has received little attention but appears to be complex. Thus, when either the *cis* or *trans* diphenylthiazetidine dioxide (**83**) was heated neat at 200°C, a mixture of products was produced among which only *E*-stilbene and benzaldehyde were identified in low yield [75BCJ480]. It seems likely that the stepwise deoxygenation of nitro compounds in basic solution by thiourea S,S-dioxide may involve extrusion of SO_2 (as sulphite) from an intermediate 1,3,2-oxathiazetidine S,S-dioxide (**84**) [84AJC2489].

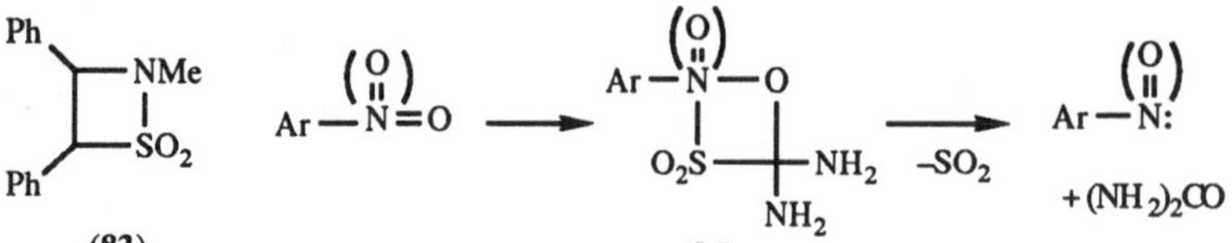

$-SO_2$; $+ (NH_2)_2CO$

(83) (84)

Extrusion of SO_2 from a number of different ring systems (**85**) has been invoked in the reaction of N-sulphinyl anilines and N-sulphinyl sulphonamides with various oxygen-containing species. The reaction was first reported by Kresze who reacted the N-sulphinyl compounds with aldehydes to

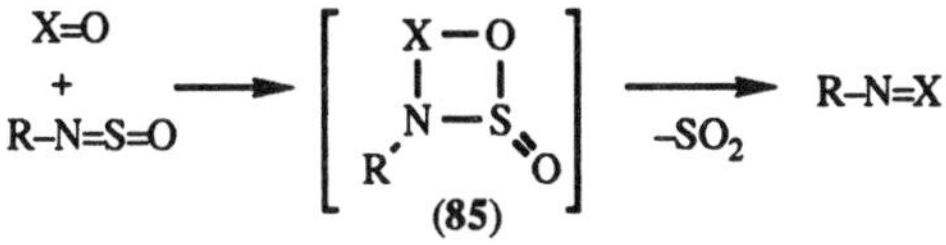

$$X{=}O + R{-}N{=}S{=}O \longrightarrow [\,(85)\,] \xrightarrow{-SO_2} R{-}N{=}X$$

(85)

give the corresponding imines [64CB483]. Formamides reacted similarly to give the formamidines *via* the intermediate (**86**). The reaction with aldehydes was further investigated by Zefirov who claimed to have isolated the intermediates (**87**) in some cases [71ZOR951], although a later study showed that the compounds isolated may have been the hydrates (**88**) [77PS333]. This reaction also takes place with certain ketones. Hexafluoroacetone reacts with N-sulphinyl anilines to form the imines $ArN{=}C(CF_3)_2$ by loss of SO_2 from the oxathiazetidine oxides (**89**) [79IZV396]. The compound (**90**) is probably an intermediate in the conversion of the ketone into the corresponding N-tosylimine by reaction with N-sulphinyl p-toluenesulphonamide [77JHC1369]. Reaction of the latter with α-diketones gives the expected α-keto-imines, again *via* an oxathiazetidine oxide [65CB601].

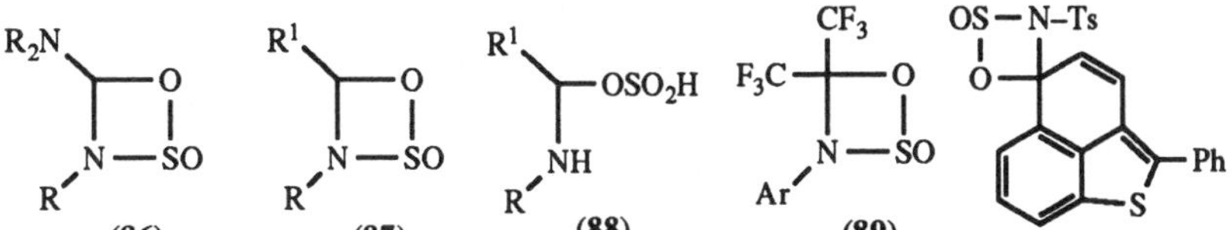

(86) (87) (88) (89) (90)

Reaction of N-sulphinyl sulphonamides with sulphoxides takes place in a similar way, with loss of SO_2 from an intermediate 1,2,4,3-oxadithiazetidine 2-oxide (**91**) to produce the sulphilimine (**92**) [63AG(E)736] which may rearrange in some cases [74JOC3412]. The related ring system (**93**) is implicated in the acid catalysed thermal disproportionation of N-sulphinyl sulphonamides to give SO_2 and sulphur diimides [66TL1671]. By treating

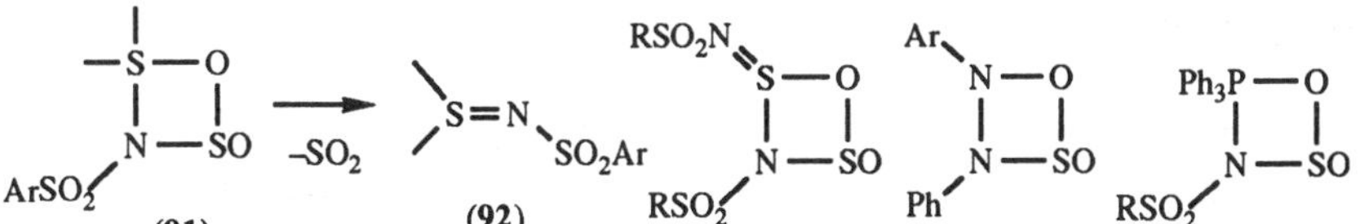

(91) $\xrightarrow{-SO_2}$ (92) (93) (94) (95)

nitrosobenzenes with N-sulphinylaniline, Zefirov obtained low yields of the azobenzenes ArN=NPh *via* the oxathiadiazetidine oxides (**94**) [72ZOR1107]. A similar reaction with triphenylphosphine oxide gives the phosphinimine product $Ph_3P{=}NSO_2R$ by SO_2 extrusion from the oxathiazaphosphetidine S-oxide (**95**) [65ACS1755].

These reactions have more recently been reported with N-sulphinyl perfluoroalkylsulphonamides, involving SO_2 extrusion from (**96**) formed by reaction with aldehydes, ketones, sulphoxides and $POCl_3$ [91CC732].

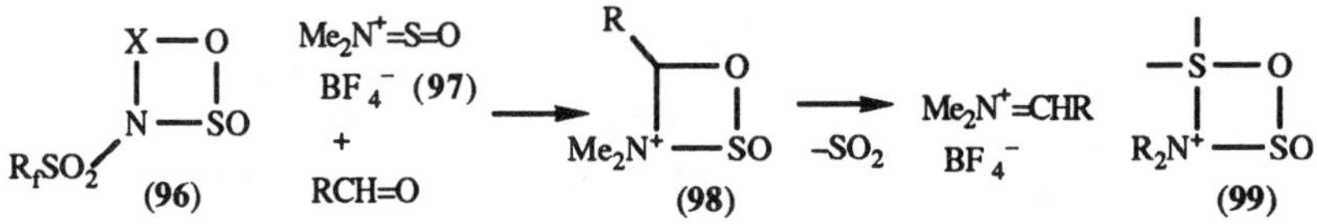

(96) (97) (98) (99)

Kresze has recently reported that reactions of this type are also successful with N-sulphinylammonium salts. Thus, for example, reaction of (**97**) with aldehydes proceeds by extrusion of SO_2 from (**98**) to afford iminium salts in over 70% yield [84S944]. The analogous reaction with sulphoxides gives aminosulphonium salts *via* intermediate (**99**) [82LA723][85LA72].

1.4 SIX-MEMBERED RINGS

1.4.1 Six-Membered Rings Containing Only Oxidised Sulphur

In general, SO_2 extrusion from saturated six-membered sulphones is not an easy process. However it can be achieved when ring strain is present. Thus, while 1,4-dithiane tetroxide (**100**) is stable at 300°C, the hexasulphone (**101**) could not be isolated from the oxidation of the corresponding hexasulphide [71JA2258]. Breakdown occurred in the manner shown to give an intermediate which underwent further fragmentation to produce two molecules of 1,3-dithiolane tetroxide. A further example of the relative stability of dithiane

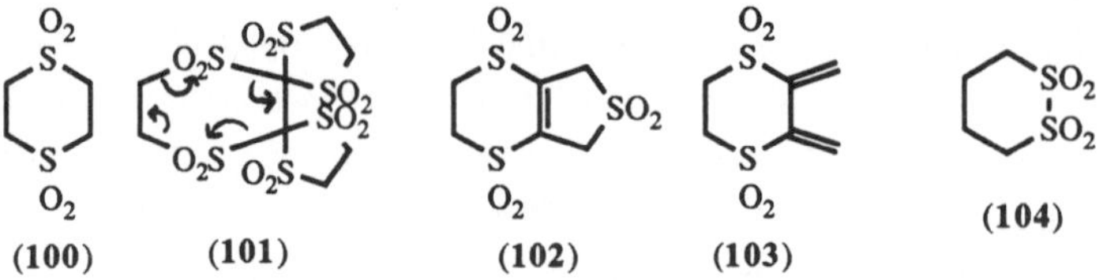

(100) (101) (102) (103) (104)

tetroxides is provided by trisulphone (**102**) which extrudes SO_2 from the five-membered ring only on solution thermolysis at 130°C, to give the useful 1,3-diene (**103**) [89CC1020]. Although 1,2-dithiane tetroxide (**104**) was recovered unchanged after heating for 1 week in boiling chlorobenzene, brief heating of the neat material to 280°C resulted in extrusion of one molecule of SO_2 to afford tetrahydrothiophene dioxide in 12% yield [69JOC1792]. Thermolysis of six-membered ring sulphones has been used preparatively in few cases. For example, Stothers obtained tetramethylacenaphthene in 20% yield by boiling the neat sulphone (**105**) at 400°C [73CJC2884]. The lithium enolates of cyclic ketones react with methylstyrylsulphone to produce bicyclic β-hydroxysulphones in good yield. Reaction of cyclohexanone, for example, gives the sulphone (**106**). When this is treated with excess sodium in ethanol, SO_2 and water are lost to give the 1,4 diene (**107**) [80JOC4789].

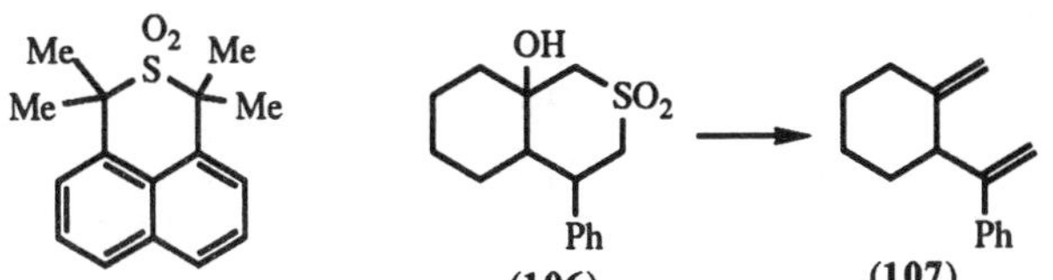

(**105**) (**106**) (**107**)

Just as was observed for the thiete dioxides, the characteristic reaction of the benzothiopyran dioxides (**108**) and (**109**) is ring expansion to a seven-membered sultine intermediate which then loses SO to give the products. Thus Smith found that on pyrolysis of (**108**), the main products were cinnamaldehyde and benzopyran (**110**), but with some indene formed *via* SO_2 loss [77TL1149]. Similarly, (**109**) gave a mixture of indene and *o*-vinylbenzaldehyde. Photolysis of (**108**) gave low yields of indene and (**110**) and three new products: the ring expanded sultine and two isomers of a five-membered sultine [74TL3633]. With (**109**), the major process on irradiation in methanol was addition of methanol

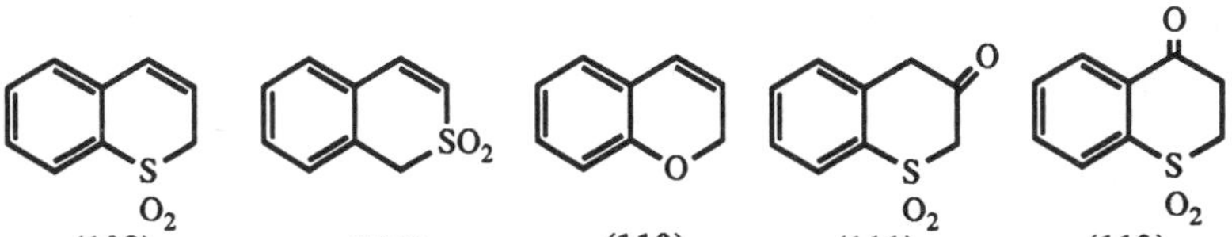

(**108**) (**109**) (**110**) (**111**) (**112**)

across the double bond, while the seven-membered sultine was obtained in 10% yield. The thiochromanone dioxides (**111**) and (**112**) have been prepared, but the sulphone group is photochemically inert [68JOC2730]. On the other hand, the α-oxosulphone (**113**), recently obtained by Schank does lose SO_2 on pyrolysis at 700°C to give fluorenone [83LA1739].

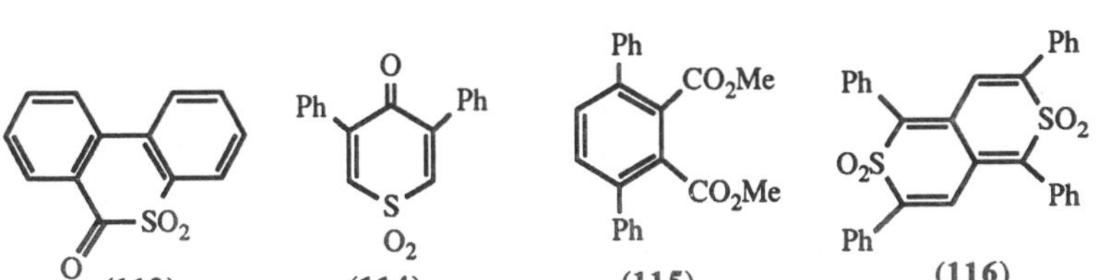

(**113**) (**114**) (**115**) (**116**)

The diphenylthiopyranone dioxide (**114**) loses SO_2 on exposure to sunlight for ten days to generate 2,5 diphenylcyclopentadienone which forms a pentacyclic trimer in 73% yield or, in the presence of a dienophile such as dimethyl acetylenedicarboxylate, a 43% yield of the aromatic adduct (**115**) [77AG(E)653]. The related compound (**116**) is reported to decompose at its melting point of 401-2°C but the products were not identified [84JOC5136].

1.4.2 1,2-Oxathiane and 1,2-Dithiane Systems

The aromatic sultines (**117**) lose SO_2 much more readily than their benzosulpholene isomers. Thus, either thermal [74JA935] or photochemical [78CJC512] decomposition produces the *o*-xylylene intermediate (**118**) which can be trapped either by reaction with a dienophile or with SO_2 to reform the

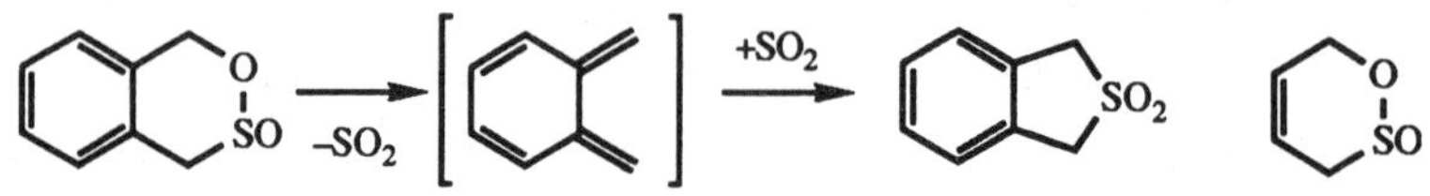

(**117**) (**118**) (**119**) (**120**)

isomeric sulphone (**119**). The thermal process proceeds rapidly in boiling benzene and the parent sultine (**120**) is even more unstable, breaking down below 0°C with quantitative evolution of butadiene and SO_2 [74JA935].

The unusually strained diene (**121**) provides a useful insight into the processes involved in these systems. Addition of SO_2 to a solution of (**121**) initially gave the sultine (**122**) [76JA2341]. This kinetic product was unstable and isomerised at room temperature, mainly to the benzosultine (**123**) but also,

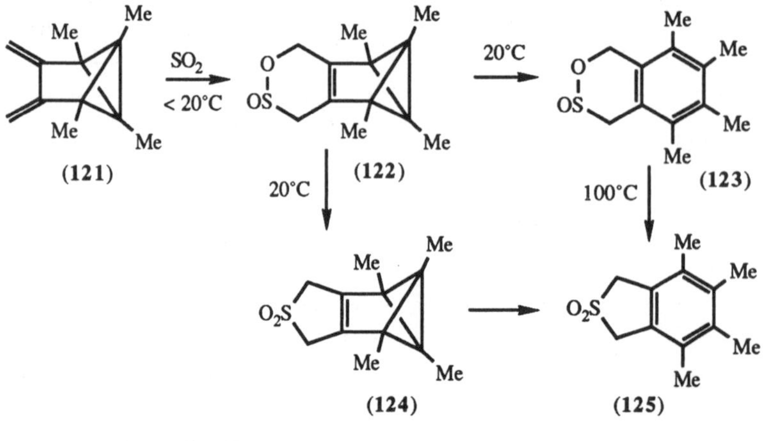

(**121**) (**122**) (**123**) (**124**) (**125**)

via loss of SO_2 and recombination, to form the thermodynamically more stable sulphone (**124**) which then rearranged to (**125**). At 100°C the sultine (**123**) isomerised, again *via* an *o*-xylylene intermediate, to (**125**).

Thermolysis of the cyclopropyl compounds (**126**) takes place in boiling chloroform to give an excellent yield of the isomerically pure 1,4-dienes (**127**) [76CC525].

The thermal extrusion of SO_2 from a six-membered ring sultone was first achieved in 1937 when Treibs heated equal weights of the sultone (**128**) and zinc oxide and obtained menthofuran (**129**) [37CB85]. Morel and Verkade later

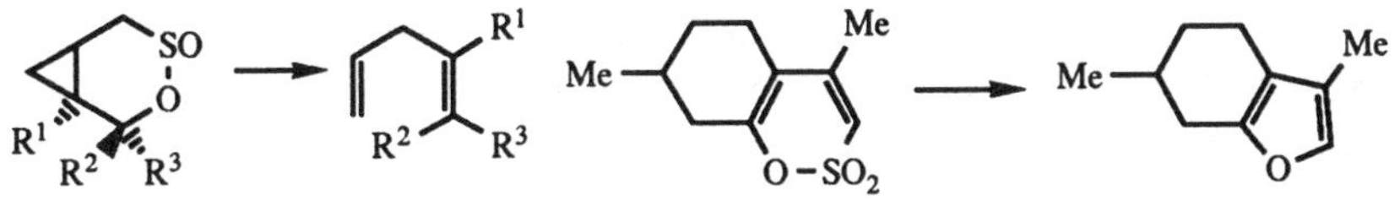

(**126**) (**127**) (**128**) (**129**)

developed this procedure into a general synthetic method for substituted furans. By heating an equal weight of the sultone (**130**) and calcium oxide in the presence of a catalytic quantity of quinoline, the substituted furan (**131**) is produced in reasonable yield [51RTC35]. The absence of calcium oxide and quinoline results in side-reactions and lower yields. Alternatively (**130**) can be heated with copper powder to give good yields of (**131**) [60LA(630)120]. In contrast, photolysis of the sultones (**130**) gave no loss of SO_2 but instead produced products derived from the reaction of the 4-oxosulphene isomer (**132**) with the solvent [63CJC100].

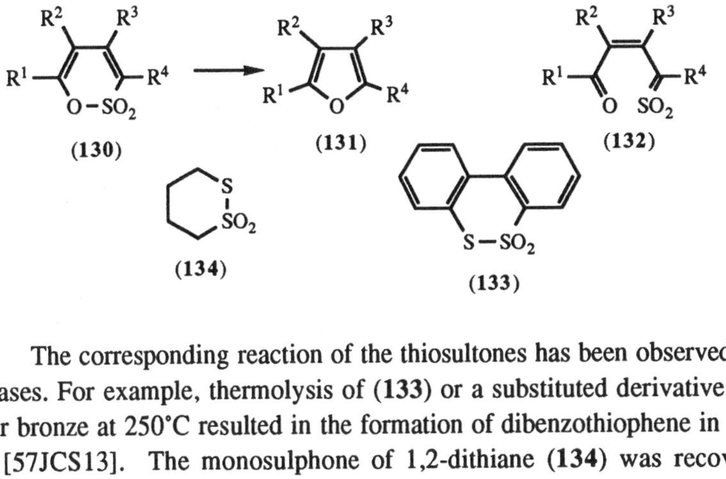

(**130**) (**131**) (**132**)

(**134**) (**133**)

The corresponding reaction of the thiosultones has been observed in a few cases. For example, thermolysis of (**133**) or a substituted derivative with copper bronze at 250°C resulted in the formation of dibenzothiophene in good yield [57JCS13]. The monosulphone of 1,2-dithiane (**134**) was recovered unchanged after heating for 1 week in boiling chlorobenzene [69JOC1792].

1.4.3 Other Systems Containing Only Oxygen and Sulphur

Thermolysis of 1,3-dithiane 1,1-dioxides (**135**) to afford the tetrahydrothiophenes (**136**) has been described [89MI117]. Attempted oxidation of 1,4-dithiins such as (**137**) leads to the loss of SO_2 and formation of

(**135**) (**136**)

2,4-diphenylthiophene [54JA4960]. This reaction probably involves elimination of SO_2 from the intermediate (**138**). In a more recent study photolysis of the isolated monsulphone of (**137**) resulted in SO_2 extrusion to give low yields of both the 2,4-diarylthiophene and its 2,5-diaryl isomer [83CL1461].

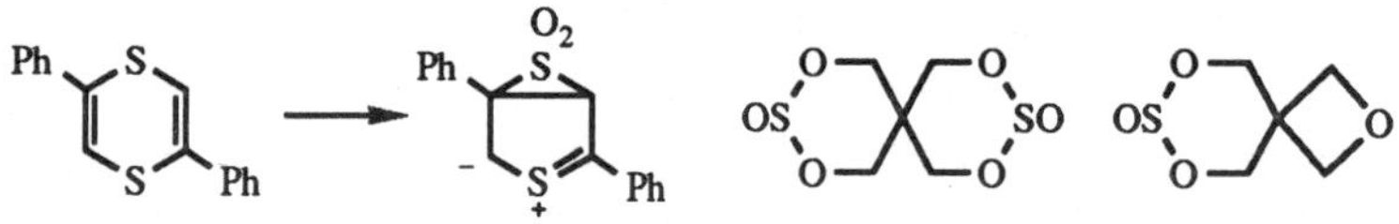

(**137**) (**138**) (**139**) (**140**)

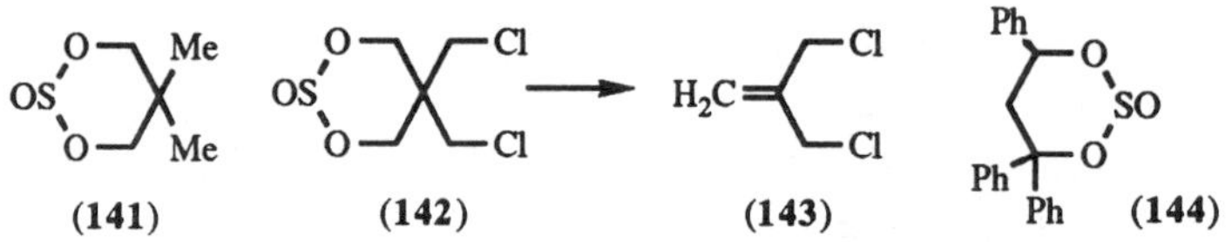

(**141**) (**142**) (**143**) (**144**)

The disulphite of pentaerythritol (**139**) lost one molecule of SO_2 on heating at 270°C to give the spiro oxetane derivative (**140**) [59JOC641]. Under the same conditions the sulphite of 2,2-dimethylpropane-1,3-diol (**141**) remained unchanged, but its dichloro derivative (**142**), readily prepared from pentaerythritol and thionyl chloride, lost SO_2 and formaldehyde on pyrolysis at 500°C and 15 mmHg to afford dichloroisobutene (**143**) in 91% yield [57JOC1723]. In the same way (**144**) decomposes in acetonitrile solution at 80°C to give benzaldehyde, $Ph_2C{=}CH_2$ and SO_2 [86JHC1099].

Salicylamides (**145**) react with N,N′-thionyldiimidazole to give the sulphites (**146**). For R=Ph this product is stable but for R=H, SO_2 is lost and the product, salicylonitrile is formed in 40% yield [79H(12)1285].

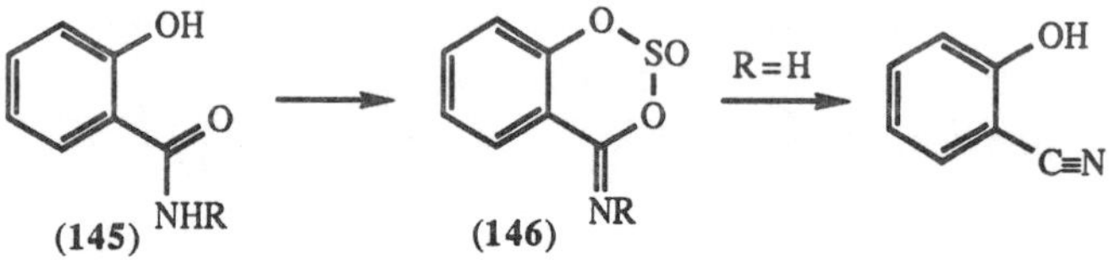

(**145**) (**146**)

A monosulphone of 5,6-diphenyl-1,2,3,4-tetrathiane is formed by decomposition of phenylsulphine in solution and it breaks down readily on heating to give sulphur, SO_2 and stilbene [68JCS(C)1612].

1.4.4 Six-Membered Rings Containing Nitrogen

Abramovitch has shown that under flash vacuum pyrolytic conditions the saturated benzosultam (**147**) loses SO_2 at 650°C to give a 75% yield of indoline with 7% of indole formed by aromatisation [75JA676]. The

N SO_2 H → N H + N H

(147)

thermolysis of unsaturated six-membered ring sultams (**148**) results in SO_2 extrusion to give the pyrroles (**149**) in good yield [61LA(646)45]. In a typical experiment the bis-sultam (**150**) was heated with lead monoxide at 280°C to give the dipyrrolylbenzene (**151**) in 78% yield [62LA(657)79]. This type of

R^3 R^1 R^2 $N-SO_2$ R^4 → $-SO_2$ R^3 R^1 N R^2 R^4

(148) (149)

O_2 S Me N Me O_2 S N Me Me → Me N Me N Me Me

(150) (151)

reaction has also been carried out photochemically [66CJC1869]. The same type of extrusion has been postulated in the reaction of 1,3-diphenyl-isobenzofuran with N-sulphinylaniline, whereby the initial adduct (**152**) rearranged to (**153**) which then lost SO_2 to give the isoindole (**154**) in 80% yield [63JOC2464].

Ph O NPh SO Ph → Ph NPh SO_2 Ph → $-SO_2$ Ph NPh Ph

(152) (153) (154)

Attempts to generate naphthalene-1,8-diyl from (**155**) by pyrolysis led only to decomposition and photolysis resulted in loss of only N_2 to give (**59**) which was however subsequently thermolysed to provide the diradical as described in Section 1.3.2 [67LA(703)96]. While treatment of the thiatriazine dioxide (**156**) with triethylamine gives 4,5-diphenyl-1,2,3-triazole by loss of SO_2 from an intermediate bicyclic thiirane dioxide (**24**), thermal decomposition of (**156**) by boiling in dichloromethane for 8h gives only *E* and *Z*-stilbene by an unknown mechanism [80JOC2604]. Sulphene reacts with substituted pyridines to form the remarkably stable 2:1 adducts (**157**). While these sublime unchanged at 160°C, they undergo stepwise loss of sulphene and SO_2 to give ions (**158**) and (**159**) in the mass spectrum [78CJC1183].

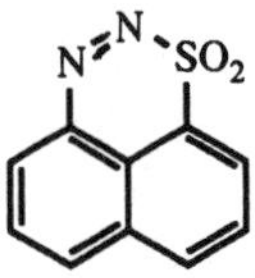

(**155**) 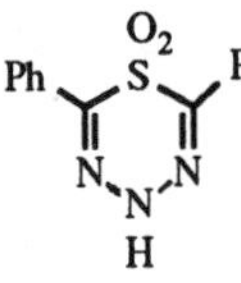 (**156**) 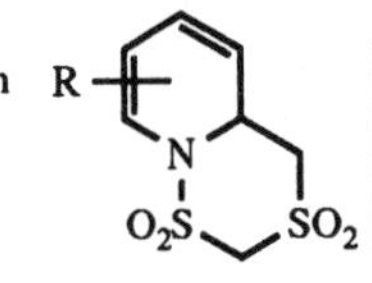 (**157**) 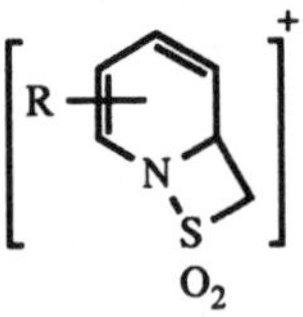 (**158**) 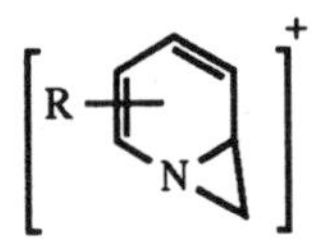 (**159**)

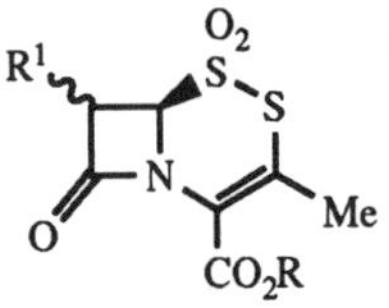

(**160**) 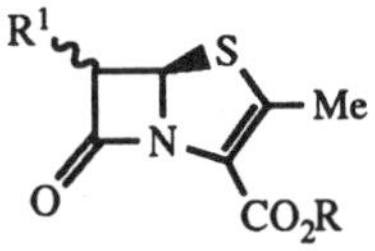 (**161**)

Solution thermolysis of 2-thiacephem 1,1-dioxides (**160**) results in stereospecific extrusion of SO_2 to give penems (**161**) [86JOC3413].

1.5 SEVEN- AND EIGHT-MEMBERED RINGS

1.5.1 Thiepine Dioxides

The first extrusion of SO_2 from seven-membered sulphone ring was observed by Truce who thermolysed benzo[d]thiepine dioxide (**162**) to produce naphthalene [56JA848]. In contrast, the isomeric benzo[b]thiepine dioxide (**163**) remained unchanged at 250°C and, on prolonged heating, it underwent polymerisation [64JOC366]. In 1967 Mock reported the first preparation of the parent thiepine dioxide (**164**) which gradually decomposed above its melting point, or in solution at 100°C, to give benzene and SO_2 [67JA1281]. The 2,4,5,7-tetraphenyl derivative behaved similarly, giving 1,2,4,5-tetraphenylbenzene in 70% yield on solution thermolysis [74JOC103].

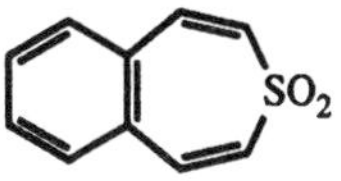

(162) 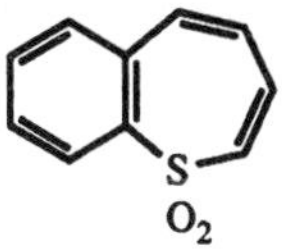 (163) 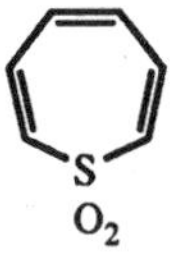 (164) 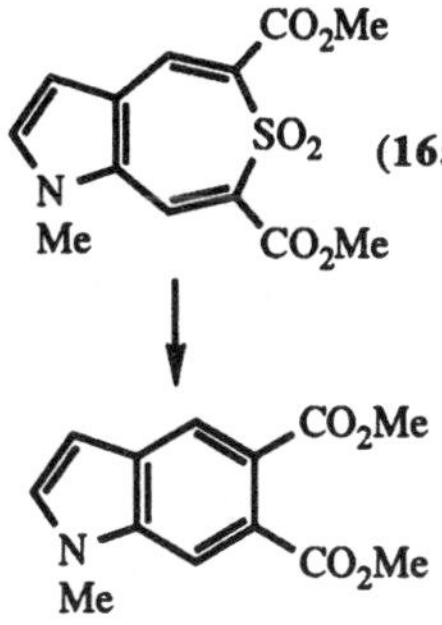 (165) (166)

The pyrrolothiepine dioxide (**165**) decomposes above 200°C with loss of SO_2 to give the substituted indole (**166**) in good yield [83JHC1191]. Heating a solution of the thienothiepine dioxide (**167**) with N-phenylmaleimide at between 90° and 240°C results in Diels-Alder reaction with concommitant loss of SO_2 to give the adduct (**168**) in 85% yield [68TL3017]. In contrast the furothiepine dioxide (**169**) reacts at 25°C to give the isolable intermediate (**170**), which then extrudes SO_2 on heating above 100°C to give (**171**) [69TL4361].

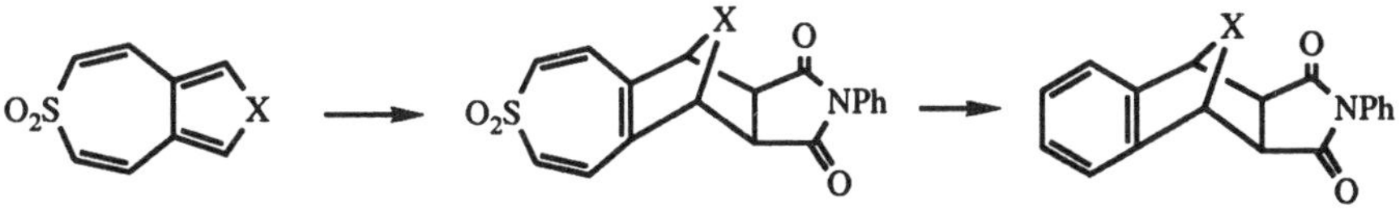

(167) X = S
(169) X = O

(170) X = O

(168) X = S
(171) X = O

The starting point for Mock's synthesis of thiepine dioxide (**164**) was the addition of SO_2 to Z-1,3,5-hexatriene. The reverse reaction, elimination of SO_2, has been the subject of detailed studies, largely because it represents a higher homologue of the sulpholene reaction of dienes. According to the Woodward-Hoffmann rules the extrusion from 2,7-dihydrothiepine dioxides (**172**) should be stereospecifically antarafacial (conrotatory). By decomposing the dimethyl compound (**172**, R = Me) and the corresponding *trans* isomer under

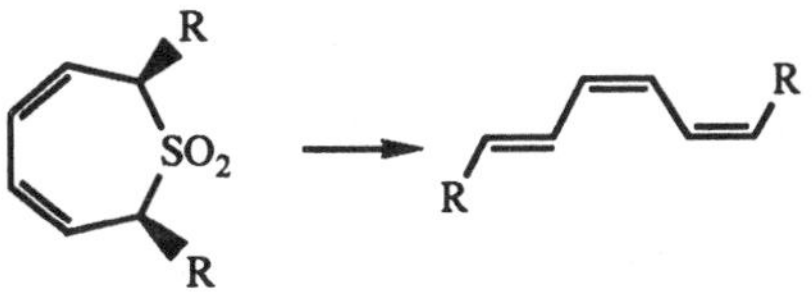

(172) 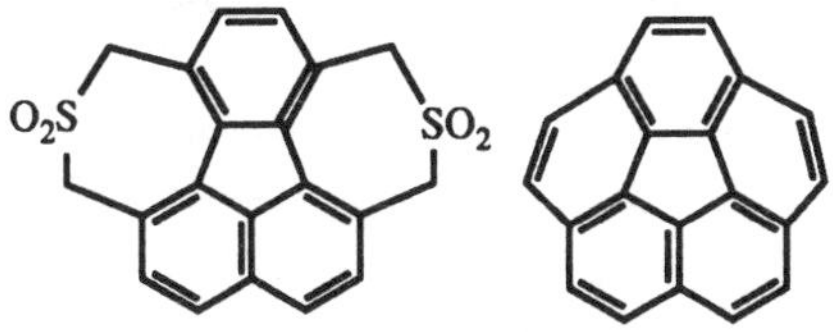 (173) (174)

g.l.c. conditions, Mock showed that this was indeed the case [69JA5682] [75JA3666]. No thermal decomposition study of the isomeric 4,5-dihydrothiepine dioxide has been reported, although it readily isomerises in the presence of base to the apparently more stable 2,7-dihydro-isomer, and the thermal decomposition might also follow this course [70CC1254].

Very recently the polycyclic disulphone (**173**) has been used to gain access to corannulene (**174**). Pyrolysis of (**173**) at 400°C gave the desired product, together with its di- and tetrahydro derivatives in low yield [92JA1922].

1.5.2 Seven- and Eight-Membered Rings Containing Oxygen

The seven-membered ring sultines (**175**) and (**176**) have already been mentioned as probable intermediates in the pyrolysis of the sulphones (**108**) and (**109**) [74TL3633]. Extrusion of SO from (**175**) and (**176**) would account for the observed products, but, although the sultines could be isolated from the photolysis of (**108**) and (**109**), their direct thermolysis has not been reported.

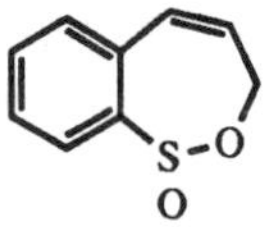

(**175**) 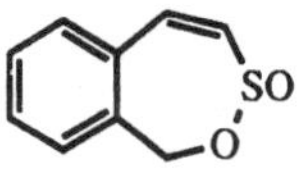 (**176**) 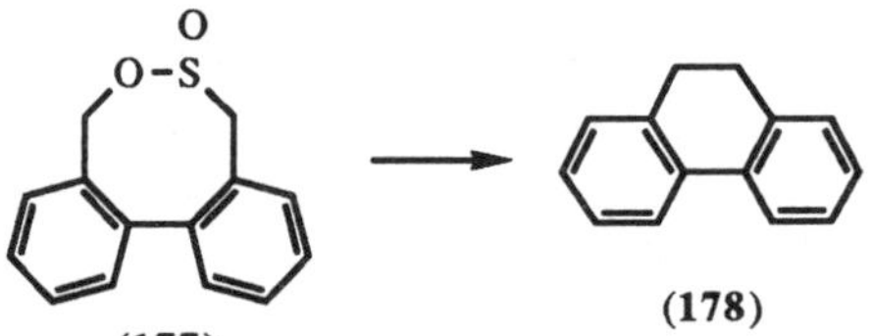 (**177**) (**178**)

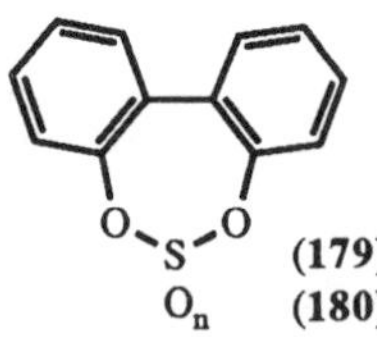

(**179**) n = 1
(**180**) n = 2
(**181**) 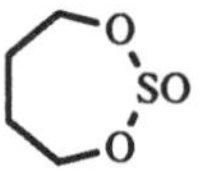 (**182**) 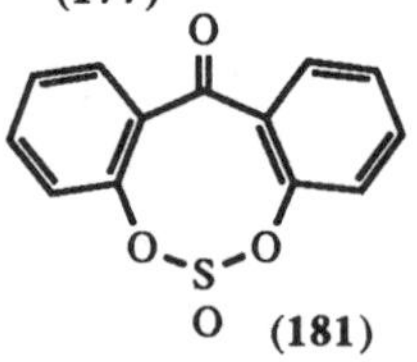

The eight-membered ring sultine (**177**) was found to lose SO_2 on irradiation in methanol to give 9,10-dihydrophenanthrene (**178**) [78CJC512]. Pyrolysis of the seven-membered ring sulphite (**179**) resulted in competitive loss of SO_2 and SO to give respectively, dibenzofuran in 20% yield and 1-hydroxydibenzofuran in 50% yield [72JOC1129]. Flash vacuum pyrolysis of the corresponding sulphate (**180**) caused exclusive loss of SO_2 to form 1-hydroxydibenzofuran (90% yield). In contrast, the closely related sulphite (**181**) underwent loss of SO_2 and partial loss of CO under the same conditions to give a mixture of dibenzofuran (30%), 9-xanthenone (10%) and 3,4-benzocoumarin (20%) [72JOC1129]. The saturated tetramethylene sulphite (**182**) was found to lose SO_2 on heating at 180°C in the presence of triethylamine to give a high yield of tetrahydrofuran [60JOC651].

Thermal extrusion of SO_2 from sulphite (**183**) is accompanied by several possible rearrangements. Thus, while heating in a sealed tube at 160°C

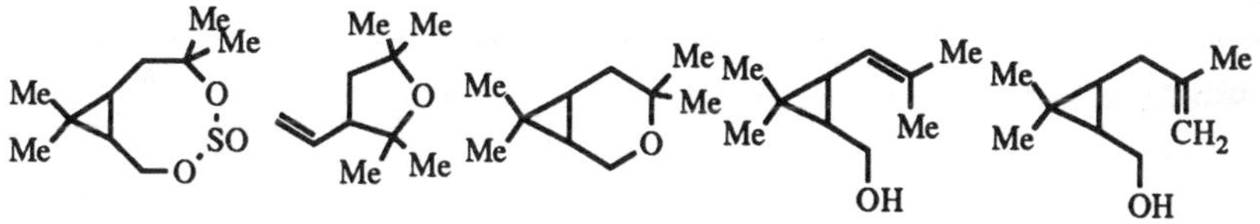

(**183**) (**184**) (**185**) (**186**) (**187**)

for 30 min. leads to quantitative conversion to (**184**), pyrolysis at 210°C in the inlet of a gas chromatography instrument gives a 3:3:2 mixture of (**185**), (**186**) and (**187**) [84JCS(P1)887].

1.5.3 Seven-Membered Rings Containing Nitrogen

Abramovitch has reported that flash vacuum pyrolysis of the benzothiazepine dioxide (**188**) at 650°C results in SO_2 extrusion to give tetrahydroquinoline (**189**) in 87% yield [84JOC3114]. The corresponding benzoxathiazepine dioxide (**190**) gives only a trace of the expected benzo-1,4-oxazine (**191**) at 450°C with mainly polymeric products being formed.

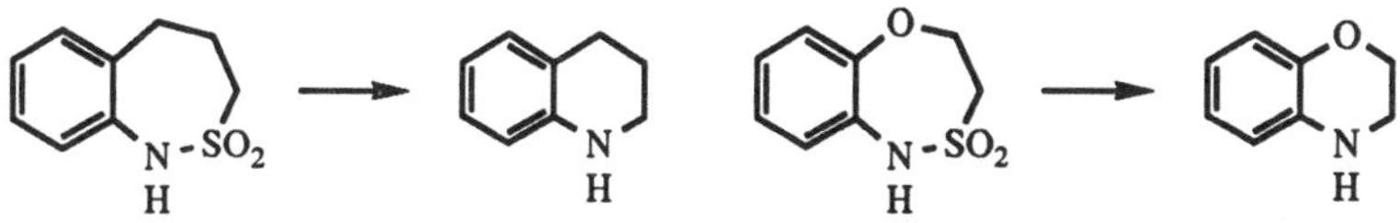

(**188**) (**189**) (**190**) (**191**)

The cyclisation of *o*-(N-sulphinylhydrazino)-benzoic acid (**192**) to benzopyrazolinone (**194**) on heating, reported in 1894, presumably involves SO_2 extrusion from benzoxathiadiazepinone S-oxide (**193**) [1894CB2549].

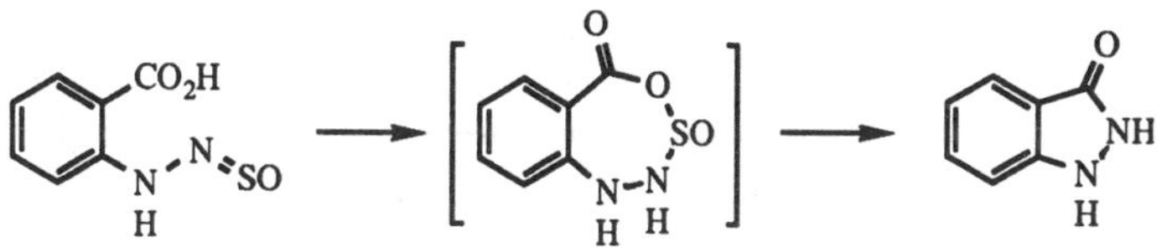

(**192**) (**193**) (**194**)

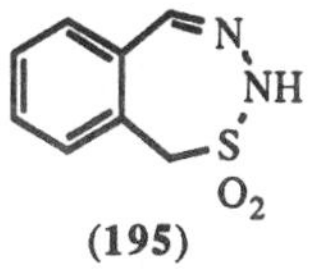

(**195**) 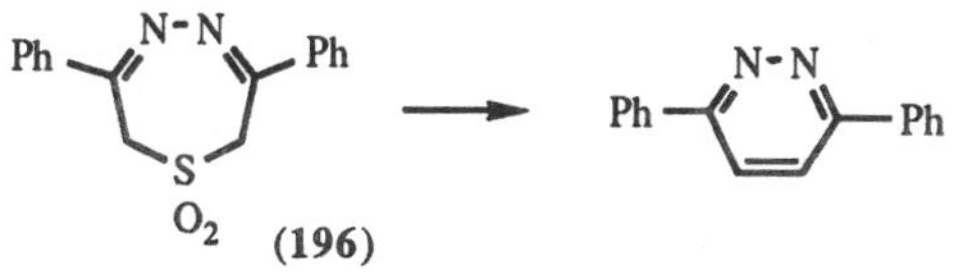 (**196**)

King found that the compound (**195**) lost only nitrogen on heating above its melting point to give benzosulpholene (**119**) in good yield [71CJC943]. However, the thiadiazepine dioxide (**196**), on heating in ethanol or acetic acid, underwent loss of "H_2SO_2" to give 3,6-diphenylpyridazine in high yield [63JCS5496]. This reaction is believed to proceed *via* a sulphinic acid intermediate.

1.6 BICYCLIC AND POLYCYCLIC MOLECULES

1.6.1 7-Thiabicyclo[2.2.1] Structures

By pyrolysing the simplest bicyclo[2.2.1] structure: 7-thiabicyclo-[2.2.1]heptane 7,7-dioxide (**197**) under flash vacuum conditions at 520°C, Corey obtained a 60% yield of 1,5-hexadiene [69JOC1233]. The carbonyl derivatives (**198**) and (**199**) decompose in solution at 220°C and 160°C to give quantitative yields of cyclohexen-3-one and hydroquinone, respectively [76TL3957].

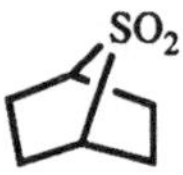

(**197**) 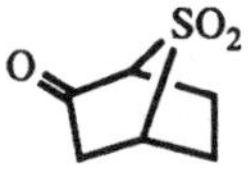 (**198**) 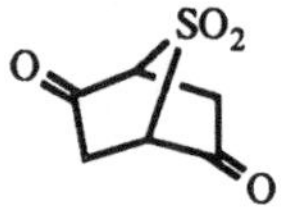 (**199**) 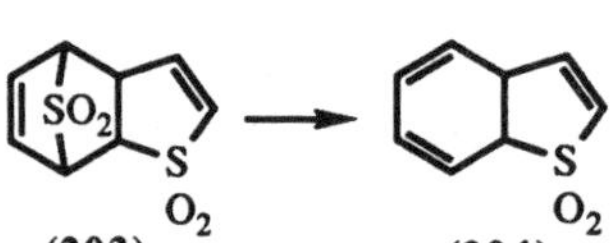 (**203**) (**204**)

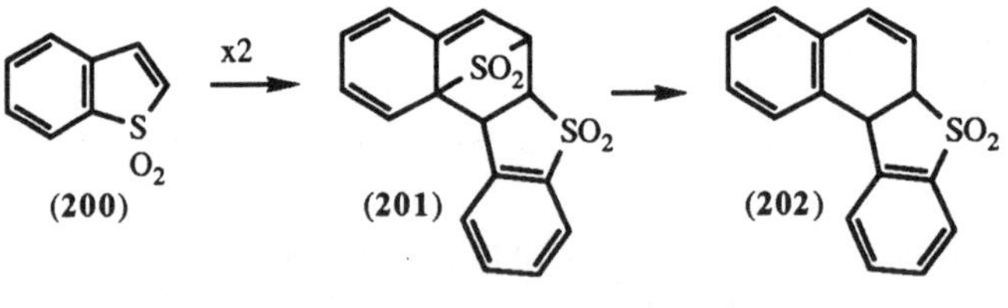

(**200**) (**201**) (**202**) 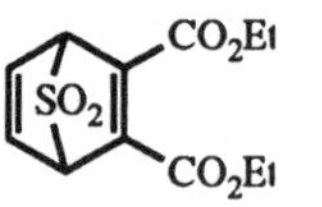 (**205**)

The vast majority of extrusions from bicyclo[2.2.1] systems however, occur in the Diels-Alder reaction of thiophene dioxides with dienophiles. This particular process was first observed when benzo[b]thiophene dioxide (**200**) was found to dimerise on heating above 200°C to give (**201**) which immediately lost SO_2 to form the benzonaphthothiophene dioxide (**202**) [51JA5566]. When thiophene dioxide itself was prepared it was found to undergo Diels-Alder reactions much more readily. Dimerisation occurred in solution below room temperature to give the dihydrobenzothiophene dioxide (**204**) *via* (**203**) [54JA1936]. The compound also reacted with other double and triple bonds with loss of SO_2 to give the expected products in low yield. Thus reaction with indene gave a 3% yield of fluorene while reaction with diethyl acetylenedicarboxylate yielded diethyl phthalate in 18% yield by SO_2 loss from the intermediate (**205**) [54JA1940]. The reaction of substituted thiophene dioxides with cyclopropenes has been used by van Tilborg and Reinhoudt to prepare the substituted cycloheptatrienes (**207**) [75RTC85]. The initially formed adducts (**206**) were unstable well below room temperature, rapidly losing SO_2 to form norcaradienes which ring expanded to give the cycloheptatrienes in quantitative yield.

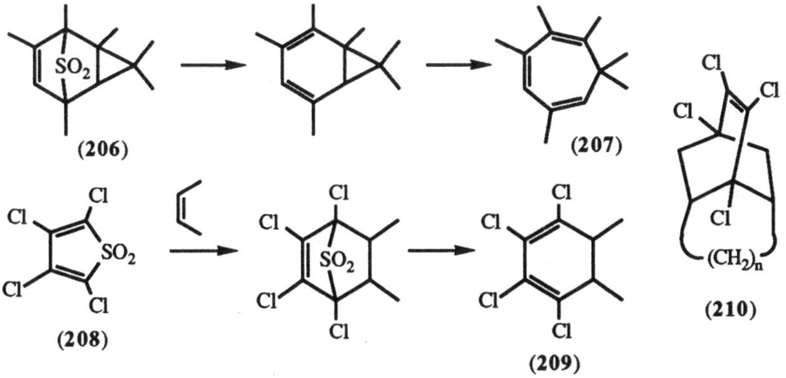
SO2
(206)
(207)
Cl
SO2
(208)
(209)
(CH2)n
(210)

Raasch has used the reaction of tetrachlorothiophene dioxide (**208**) with a large range of unsaturated systems to prepare annelated products in good to excellent yield [80JOC856]. The reactions proceed readily in solution between 0°C and 100°C with loss of SO_2 to give the product (**209**). Reaction with α,ω-dienes provides a convenient route to "isotwistenes" (**210**) *via* an intramolecular Diels-Alder reaction of the initial adduct.

In recent work, cycloaddition of 3,4-diadamantylthiophene dioxide (**211**) with phenyl vinyl sulphone and N-phenyltriazolinedione has been reported

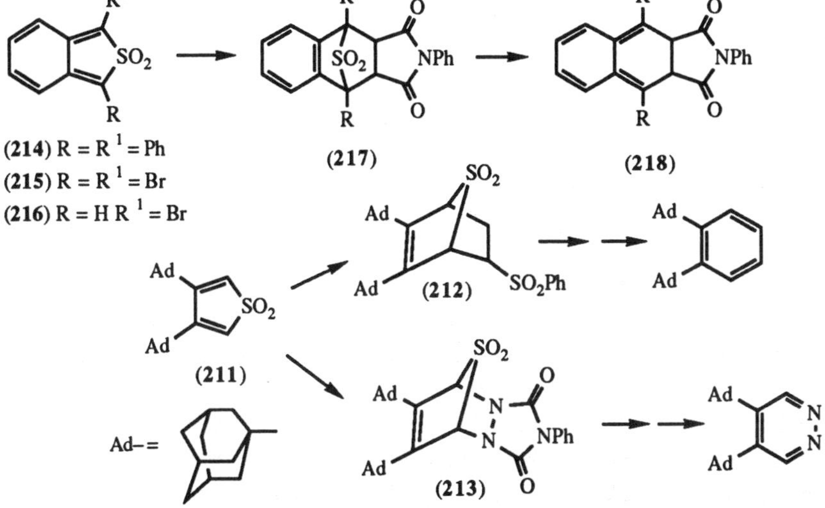
R1
SO2
R
(214) R = R1 = Ph
(215) R = R1 = Br
(216) R = H R1 = Br
O
NPh
(217)
(218)
Ad
(211)
Ad–=
(212)
SO2Ph
N
(213)

to lead *via* SO_2 extrusion from the initial adducts (**212**) and (**213**) and subsequent reactions, to 1,2-diadamantylbenzene and 4,5-diadamantylpyridazine, respectively [88JA6598][89H(29)1241][90JA5654].

Cava has reported the Diels-Alder reaction of the unstable benzo[c]thiophene dioxides (**214-6**). These react immediately with N-phenylmaleimide at room temperature to give a high yield of the products (**218**) arising by elimination of SO_2 from (**217**) [75JOC72].

1.6.2 2-Thiabicyclo[2.2.1] Structures

Some time ago Block reported that sulphene, generated by a novel method, reacts with diphenylisobenzofuran to form the product (**220**) in 92% yield, presumably by loss of SO_2 from the initial Diels-Alder adduct (**219**) [82TL4203].

1.6.3 2-Thiabicyclo[2.2.2] structures

King has shown that the bicyclic compounds (**221**) generally fragment to a sulphene and an aromatic ring on thermolysis [73CJC3044]. However, Fischer discovered that in the case of the diester (**222**), the loss of sulphene is only a minor pathway and that thermolysis in solution at 200°C results mainly

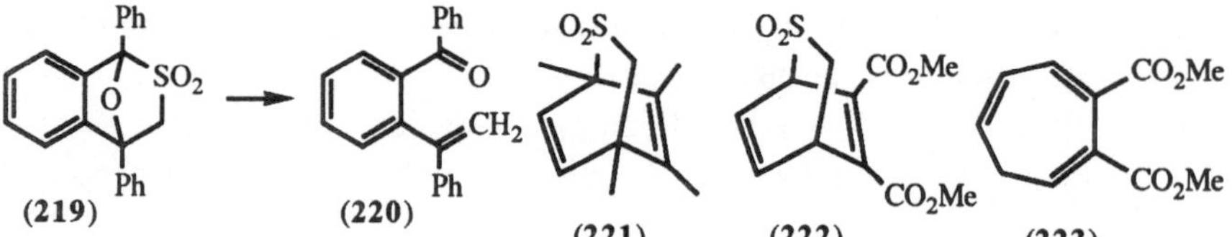

(**219**) (**220**) (**221**) (**222**) (**223**)

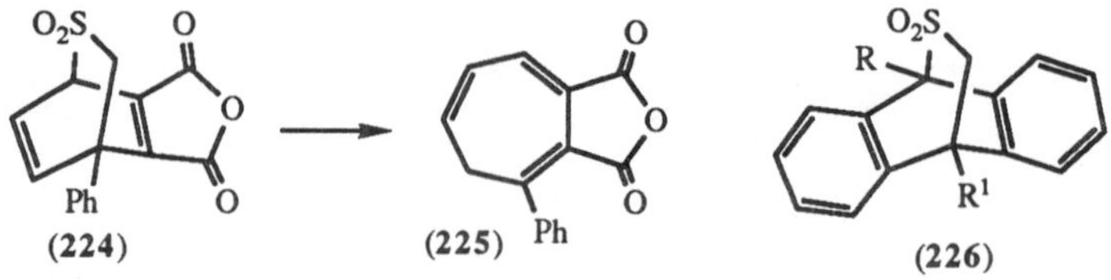

(**224**) (**225**) (**226**)

in elimination of SO_2 and rearrangement to give the cycloheptatriene diester (**223**) and its isomers [73JOC3073]. King also later observed these competing processes in the case of the anhydride (**224**) which extruded sulphene and SO_2 to an equal extent to give, respectively, 3-phenylphthalic anhydride and the cycloheptatriene (**225**) [74CJC2409]. Smith has also shown that the dibenzo compounds (**226**) undergo exclusive loss of SO_2 to give either the dibenzocycloheptatriene or the 9-methylanthracene depending on the conditions used [76CC981].

1.6.4 8-Thiabicyclo[3.2.1] Structures

In 1963 Cava obtained the resonance-stabilised o-xylylene compound pleiadene (**228**), by heating the corresponding bridged sulphone (**227**) at 210°C [63JA835]. The product could be isolated as the dimer or trapped as an adduct with N-phenylmaleimide. A year later he reported the realisation of the same reaction photochemically and this was indeed the first example of photochemical extrusion of SO_2 from a sulphone [64JA3173].

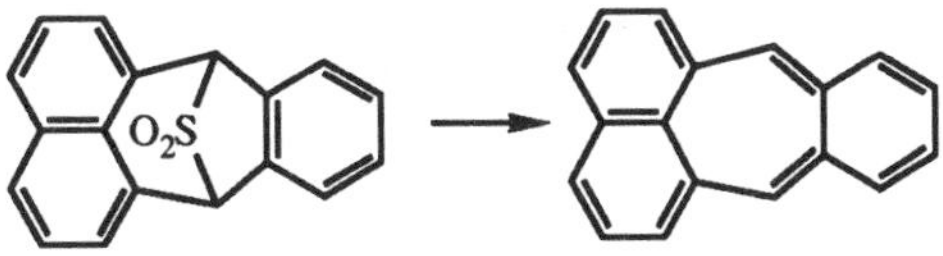

(227) (228) 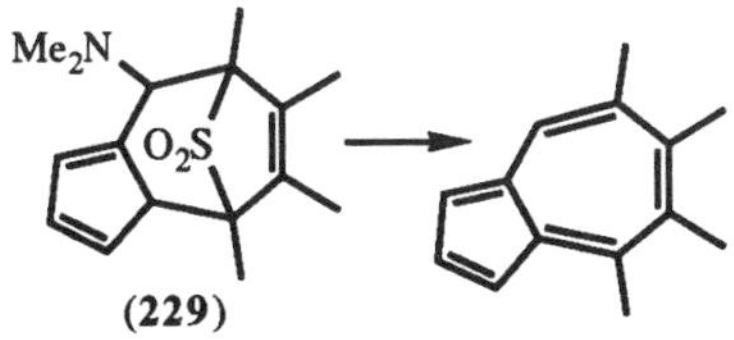 (229) (230)

(231) (232)  (233)

Extrusion of a bridging SO_2 group is the key step in the azulene synthesis reported simultaneously by both Leaver and Houk in 1977 [77TL639][77JA4199]. Reaction of a thiophene dioxide with dimethylamino-fulvene gave the adduct (**229**) which readily lost SO_2 and dimethylamine to give the azulene (**230**) in low yield. A similar approach was later used by Kanematsu to prepare 5,6-diaza-azulenes from a substituted thiadiazole dioxide [80CC873]. Flash vacuum pyrolysis of (**231**) results in loss of SO_2 followed by a [1,5]-hydrogen shift to give (**232**) [81JCS(P1)1846]. At 550°C the yield is quantitative and this is the best synthesis of (**232**). In contrast the cyclopropyl compound (**233**) does not lose SO_2 even at 300°C [70JA6918].

1.6.5 9-Thiabicyclo[3.3.1] Structures

Corey prepared the saturated bicyclo[3.3.1] system (**234**) and found that it loses SO_2 on flash vacuum pyrolysis at 710°C to give a 50% yield of bicyclo[3.3.0]octane in addition to a small quantity of cyclooctene [69JOC1233].

1.6.6 9-Thiabicyclo[4.2.1] Structures

In contrast to the radical process involved in the decomposition of (**234**), extrusion of SO_2 from unsaturated compounds of the isomeric [4.2.1] structure can in some cases be concerted and proceed more readily. For example, Paquette has prepared the compounds (**235-8**) from an unusual rearrangement of α-chlorosulphones under Ramberg-Bäcklund conditions and found that they readily lose SO_2, either on flash vacuum pyrolysis at 400°C, or by photolysis in acetone, to give the cyclooctatetraenes [71JA1047]. Hydrogenation of **(236)** occurred only at the two conjugated double bonds and the product - effectively a bridged sulpholene - again readily lost SO_2 on pyrolysis.

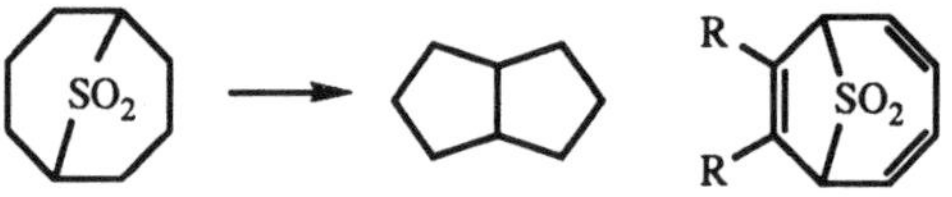

(234)

(235) R = Me
(236) $R_2 = (CH_2)_4$
(237) $R_2 = (CH_2)_5$
(238) $R_2 =$

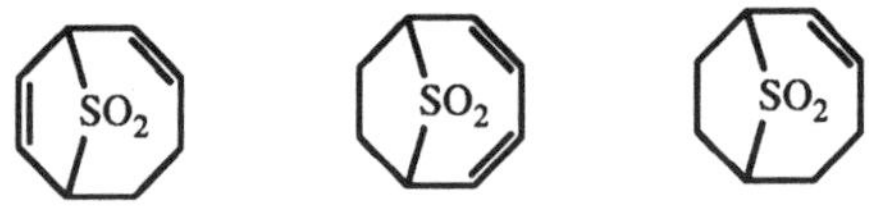

(239) **(240)** **(241)**

The thermal decomposition of the bridged sulphones (**239-241**) has been studied in detail by Mock in relation to the orbital symmetry constraints which apply to these systems [70JA3807]. Compound (**239**) dissociates readily above 100°C by an allowed 1,4-extrusion of SO_2 to give 1,3,5-cyclooctatriene. The isomeric sulphone (**240**) on the other hand only extrudes SO_2 above 200°C and the rate is 60,000 times slower than for (**239**). This was attributed to the fact that the symmetry-allowed antarafacial (conrotatory) 1,6-extrusion is sterically impossible. The compound (**241**) with no possible concerted fragmentation pathway is even more stable.

1.6.7 Propellanes

Extrusion of SO_2 from tricyclo[l,m,n,0] systems, using the Ramberg-Bäcklund reaction provides a convenient route to propellanes i.e. tricyclo[l,m,n,0] hydrocarbons [75MI1]. In a few cases, however, unsuccessful attempts have been made to prepare propellanes by direct extrusion of SO_2 from the parent sulphone. For example, Weinges found that the highly strained sulphone (**242**) could not be isolated from oxidation of the corresponding sulphide since it immediately lost one molecule of SO_2 under the conditions used to give the monosulphone (**243**) [77CB2978].

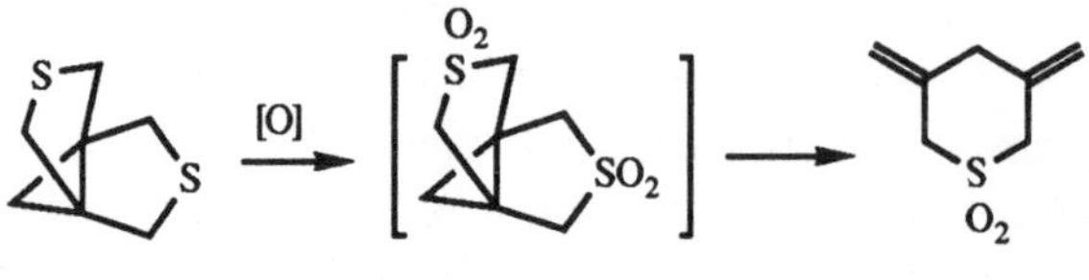

(242) **(243)**

Likewise, an attempt by Ginsburg to prepare [2.2.2] propellane proved unsuccessful. Although the sulphones (**244**) and (**245**) could not be prepared from the corresponding sulphides under normal conditions, they were obtained by low temperature ozonolysis. On warming to room temperature, loss of one molecule of SO_2 occurred to give 1,4-dimethylenecyclohexane from (**244**) and the monosulphone (**246**) from (**245**) [72T2507].

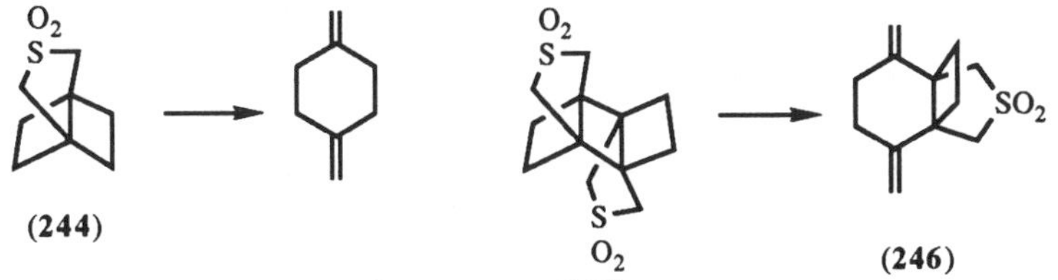

(**244**) (**245**) (**246**)

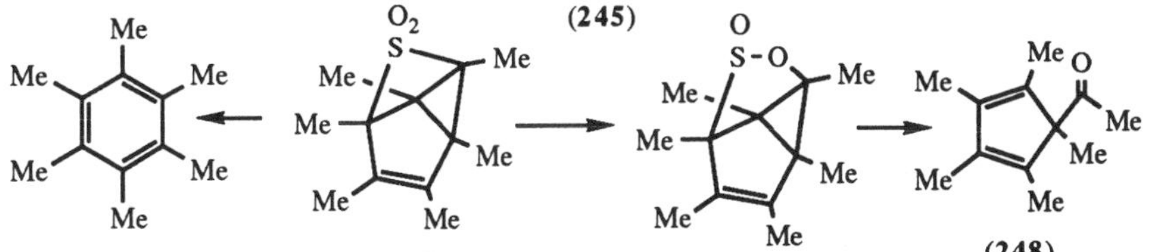

(**247**) (**248**)

1.6.8 Other Polycyclic Systems

The stable tricyclic adduct (**247**) of SO_2 with hexamethyldewarbenzene loses SO_2 to give hexamethylbenzene on flash vacuum pyrolysis at 600°C, but at lower temperatures a competing pathway also gives the acetyl cyclopentadiene (**248**) by loss of SO [75TL1795][82RTC365].

1.7 CYCLOPHANES

In recent years there has been intense interest in the chemistry of cyclophanes, i.e. molecules in which two or more aromatic nuclei are joined by two or more saturated carbon chains. Many of the early cyclophane-forming reactions relied on elimination of halogen atoms to form the bridges, but more recently extrusion of SO_2 from the sulphones of thiacyclophanes has emerged as an important method for their preparation. The synthesis of cyclophanes has been the subject of several reviews [73S85][80ACR65] with an excellent review of the use of sulphones as precursors available [79AG(E)515].

Vögtle was the first to prepare a cyclophane by sulphone thermolysis when, in 1969, he obtained a 20% yield of [2.2]-metacyclophane (**250**) by heating the disulphone (**249**) at 350°C [69AG(E)274]. It was later discovered by

Staab that photolytic removal of SO_2 could also be used to prepare cyclophanes. The disulphone (**251**), for example, could be converted into the cyclophane (**252**) either by flash vacuum pyrolysis at 440°C or by photolysis in benzene [73AG(E)776].

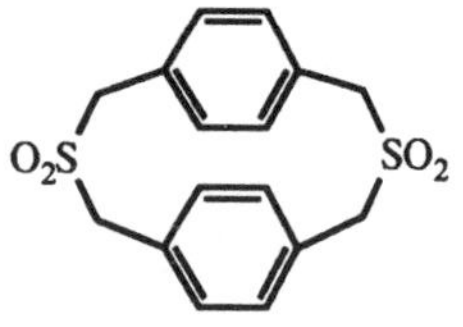

(**249**) 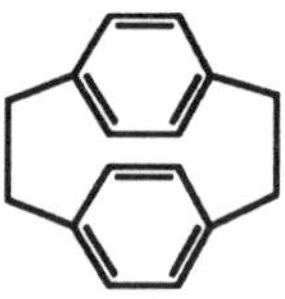 (**250**) 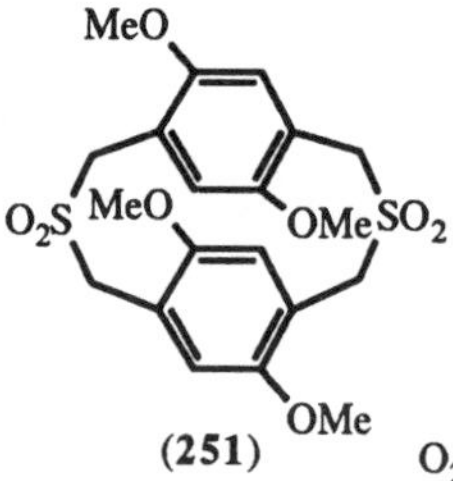 (**251**) 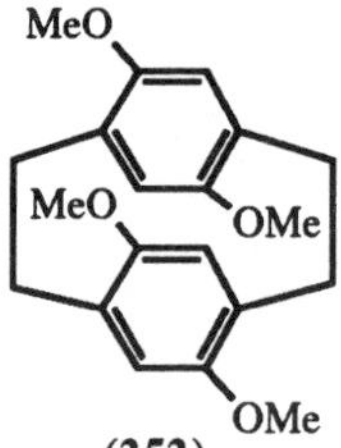 (**252**)

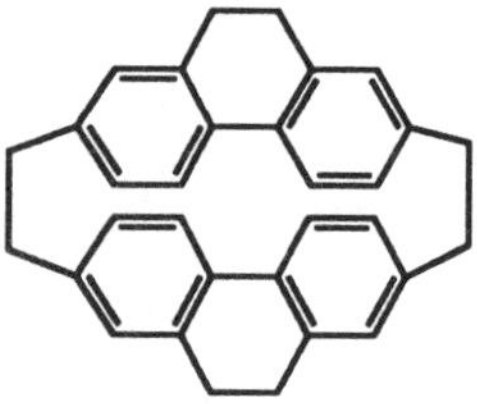

(**253**) 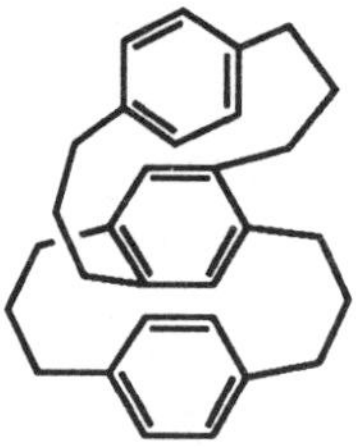 (**254**) 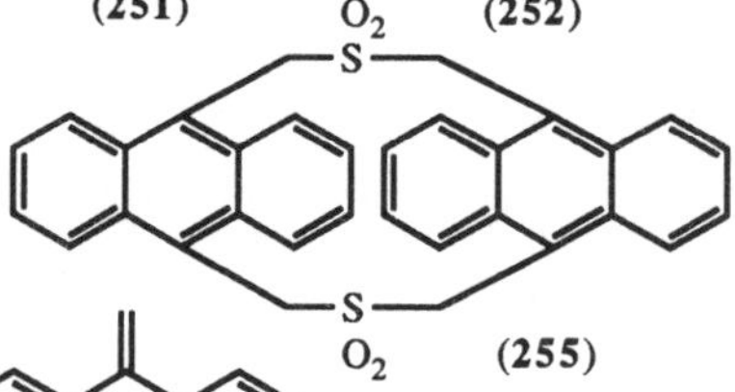 (**255**) 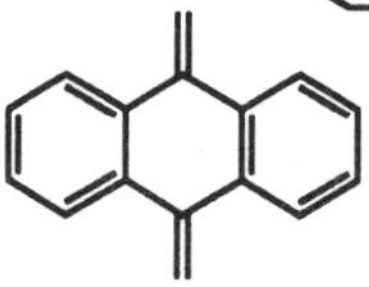 (**256**)

Sulphone pyrolysis has since been used to prepare a huge number of cyclophanes. An advantage is that the thiacyclophanes are readily accessible in good yields, either by reaction of two bromoalkylaromatic fragments with sodium sulphide or by reaction of a bromoalkylaromatic with a thioalkylaromatic. Some typical examples of cyclophanes which have been prepared by vapour phase pyrolysis of disulphones are (**253**) [70TL3585], and the triple-layered compound (**254**), which was obtained by Otsubo in a reaction sequence involving initial sulphone pyrolysis to form the first ring, followed by a further sequence of thiacyclophane building, oxidation and pyrolysis to attach the second cyclophane ring [78TL2507].

Finally, it should be noted that when the cyclophane to be formed is too strained, the pyrolysis may lead to products from the two separate diradicals. As long ago as 1964 Millar and Wilson thermolysed the disulphone (**255**) in diethyl phthalate and obtained 9,10-anthraquinodimethane (**256**) which could be trapped as its maleic anhydride adduct in 70% yield [64JCS2121].

1.8 REFERENCES

1894CB2549 J. Klieeisen, *Ber. Dtsch. Chem. Ges.*, 1894, **27**, 2549.
16CB1941 H. Staudinger and F. Pfenninger, *Ber. Dtsch. Chem. Ges.*, 1916, **49**, 1941.
37CB85 W. Treibs, *Ber. Dtsch. Chem. Ges.*, 1937, **70**, 85.
51JA5566 F. G. Bordwell, W. H. McKellin, and D. Babcock, *J. Am. Chem. Soc.*, 1951, **73**, 5566.
51RTC35 T. Morel and P. E. Verkade, *Recl. Trav. Chim. Pays Bas*, 1951, **70**, 35.
54JA1936 W. J. Bailey and E. W. Cummins, *J. Am. Chem. Soc.*, 1954, **76**, 1936.
54JA1940 W. J. Bailey and E. W. Cummins, *J. Am. Chem. Soc.*, 1954, **76**, 1940.
54JA4960 W. E. Parham and V. J. Traynelis, *J. Am. Chem. Soc.*, 1954, **76**, 4960.
56JA848 W. E. Truce and F. J. Lotspeich, *J. Am. Chem. Soc.*, 1956, **78**, 848.
57JCS13 W. L. F. Armarego and E. E. Turner, *J. Chem. Soc.*, 1957, 13.
57JOC1723 A. S. Matlock and D. S. Breslow, *J. Org. Chem.*, 1957, **22**, 1723.
59JOC641 S. Wawzonek and J. T. Loft, *J. Org. Chem.*, 1959, **24**, 641.
60JOC651 R. G. Gillis, *J. Org. Chem.*, 1960, **25**, 651.
60LA(630)120 W. Treibs, *Liebigs Ann. Chem.*, 1960, **630**, 120.
61LA(646)45 B. Helferich, R. Dhein, K. Geist, H. Jünger, and D. Wiehle, *Liebigs Ann. Chem.*, 1961, **646**, 45.
62LA(657)79 B. Helferich and W. Klebert, *Liebigs Ann. Chem.*, 1962, **657**, 79.
63AG(E)736 G. Schulz and G. Kresze, *Angew. Chem., Int. Ed. Engl.*, 1963, **2**, 736.
63CI(L)450 R. M. Dodson and G. Klose, *Chem. Ind. (London)*, 1963, **41**, 450.
63CJC100 J. F. King, P. de Mayo, E. Morkved, A. B. M. A. Sattar, and A. Stoessl, *Can. J. Chem.*, 1963, **41**, 100.
63JA835 M. P. Cava and R. H. Schlessinger, *J. Am. Chem. Soc.*, 1963, **85**, 835.
63JA3236 W. E. Truce and J. R. Norell, *J. Am. Chem. Soc.*, 1963, **85**, 3236.
63JCS5496 J. D. Loudon and L. B. Young, *J. Chem. Soc.*, 1963, 5496.
63JOC2464 M. P. Cava and R. H. Schlessinger, *J. Org. Chem.*, 1963, **28**, 2464.
63USP3136781 W. J. Middleton, *U. S. Pat.*, 3 136 781 (1963) [*Chem. Abstr.*, 1964, **61**, 5612].
64CB483 R. Albrecht, G. Kresze, and B. Mlakar, *Chem. Ber.*, 1964, **97**, 483.
64JA3173 M. P. Cava, R. H. Schlessinger, and J. P. van Meter, *J. Am. Chem. Soc.*, 1964, **86**, 3173.
64JCS2121 I. T. Millar and K. V. Wilson, *J. Chem. Soc.*, 1964, 2121.
64JOC366 V. J. Traynelis and R. F. Love, *J. Org. Chem.*, 1964, **29**, 366.
65ACS1755 A. Senning, *Acta Chem. Scand.*, 1965, **19**, 1755.
65AG(E)70 G. Opitz and K. Fischer, *Angew. Chem., Int. Ed. Engl.*, 1965, **4**, 70.
65AG(E)433 R. Ohme and E. Schmitz, *Angew. Chem., Int. Ed. Engl.*, 1965, **4**, 433.
65CB601 G. Kresze, D. Sommerfeld, and R. Albrecht, *Chem. Ber.*, 1965, **98**, 601.
65JA5804 L. A. Carpino and L. V. McAdams, *J. Am. Chem. Soc.*, 1965, **87**, 5804.
66CJC1869 T. Durst and J. F. King, *Can. J. Chem.*, 1966, **44**, 1869.
66TL1671 W. Wucherpfennig and G. Kresze, *Tetrahedron Lett.*, 1966, **7**, 1671.
66TL3929 S. Matsumura, T. Nagai, and N. Tokura, *Tetrahedron Lett.*, 1966, **7**, 3929.
67JA1281 W. L. Mock, *J. Am. Chem. Soc.*, 1967, **89**, 1281.
67JA2781 R. Kalish and W. H. Pirkle, *J. Am. Chem. Soc.*, 1967, **89**, 2781.
67LA(703)96 R. W. Hoffmann and W. Sieber, *Liebigs Ann. Chem.*, 1967, **703**, 96.
68JCS(C)1612 A. M. Hamid and S. Trippett, *J. Chem. Soc. (C)*, 1968, 1612.
68JOC2730 I. W. J. Still and M. T. Thomas, *J. Org. Chem.*, 1968, **33**, 2730.
68TL3017 R. H. Schlessinger and G. S. Ponticello, *Tetrahedron Lett.*, 1968, **9**, 3017.
69AG(E)274 F. Vögtle, *Angew. Chem., Int. Ed. Engl.*, 1969, **8**, 274.
69CC32 C. L. McIntosh and P. de Mayo, *J. Chem. Soc., Chem. Commun.*, 1969, 32.
69JA2097 D. C. Dittmer, G. C. Levy, and G. E. Kuhlmann, *J. Am. Chem. Soc.*, 1969, **91**, 2097.
69JA5682 W. L. Mock, *J. Am. Chem. Soc.*, 1969, **91**, 5682.
69JOC1233 E. J. Corey and E. Block, *J. Org. Chem.*, 1969, **34**, 1233.
69JOC1310 D. C. Dittmer, R. S. Henion, and N. Takashina, *J. Org. Chem.*, 1969, **34**, 1310.
69JOC1792 L. Field and R. B. Barbee, *J. Org. Chem.*, 1969, **34**, 1792.
69TL4361 R. H. Schlessinger and G. S. Ponticello, *Tetrahedron Lett.*, 1969, **10**, 4361.
70CC1254 W. L. Mock, *J. Chem. Soc., Chem. Commun.*, 1970, 1254.

70CJC3704	J. F. King, P. de Mayo, C. L. McIntosh, K. Piers, and D. J. H. Smith, *Can. J. Chem.*, 1970, **48**, 3704.
70JA3807	W. L. Mock, *J. Am. Chem. Soc.*, 1970, **92**, 3807.
70JA6918	W. L. Mock, *J. Am. Chem. Soc.*, 1970, **92**, 6918.
70S393	N. H. Fischer, *Synthesis*, 1970, 393.
70TL3585	M. Hänel and H. A. Staab, *Tetrahedron Lett.*, 1970, **11**, 3585.
70TL4683	B. Zwanenburg, A. Wagenaar, and J. Strating, *Tetrahedron Lett.*, 1970, **11**, 4683.
71CC22	J. C. Philips, J. V. Swisher, D. Haidukewych, and O. Morales, *J. Chem. Soc., Chem. Commun.*, 1971, 22.
71CJC943	J. F. King, A. Hawson, B. L. Huston, L. J. Danks, and J. Komery, *Can. J. Chem.*, 1971, **49**, 943.
71JA476	L. A. Carpino, L. V. McAdams, R. H. Rhynbrandt, and J. W. Spiewak, *J. Am. Chem. Soc.*, 1971, **93**, 476.
71JA676	B. M. Trost, W. L. Schinski, F. Chen, and I. B. Mantz, *J. Am. Chem. Soc.*, 1971, **93**, 676.
71JA1047	L. A. Paquette, R. E. Wingard, and R. H. Meisinger, *J. Am. Chem. Soc.*, 1971, **93**, 1047.
71JA2258	D. L. Coffen, J. Q. Chambers, D. R. Williams, P. E. Garrett, and N. D. Canfield, *J. Am. Chem. Soc.*, 1971, **93**, 2258.
71ZOR951	N. S. Zefirov and T. M. Pozdnyakova, *Zh. Org. Khim.*, 1971, **7**, 951.
72CJC866	J. F. King, E. G. Lewars, and L. J. Danks, *Can. J. Chem.*, 1972, **50**, 866.
72JOC225	D. C. Dittmer, K. Ikura, J. M. Balquist, and N. Takashina, *J. Org. Chem.*, 1972, **37**, 225.
72JOC1129	D. C. De Jongh and R. Y. Van Fossen, *J. Org. Chem.*, 1972, **37**, 1129.
72T2507	I. Lantos and D. Ginsburg, *Tetrahedron*, 1972, **28**, 2507.
72TL4781	R. F. J. Langendries and F. C. De Schryver, *Tetrahedron Lett.*, 1972, **13**, 4781.
72ZOR1107	T. M. Pozdnyakova and N. S. Zefirov, *Zh. Org. Khim.*, 1972, **8**, 1107.
73AG(E)776	W. Rebafka and H. Staab, *Angew. Chem., Int. Ed. Engl.*, 1973, **12**, 776.
73CJC2884	D. H. Hunter and J. B. Stothers, *Can. J. Chem.*, 1973, **51**, 2884.
73CJC3044	J. F. King and E. G. Lewars, *Can. J. Chem.*, 1973, **51**, 3044.
73JA634	J. W. Timberlake and M. L. Hodges, *J. Am. Chem. Soc.*, 1973, **95**, 634.
73JA3420	F. Jung, N. K. Sharma, and T. Durst, *J. Am. Chem. Soc.*, 1973, **95**, 3420.
73JOC2652	J,. M. Bohen and M. M. Joullié, *J. Org. Chem.*, 1973, **38**, 2652.
73JOC3073	N. H. Fischer and H.-N. Lin, *J. Org. Chem.*, 1973, **38**, 3073.
73S85	F. Vögtle and P. Neumann, *Synthesis*, 1973, 85.
74AG(E)742	H. Quast and F. Kees, *Angew. Chem., Int. Ed. Engl.*, 1974, **13**, 742.
74CJC2409	J. F. King, R. M. Enanoza, and E. G. Lewars, *Can. J. Chem.*, 1974, **52**, 2409.
74JA935	F. Jung, M. Molin, R. van der Elzen, and T. Durst, *J. Am. Chem. Soc.*, 1974, **96**, 935.
74JA2964	R. Langendries, F. C. De Schryver, P. de Mayo, R. A. Marty, and J. Schutyser, *J. Am. Chem. Soc.*, 1974, **96**, 2964.
74JOC103	N. Ishibe, K. Hashimoto, and M. Sunami, *J. Org. Chem.*, 1974, **39**, 103.
74JOC1817	J. C. Sheehan, U. Zoller, and D. Ben-Ishai, *J. Org. Chem.*, 1974, **39**, 1817.
74JOC2722	U. Jacobsson, T. Kempe, and T. Norin, *J. Org. Chem.*, 1974, **39**, 2722.
74JOC3412	T. Minami, Y. Tsumori, K. Yoshida, and T. Agawa, *J. Org. Chem.*, 1974, **39**, 3412.
74JOC3777	P. Vouros and L. A. Carpino, *J. Org. Chem.*, 1974, **39**, 3777.
74TL3633	C. R. Hall and D. J. H. Smith, *Tetrahedron Lett.*, 1974, **15**, 3633.
75BCJ480	T. Hiraoka and T. Kobayashi, *Bull. Chem. Soc. Jpn.*, 1975, **48**, 480.
75CC724	T. Durst and B. P. Gimbarzevsky, *J. Chem. Soc., Chem. Commun.*, 1975, 724.
75JA676	R. A. Abramovitch and W. D. Holcomb, *J. Am. Chem. Soc.*, 1975, **97**, 676.
75JA3666	W. L. Mock, *J. Am. Chem. Soc.*, 1975, **97**, 3666.
75JCS(P1)2513	R. J. Bushby, *J. Chem. Soc., Perkin Trans. 1*, 1975, 2513.
75JHC21	P. Vouros, *J. Heterocycl. Chem.*, 1975, **12**, 21.
75JOC72	M. P. Cava and J. McGrady, *J. Org. Chem.*, 1975, **40**, 72.
75MI1	D. Ginsburg, "Propellanes", Verlag Chemie, Weinheim, 1975.

75RTC85	W. J. M. van Tilborg, P. Smael, J. P. Visser, C. G. Kouwenhoven, and D. N. Reinhoudt, *Recl. Trav. Chim. Pays Bas*, 1975, **94**, 85.
75TL1795	H. Hogeveen, H. Joritsma, and P. W. Kwant, *Tetrahedron Lett.*, 1975, **16**, 1795.
75TL2179	J. Nokami, N. Kunieda, and M. Kinoshita, *Tetrahedron Lett.*, 1975, **16**, 2179.
76CC525	F. Jung, *J. Chem. Soc., Chem. Commun.*, 1976, 525.
76CC981	N. J. Hales, D. J. H. Smith, and M. E. Swindles, *J. Chem. Soc., Chem. Commun.*, 1976, 981.
76JA2341	R. F. Heldeweg and H. Hogeveen, *J. Am. Chem. Soc.*, 1976, **98**, 2341.
76JOC3044	D. C. Dittmer and T. R. Nelsen, *J. Org. Chem.*, 1976, **41**, 3044.
76TL3957	I. Tabuchi, Y. Tamura, and Z. Yoshida, *Tetrahedron Lett.*, 1976, **17**, 3957.
76TL4199	E. E. Nunn, *Tetrahedron Lett.*, 1976, **17**, 4199.
77AG(E)653	W. Ried and H. Bopp, *Angew. Chem., Int. Ed. Engl.*, 1977, **16**, 653.
77CB1780	H. Quast and F. Kees, *Chem. Ber.*, 1977, **110**, 1780.
77CB2978	K. Weinges, H. Baake, H. Distler, K. Klessing, R. Kolb, and G. Schilling, *Chem. Ber.*, 1977, **110**, 2978.
77CC713	J. C. Philips and O. Morales, *J. Chem. Soc., Chem. Commun.*, 1977, 713.
77JA4199	S. E. Reiter, L. C. Dunn, and K. N. Houk, *J. Am. Chem. Soc.*, 1977, **99**, 4199.
77JHC1369	R. Neidlein and A. D. Kraemer, *J. Heterocycl. Chem.*, 1977, **14**, 1369.
77JOC3506	J. E. Coates and F. S. Abbott, *J. Org. Chem.*, 1977, **42**, 3506.
77OR1	L. A. Paquette, *Org. React.*, 1977, **25**, 1.
77PS333	J. Mirek and S. Rachwal, *Phosphorus and Sulfur*, 1977, **3**, 333.
77TL639	D. Copland, D. Leaver, and W. B. Menzies, *Tetrahedron Lett.*, 1977, **18**, 639.
77TL1149	J. D. Finlay, C. R. Hall, and D. J. H. Smith, *Tetrahedron Lett.*, 1977, **18**, 1149.
78CJC512	T. Durst, J. C. Huang, N. K. Sharma, and D. J. H. Smith, *Can. J. Chem.*, 1978, **56**, 512.
78CJC1183	J. S. Grossert, M. M. Bharadwaj, R. F. Langler, T. S. Cameron, and R. E. Cordes, *Can. J. Chem.*, 1978, **56**, 1183.
78S579	J. D. Finlay, D. J. H. Smith, and T. Durst, *Synthesis*, 1978, 579.
78TL2507	T. Otsubo, T. Kohda, and S. Misumi, *Tetrahedron Lett.*, 1978, **19**, 2507.
79AG(E)515	F. Vögtle and L. Rossa, *Angew. Chem., Int. Ed. Engl.*, 1979, **18**, 515.
79H(12)1285	M. Ogata, H. Matsumoto, and S. Kida, *Heterocycles*, 1979, **12**, 1285.
79IZV396	Yu. V. Zeifman, E. G. Ter-Gabrielyan, D. P. Del'tsova, and N. P. Gambaryan, *Izv. Akad. Nauk SSSR, Ser. Khim.*, 1979, **28**, 396.
80ACR65	V. Boekelheide, *Acc. Chem. Res.*, 1980, **13**, 65.
80AG(E)203	R. Seelinger and W. Sundermeyer, *Angew. Chem., Int. Ed. Engl.*, 1980, **19**, 203.
80CB1632	M. Regitz and B. Mathieu, *Chem. Ber.*, 1980, **113**, 1632.
80CC873	M. Mori and K. Kanematsu, *J. Chem. Soc., Chem. Commun.*, 1980, 873.
80H1677	H. Quast, *Heterocycles*, 1980, **14**, 1677.
80JOC856	M. S. Raasch, *J. Org. Chem.*, 1980, **45**, 856.
80JOC2604	B. B. Jarvis and G. P. Stahly, *J. Org. Chem.*, 1980, **45**, 2604.
80JOC3646	P. G. De Benedetti, C. De Micheli, R. Gandolfi, P. Gariboldi, and A. Rastelli, *J. Org. Chem.*, 1980, **45**, 3646.
80JOC4789	K. Takaki, K. Nakagawa, and K. Negoro, *J. Org. Chem.*, 1980, **45**, 4789.
81CB774	H. Quast and F. Kees, *Chem. Ber.*, 1981, **114**, 774.
81JCS(P1)1846	T. Durst, M. Lancaster, and D. J. H. Smith, *J. Chem. Soc., Perkin Trans. 1*, 1981, 1846.
82JA3119	E. Block, E. R. Corey, R. E. Penn, T. L. Renken, P. F. Sherwin, H. Bock, T. Hirabayashi, S. Mohmand, and B. Solouki, *J. Am. Chem. Soc.*, 1982, **104**, 3119.
82LA723	A. Schwöbel, M. A. Pérez, M. Rössert, and G. Kresze, *Liebigs Ann. Chem.*, 1982, 723.
82RTC365	U. E. Wiersum, *Recl. Trav. Chim. Pays Bas*, 1982, **101**, 365.
82TL4203	E. Block and M. Aslam, *Tetrahedron Lett.*, 1982, **23**, 4203.
83CL1461	K. Kobayashi and K. Mutai, *Chem. Lett.*, 1983, 1461.

83CPB2957	T. Aoyama, S. Toyama, N. Tamaki, and T. Shioiri, *Chem. Pharm. Bull.*, 1983, **31**, 2957.
83JHC1191	J. Duflos, G. Dupas, and G. Quéguiner, *J. Heterocycl. Chem.*, 1983, **20**, 1191.
83JOC60	J. Nakayama, E. Oshima, A. Ishii, and M. Hoshino, *J. Org. Chem.*, 1983, **48**, 60.
83LA1739	K. Schank and F. Werner, *Liebigs Ann. Chem.*, 1983, 1739.
84AJC2489	J. Rosevear and J. F. K. Wilshire, *Aust. J. Chem.*, 1984, **37**, 2489.
84JCS(P1)887	B. Crammer, Z. Goldschmidt, R. Ikan, and S. Cohen, *J. Chem. Soc., Perkin Trans. 1*, 1984, 887.
84JCS(P1)2457	R. J. Bushby, S. Mann, and M. V. Jesudason, *J. Chem. Soc., Perkin Trans. 1*, 1984, 2457.
84JOC1300	M. Komatsu, Y. Yoshida, M. Uesaka, Y. Ohshiro, and T. Agawa, *J. Org. Chem.*, 1984, **49**, 1300.
84JOC3114	R. A. Abramovitch, A. O. Kress, S. P. McManus, and M. R. Smith, *J. Org. Chem.*, 1984, **49**, 3114.
84JOC5136	C. H. Chen, G. A.Reynolds, D. L. Smith, and J. L. Fox, *J. Org. Chem.*, 1984, **49**, 5136.
84S944	A. Schwöbel and G. Kresze, *Synthesis*, 1984, 944.
84TL4553	J. Takayama, M. Tanuma, Y. Honda, and M. Hoshino, *Tetrahedron Lett.*, 1984, **25**, 4553.
85AG(E)855	R. W. Saalfrank and W. Rost, *Angew. Chem., Int. Ed. Engl.*, 1985, **24**, 855.
85CB1415	R. Schork and W. Sundermeyer, *Chem. Ber.*, 1985, **118**, 1415.
85CB2852	U. Rheude, R. Schork, and W. Sundermeyer, *Chem. Ber.*, 1985, **118**, 2852.
85CB4553	A. Elsässer and W. Sundermeyer, *Chem. Ber.*, 1985, **118**, 4553.
85CJC1	J. F. King, J. H. Hillhouse, and K. C. Khemani, *Can. J. Chem.*, 1985, **63**, 1.
85LA72	A. Schwöbel, G. Kresze, and M. A. Pérez, *Liebigs Ann. Chem.*, 1985, 72.
86JHC1099	R. J. Olsen, T. A. Lewis, M. A. Mehta, and J. G. Stack, *J. Heterocycl. Chem.*, 1986, **23**, 1099.
86JOC3413	E. Perrone, M. Alpegiani, A. Bedeschi, D. Borghi, F. Giudici, and G. Franceschi, *J. Org. Chem.*, 1986, **51**, 3413.
87CB1499	R. Henn, W. Sundermeyer, and H. Pritzkow, *Chem. Ber.*, 1987, **120**, 1499.
88JA6598	J. Nakayama, S. Yamaoka, T. Nakanishi, and M. Hoshino, *J. Am. Chem. Soc.*, 1988, **110**, 6598.
89CC1020	S.-J. Lee, J.-C. Lee, M.-L. Peng and T. Chou, *J. Chem. Soc., Chem. Commun.*, 1989, 1020.
89H(29)1241	J. Nakayama and A. Hirashima, *Heterocycles*, 1989, **29**, 1241.
89MI117	K. Kabzinska and R. Kawecki, *Bull. Pol. Acad. Sci., Chem.*, 1989, **37**, 117 [*Chem. Abstr.*, 1990, **112**, 198257].
89TL3267	A. G. Sutherland and R. J. K. Taylor, *Tetrahedron Lett.*, 1989, **30**, 3267.
90CB1989	G. Opitz, T. Ehlis, and K. Reith, *Chem. Ber.*, 1990, **123**, 1989.
90JA5654	J. Nakayama and R. Hasemi, *J. Am. Chem. Soc.*, 1990, **112**, 5654.
91CC732	S.-Z. Zhu and Q.-Y. Chen, *J. Chem. Soc., Chem. Commun.*, 1991, 732.
92JA1921	A. Borchardt, A. Fuchicello, K. V. Kilway, K. K. Baldridge, and J. Siegel, *J. Am. Chem. Soc.*, 1992, **114**, 1921.

CHAPTER 2

2-Isoxazolines

SH. SHATZMILLER
Tel-Aviv University, Israel

Owing to unforeseen circumstances the author was unable to submit this chapter.

CHAPTER 3

Three-Membered Ring Systems

ALBERT PADWA AND SHAUN MURPHREE
Emory University, Atlanta, GA, USA

3.1 INTRODUCTION

Small ring heterocycles, particularly epoxides and aziridines, continue to command interest among the ranks of synthetic chemists. These compounds are not only useful synthetic intermediates, but also hold interest as end-products in their own right. Each year reveals yet another facet in the evolution of their chemistry. Herein lies the view of the past year's increment from one perspective.

3.2 EPOXIDES

3.2.1 Preparation

Since epoxides are extremely useful precursors to biologically significant compounds, the primary focus of most preparative methods is the concern for the introduction and control of stereocenters. The asymmetric epoxidation of functionalized alkenes, now firmly entrenched in common practice, has been the subject of much investigation since Sharpless laid the groundwork for this most important protocol. Sharpless himself has recently issued a mechanistic study of the titanium tartrate-mediated asymmetric epoxidation of allylic alcohols, wherein it was shown that the reaction proceeded with first-order kinetics with respect to substrate and oxidant [91JA106].

While the epoxidation of functionalized alkenes continues to hold its due prominence, another major thrust in this area is the enantioselective epoxidation of simple alkenes. This has been the topic of at least one short review during the past year [91ANG403]. Such epoxidations are usually accomplished using a chiral salen (Mn) complex as a catalyst, in the presence of some other stoichiometric oxidant such as sodium hypochlorite. The research groups of Jacobsen [91TL5055, 6533; 91JOC2296, 6497], as well as Katsuki and co-workers [91SL691] have been extremely active in

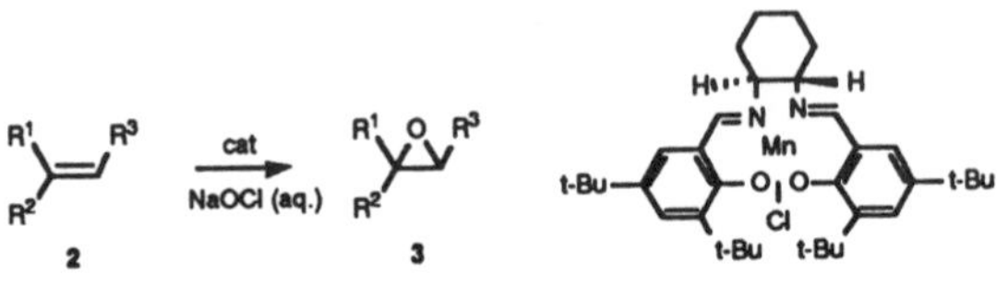

this area. For example, using the catalyst 1, trisubstituted olefins 2 could be epoxidized with very good enantioselectivity [91JOC6497]. It has been surmised from mechanistic evidence that such epoxidations occur *via* a concerted process, except in the case of aryl olefins. Besides applying this methodology to the preparation of vinyl and alkynyl epoxides [91TL6533], Jacobsen has also reported the first successful example of the asymmetric epoxidation of chromones (e.g., 4) by this method [91TL5055].

Aside from the asymmetric epoxidation protocol, chiral epoxides have also been accessed by enzymatic means. Thus, Schneider and Goergens [91CC1064] fashioned the useful chiral building block 8 in straightforward manner from the pre cursor 6, which is available chemoenzymatically in enantiomerically pure form.

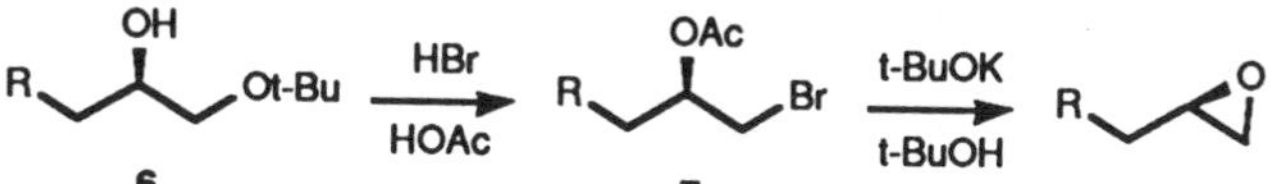

Introduction of chirality is not the only challenge in the preparation of epoxides. Equally demanding is the epoxidation of sensitive substrates that require the mildest of conditions. In the scrutiny of such obstacles, one reagent has been brought to the fore. Dimethyldioxirane (9), also useful as an oxidant for other purposes (*vide infra*), has been shown to be an ideal reagent for the epoxidation of recalcitrant or exotic substrates, often yielding products not available by classical routes. For example, Adam and co-workers [91JA8005] have reported the first benzofuran epoxide synthesis (i.e., 1 1) using dimethyldioxirane in their study on the mutagenesis of benzofuran dioxetanes. In addition, the epoxidation of other challenging substrates has been studied. Various enol-type alkenes 1 2 were found to be amenable to this protocol, producing labile epoxides 1 3 [91TL1295]. In a similar vein, the previously unreported flavone epoxides 1 5 were prepared in excellent yield from the flavone precursors 1 4 at subambient temperatures [91TL1041]. Furthermore, Adam has shown that dimethyldioxirane efficiently epoxidizes electron-poor alkenes, thus providing a milder alternative to the Weitz-Scheffer epoxidation [91CB227].

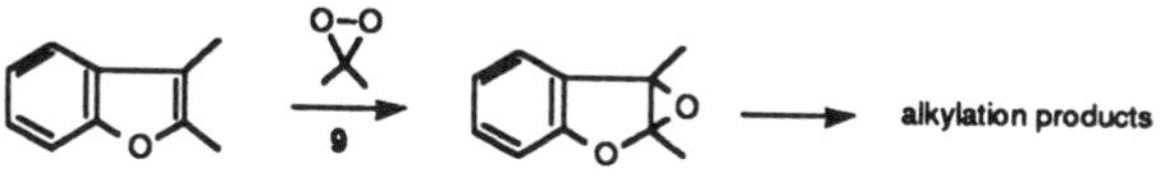

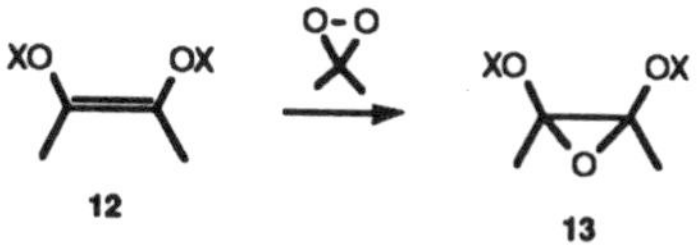

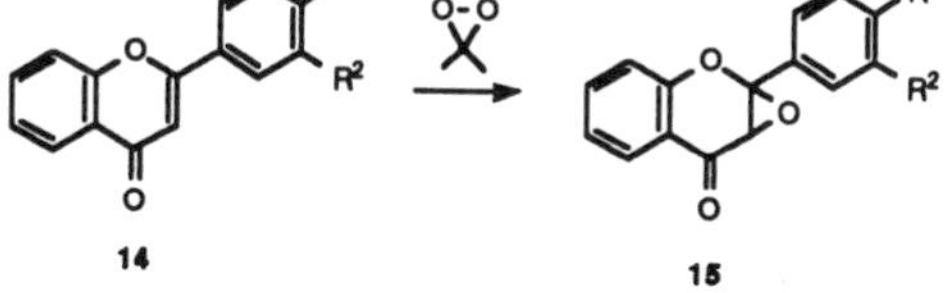

Crandall and co-workers [91JOC1153] have studied the dimethyldioxirane-mediated epoxidation of various allenes **16**. The corresponding 1,4-dioxaspiro[2.2]pentanes **17** were produced in good yields, the mono- and di-substituted allenes giving anti diastereomers with good stereoselectivity. These spirodiepoxides then underwent nucleophilic cleavage under buffered conditions to give highly functionalized products of type **18**.

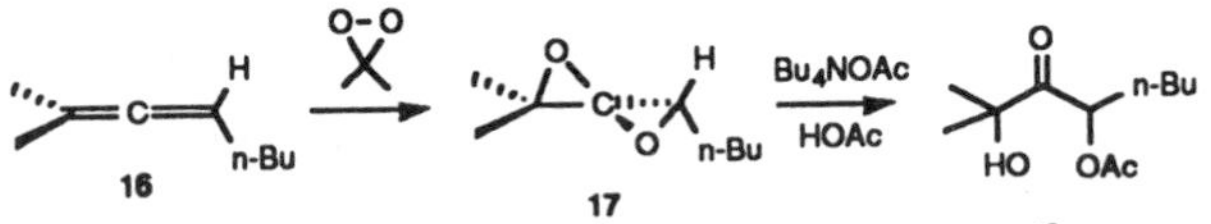

In the area of natural products, Marples and co-workers [91TL533] have used dioxiranes generated *in situ* from a range of ketones to effect 5,6-epoxidation of cholesterol or its acetate in high yield. Messeguer and co-workers [91TET1291] have employed dimethyldioxirane to prepare a proposed epoxide-containing metabolic intermediate (e.g., **20**) of juvenile hormone III.

3.2.2 Reactions of Epoxides

3.2.2.1 Nucleophilic Ring Opening

Probably the most familiar reaction of epoxides is their ring-cleavage by various nucleophiles. Interesting chemistry continues to be elaborated on in this area. For example, Oguni and co-workers [91SL774] have utilized chiral titanium complexes to effect the asymmetric ring opening of cyclohexene oxide (**21**) with trimethylsilyl azide (**22**) to provide the azidocyclohexanol **23** in up to 63% optical yield. In a similar vein, Saito and Moriwake [91TL667] have used the Sharpless methodology to cleave the 2,3-epoxyester **24** with high regio- and stereoselectivity using a tertiary ammonium azide salt as a novel nucleophilic reagent. This methodology allows for the facile preparation of primary β-hydroxy-α-amino acids.

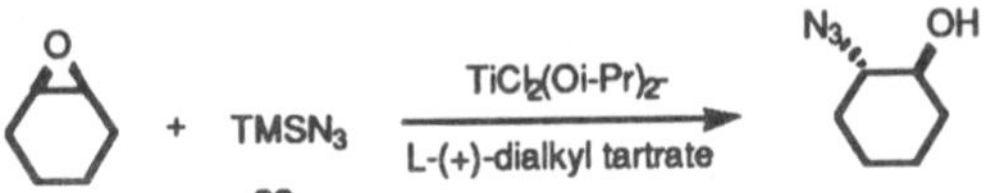

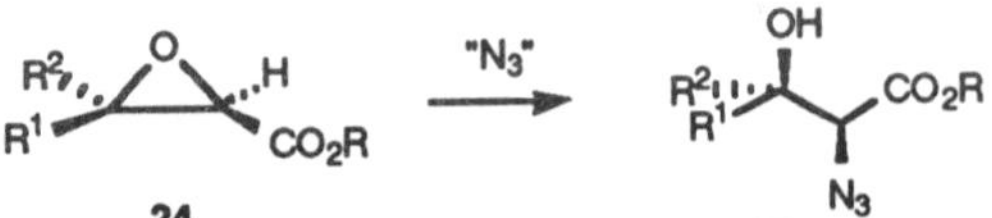

Pericás and Riera [91TL6931] have re-examined Sharpless' study on the regioselective ring-opening of chiral epoxyalcohols with various amines. These investigators found that primary amines could also be used for the reaction, obviating the need to utilize the indirect azide route. This protocol proceeds with the established propensity for attack at C-3, giving aminodiols 27.

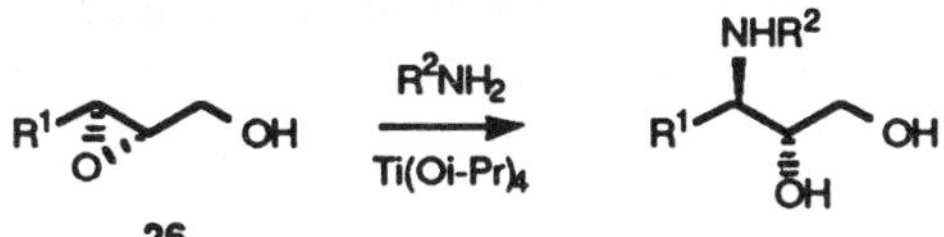

Otera and co-workers [91SL97] have elaborated a practical method for transforming (R)-2-phenyloxirane (28), available from microbial procedures, into the more difficultly accessible antipode. The procedure centers around the ring-opening of 28 with benzyl alcohol under catalysis using organotin phosphates, a reaction which proceeds with complete inversion of stereochemistry.

Halohydrins are also available from the cleavage of epoxides with halides. Thus, Bonini and co-workers [91JOC6206] have treated epoxyalcohols (e.g., 31) with magnesium iodide at low temperature. The resulting iodohydrins (e.g., 32) were subsequently reduced with tin hydride to the simple diol (e.g., 33). The high regioselectivity of the reaction, which proceeds by exclusive attack at C-3, has been rationalized by invoking a metal-centered chelate throughout the course of the reaction. Similarly, Bajwa and co-workers [91TL3021] have reported the conversion of epoxides 34 to halohydrins 35 using lithium halides in the presence of an acid.

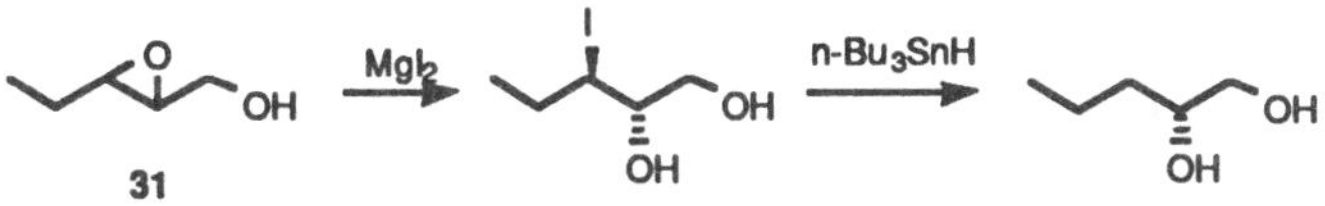

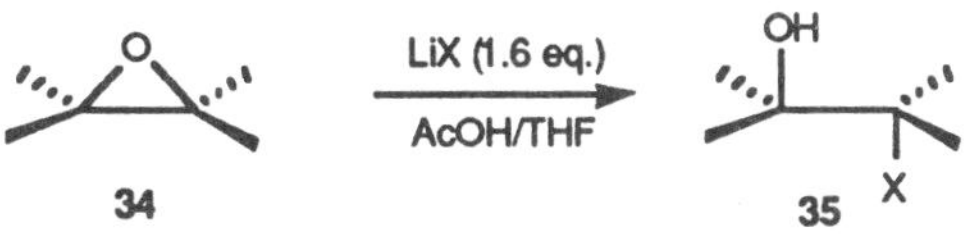

By extending such cleavage reactions into the realm of intramolecular attack, the epoxide functionality can be used to advantage for the preparation of a variety of heterocyles. For example, Fotsch and Chamberlin [91JOC4141] have studied the intramolecular cleavage of epoxides (e.g., 36) by neighboring carbonyl groups under

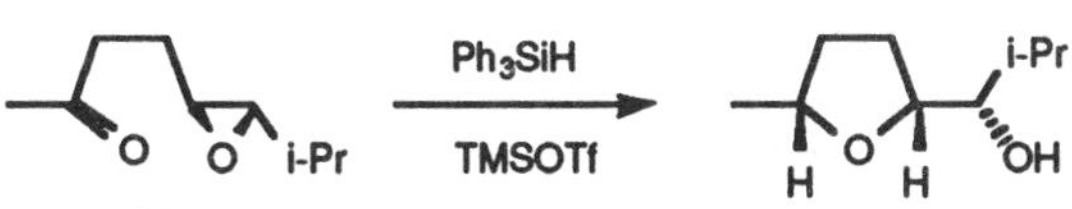

Lewis acid conditions. The intermediate oxocarbenium ion formed was trapped with high stereochemical control to give 2,5-disubstituted tetrahydrofurans (e.g., 37).

Such reactions are more commonly encountered with an amine acting as the nucleophile. For example, Nuhrich and Moulines [91TET3075] have studied the cyclization of N-tosyl oxiranepropyl amines 39, finding that in most cases, a regiospecific 5-*exo-tet* ring closure occurs to give N-tosyl-2-pyrrolidine methanols 40 in high yields.

NHTs MCPBA NHTs NaOH NTs CrO_3 NTs CO_2H OH O

38 39 40 41

Pearson and Bergmeier [91JOC1976] have used a similar cyclization reaction in their approach to (-)-slaframine. Thus, catalytic reduction of the optically pure epoxy azide 43 resulted in spontaneous double cyclization to the indolizidine 44, which was carried through to (-)-slaframine. Similarly, Husson and co-workers [91HET879] have used this protocol in their synthesis of swainsonine analogs. The key step involves a ring opening reaction of a tethered pyrrolidine (i.e., 45). Only the [5+6] product 46 was formed, in spite of the known tendency for such cyclizations to produce the smaller ring. This was rationalized in terms of the high ring strain associated with the alternative [4+6] product.

OH MCPBA OH Pd/C H_2 H OH

R^1R^2N N_3 R^1R^2N N_3 O R^1R^2N N

42 43 44

H O RO H N H → HO RO N H

45 46

Such ring-opening reactions are not limited to heteroatoms. Shimizu and co-workers [91TET2991] have published a facile synthesis of (-)-serricornin which has as the key step a palladium-catalyzed hydrogenolysis reaction of alkenyloxirane 47 to the homoallylic alcohol 48, using formic acid as a hydride equivalent.

Et O CO_2Et Pd-cat HCO_2H Et OH CO_2Et → Et OH COEt

47 48 49

Nucleophilic ring-opening of vinyl epoxides is a frequently encountered method for the formation of carbon-carbon bonds. To this end, a wide spectrum of carbon nucleophiles have been employed. Representing perhaps the simplest variety is the regio- and stereospecific methylation of epoxy alkenes 50 and 52, studied by Marshall [91JOC2225] and Miyashita [91JOC6483], respectively. Marshall's method employs a cuprate, whereas Miyashita's protocol involves the use of organoaluminum reagents.

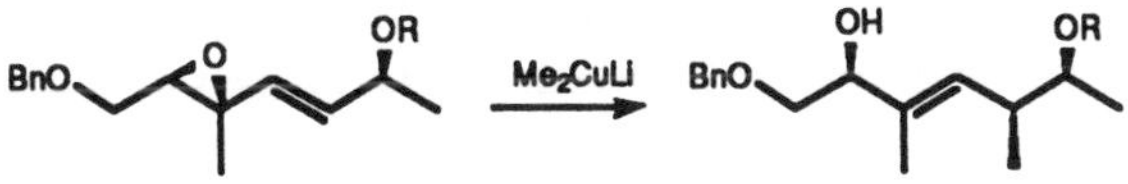

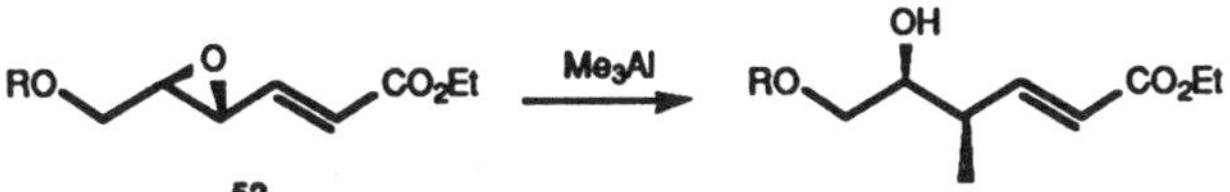

Anions derived from alkynes also take part in the intermolecular cleavage of epoxides, often proceeding with high regio- and stereoselectivity. For example, Crotti and co-workers [91TL6617] have examined the metal salt-catalyzed alkynylation of 1,2-epoxides 54 which results in the formation of β-hydroxyacetylenes 55. The alkynyl anion attacks exclusively on the less substituted carbon atom, while the stereoselectivity arises from attack on the β-face. Similarly, Matsubara and co-workers [91BSJ1810] found that the addition of lithioalkynes 56 to epoxides 57 proceeds with excellent regioselectivity using trimethylgallium as a catalyst, which presumably activates the epoxide ring by coordination. In this case, the low acidity of the catalyst is desirable so as to prevent isomerization of the epoxide.

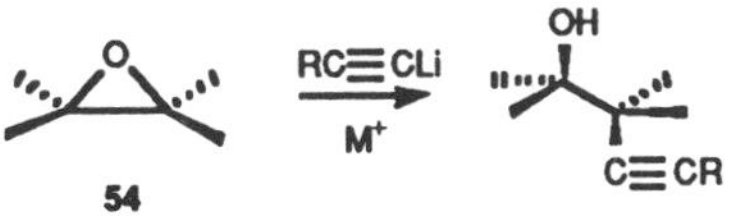

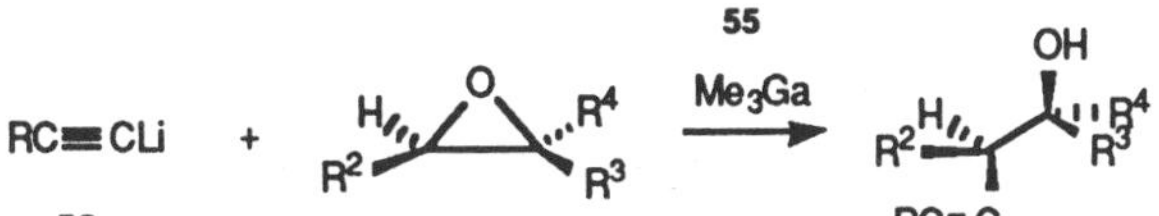

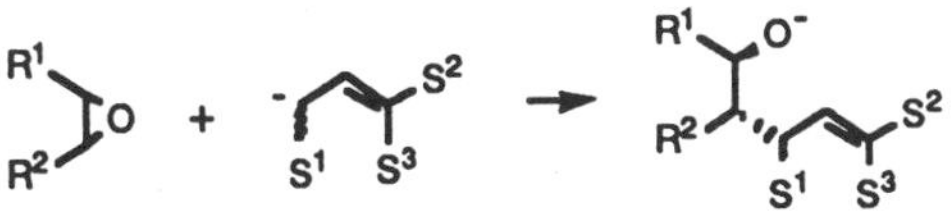

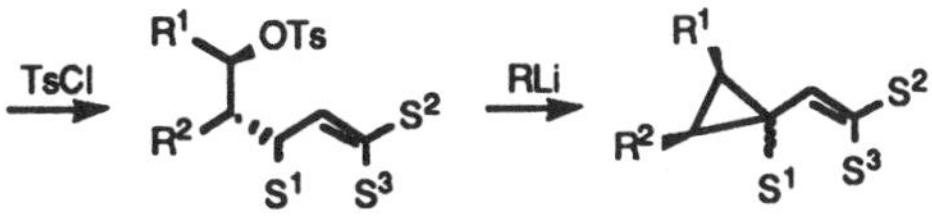

More exotic carbon nucleophiles have also been employed. A protocol for the synthesis of vinylcyclopropanes from epoxides was reported by Schaumann and co-workers [91JOC717]. In this procedure, an epoxide (59) is attacked by a het eroatom-substituted allyl anion (60) and this is followed by an *in situ* tosylation of the resultant alkoxide. Subsequent deprotonation leads to ring closure. This method also represents an interesting route for the annulation of vinylcyclopropanes onto existing carbocycles.

While certain regiochemical trends are evident in such displacement reactions, the size of the nucleophile itself can play a role in the site selectivity. For example, Wicha and co-workers [91CC297] discovered an anomolous result in the reaction of epoxysilanes 64 with hindered α-phenylsulfonyl carbanions. Although exclusive α-attack is usually the rule with such epoxides, in the case of very bulky anions (e.g. 65) a regiochemical crossover was observed due to steric interactions.

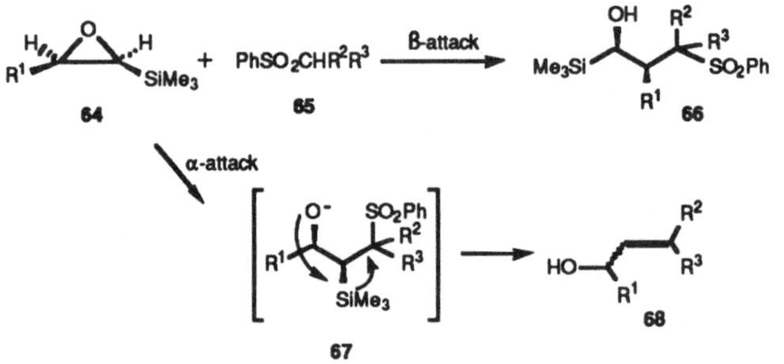

As in the case with heteronucleophiles, carbon-centered nucleophiles have also been utilized intramolecularly, providing novel routes to carbocycles. For example, Biellmann and co-workers [91TL2637] have published on an enantiospecific synthesis of a chrysanthemic acid precursor which centered around the Stork cyclization of epoxynitrile 70 which produced the kinetically favored cyclopropane 71.

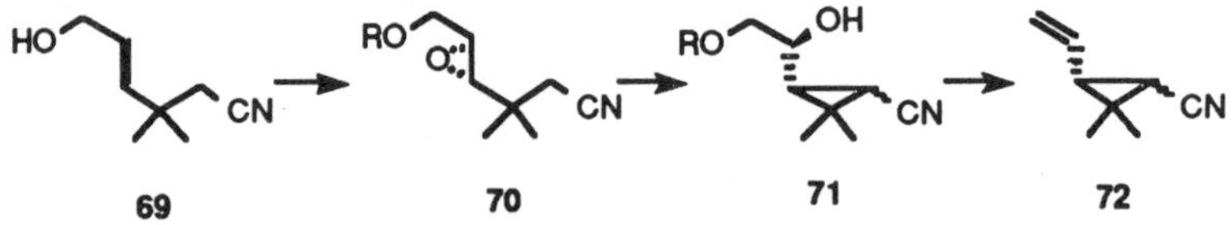

During the past year Whiting and Mallayon [91PTI2277] utilized a phenolic residue as a latent nucleophile. Epoxyphenol 73 was found to undergo cyclization to the tricyclospirodienone 74 *via* a 7-*endo* type of ring closure, in contrast to the corresponding phenolic iodides, which cyclize primarily in the 5- and 6-*exo* sense.

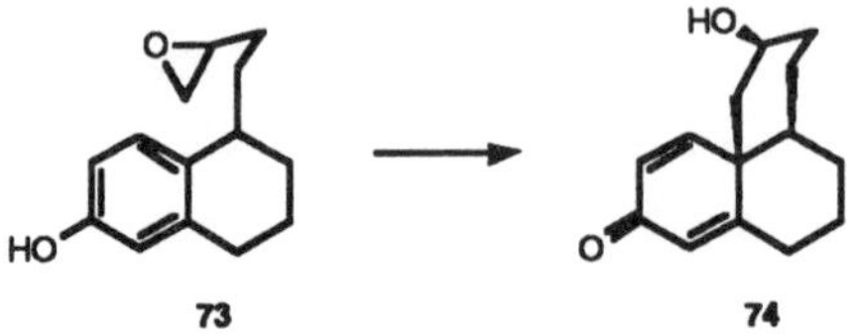

In a study of the transannular epoxide opening reactions of medium-size ring epoxy sulfones, Sandri and Fava [91JOC4513] found that no clear-cut regiochemical bias was operative. Thus, treatment of epoxysulfone 75 with base generally led to mixtures of regioisomeric products, the distribution of which was assumed to be related to temperature-dependent proton exchange, as well as aggregation phenomena.

3.2.2.2 Rearrangements of Epoxides

The epoxide nucleus also represents a useful locus for the initiation of rearrangement reactions. One of the more interesting examples involves radical addition of tri-n-butyl stannane onto a vinyl epoxide followed by a subsequent rearrangement. Thus, in the case of spiro-epoxide **76**, addition of tin produces an alkoxy radical which fragments and cyclizes to give cycloalkanone **79** as the major product [91TL6575]. In the case of cyclohexene oxide **80**, the initially produced alkoxy radical **81** initiates an internal hydrogen atom abstraction, leading to a carbon centered radical (**82**) which is set up to undergo cyclization to bicycloalkanol **83** [91JA5106].

Building upon the well known acid-catalyzed rearrangement of epoxides to carbonyl compounds, Yamamoto [91SL491] developed an interesting selective rearrangement of trisubstituted epoxides to either aldehydes or ketones. Whereas treatment of epoxide **84** with antimony pentafluoride led mostly to ketone **85**, the same epoxide in the presence of methylaluminum bis(4-bromo-2,6-tert-butylphenoxide) was converted to aldehyde **86**. This remarkable crossover was rationalized in terms of the latter catalyst's extreme steric bulk and oxygenophilicity.

	85		86
MABR	0	:	1C
SbF_5	83	:	1'

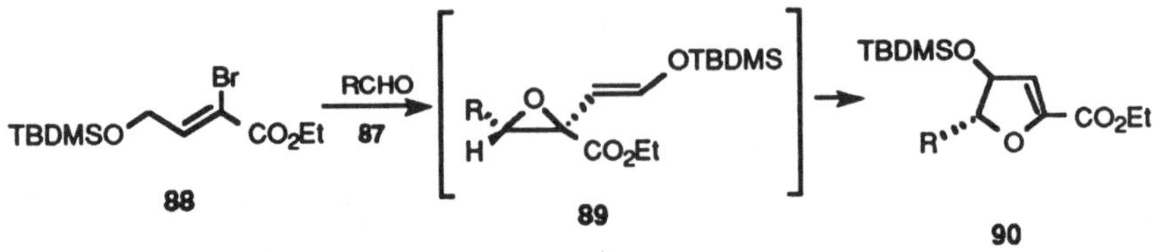

Hudlicky and Barbieri [91JOC4598] have reported a unique low-temperature, 1,3-sigmatropic rearrangement of silyl enol ether terminated epoxides of type **89** in their approach to some interestingly functionalized dihydrofurans **90**.

Alkynyl substituted epoxides such as **91** undergo base-catalyzed rearrangement to form furans **93** through the intermediacy of a cumulene alkoxide (**92**) [91JOC1685].

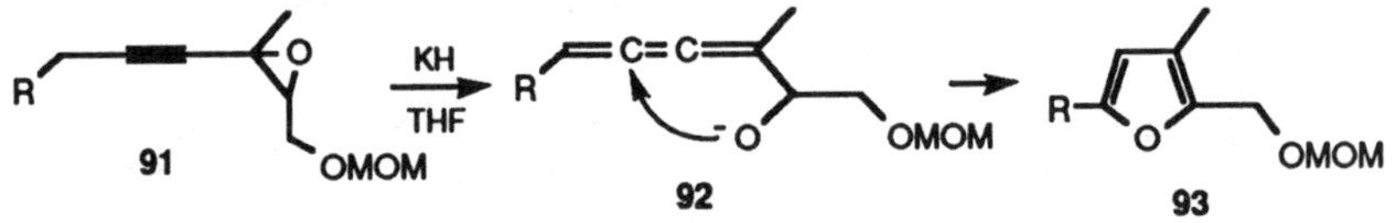

Finally, Abad and Arnó [91SL787] described a silica-gel variant of the Eschenmoser fragmentation reaction, providing an alternative to conventional methods.

3.2.2.3 Additions of Epoxides to Electrophiles

α-Phenylsulfonyl epoxides **94** have been lithiated with n-butyllithium at low temperatures. The α-sulfonyl anions so formed were captured with electrophiles such as allyl halides and ketones to give adducts **95** in good yield. Subsequent treatment with magnesium bromide etherate produced bromoketones **96** by a bromide ring-opening/phenylsulfonyl ejection sequence. The entire reaction sequence proceeds with complete regiochemical control [91PTI897]. Lithiated epoxides such as **98** also can be accessed from the organotin precursors **97** by tin-lithium exchange. In the presence of TMEDA, these anions react with aldehydes and ketones to give epoxy alcohols **99** in good yield [91TL615].

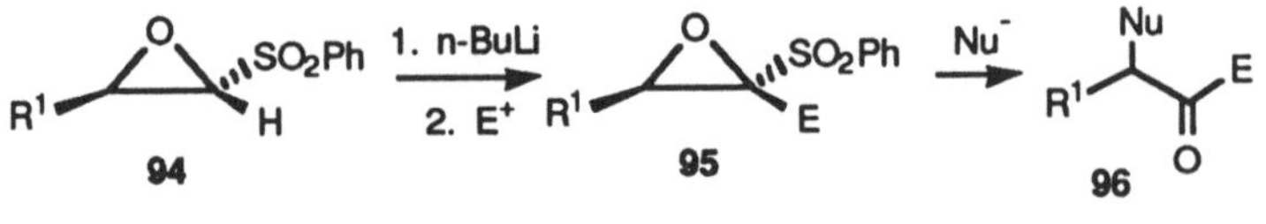

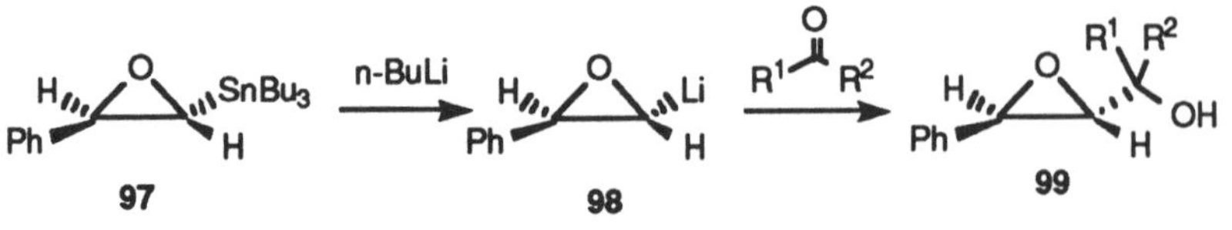

3.2.2.4 Deoxygenation of Epoxides

The net reversal of the epoxidation reaction, namely the eliminative deoxygenation of epoxides, has been carried out in a variety of ways. For example, tungsten reagents react with epoxides **100** to form tungsten (IV)-oxo complexes which ultimately lead to the corresponding olefins **101** with predominant retention of configuration [91JA870]. Silyl epoxides **102** are converted to vinyl silanes **104** upon treatment with excess organolithium reagents. The reaction is observed to be a predominantly Z-selective process for simple systems, but produces the E isomer in the cases of sterically demanding substrates [91TL3457]. Similar behavior was observed for the triphenylsilyl epoxide **105**. In this case, the epoxide is deprotonated with a stoichiometric amount of sec-butyllithium. Subsequent reaction with trimethylaluminum leads to deoxygenation with concomitant introduction of a methyl group, eventually affording vinyl silane **107** [91TL2783].

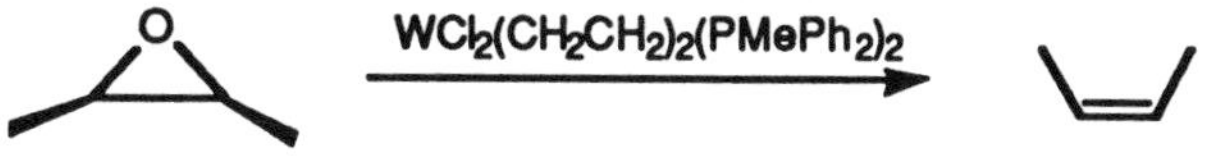

$WCl_2(CH_2CH_2)_2(PMePh_2)_2$

100 101

LiR² LiR²

102 103 104

s-BuLi Me_3Al

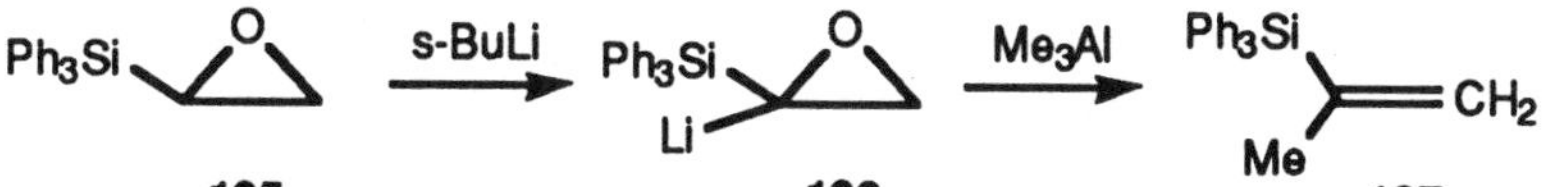

105 106 107

3.3 DIOXIRANES

Aside from the aforementioned utility of dioxiranes for the preparation of epoxides, these compounds are versatile oxidizing reagents for other transformations as well. For example, Guertin and Chan have used dimethyldioxirane to prepare α-hydroxy carbonyl compounds **110** in excellent yields by the direct oxidation of enolates [91TL715]. In addition, Maynard and Paquette have observed a remarkably facile conversion of the 3-methyl group of cyclopropene **111** to a carboxylic acid functionality in the presence of dimethyldioxirane [91JOC5480].

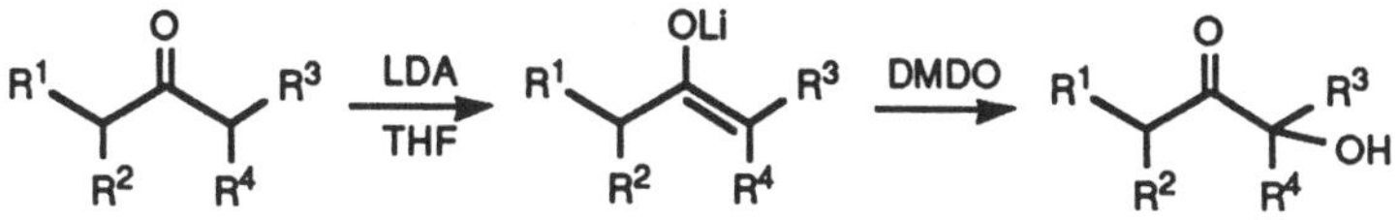

LDA THF DMDO

108 109 110

DMDO

111 112

3.4 OXAZIRIDINES

Chiral oxaziridines have found great utility as versatile asymmetric oxidizing agents. Davis and co-workers [91JOC809, 91JOC1143; 91TL867] have prepared chiral oxaziridines and demostrated their ability to effect the asymmetric oxidation of sulfides to sulfoxides (11-52% ee), enolates to α-hydroxy carbonyl compounds (11-81%), as well as the asymmetric epoxidation of non-functionalized alkenes (17-61%). An illustrative example involves the preparation of the rhodomycinone AB-synthon 115 (94% ee) via the asymmetric hydroxylation of tetralone **113** with the enantiomerically pure (+)-[(8,8-dimethoxycamphoryl)sulfonyl]oxaziridine **116**.

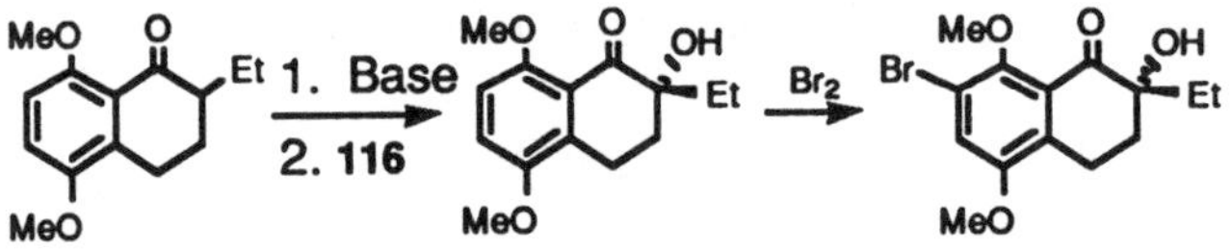

113 114 115

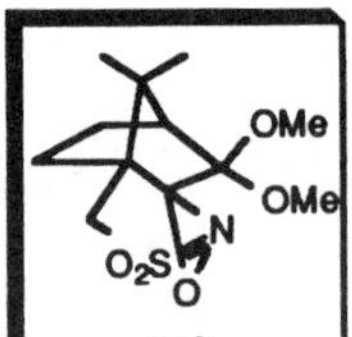

116

Oxaziridines also represent useful synthetic intermediates. Toward this end, Aubé and co-workers [91JOC499] have developed useful alternatives to the Beckmann rearrangement and Schmidt reaction through the intermediacy of spirocyclic oxaziridines of type 118. Photochemical rearrangement leads to the ring enlarged products (i.e., **119**). Optically active ketones can be converted to either lactam isomer by the proper choice of the chiral amine auxiliary.

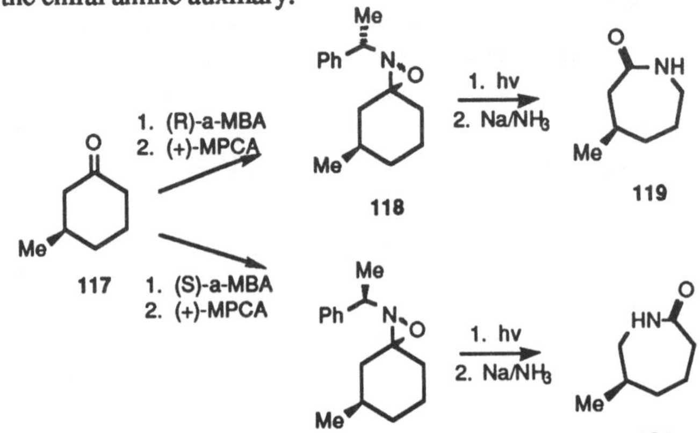

117 118 119 120 121

3.5 AZIRIDINES

3.5.1 Preparation of Aziridines

Both electron-rich and electron-poor olefins may be induced to undergo aziridination in the presence of copper (I) or copper (II) catalysts, in good to excellent yield [91JOC6744]. Aziridines are also available through the aminolysis of epoxides, followed by stereospecific ring closure of the resulting aminoalcohols **123** [91TL6935]. Finally, the addition of lithium enolates of 2-halo carboxylic esters **126** to N-trimethylsilyl imines **125** results in the formation of aziridine derivatives of type **127** with high *cis*-selectivity [91TL121].

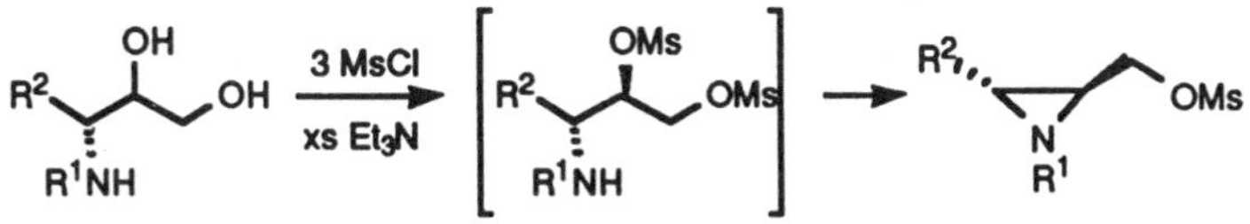

122 123 124

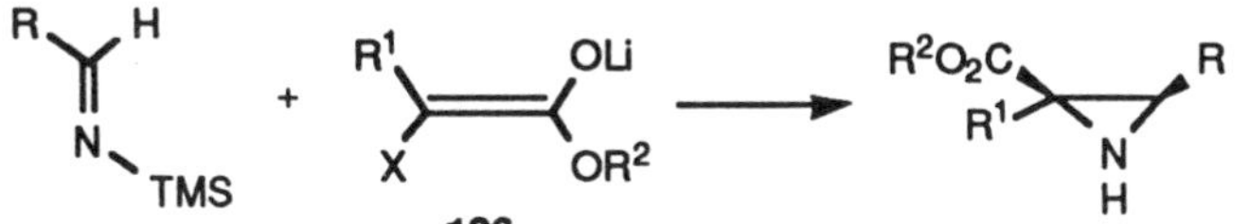

125 126 127

3.5.2 Reactions of Aziridines

Aziridines are known to undergo isomerization to the corresponding azomethine ylides, which have been utilized for further chemical transformations. Garner and co-workers [91JOC5893] have taken advantage of this behavior in their asymmetric approach to the 3,8-diazabicyclo[3.2.1]octane moiety of quinocarcin. The key step involves photolysis of bicyclic aziridine **128** to form the azomethine ylide **129**, which is then allowed to undergo 1,3-dipolar cycloaddition across a chiral acryloyl sultam to provide cycloadduct **130**, containing three new chiral centers.

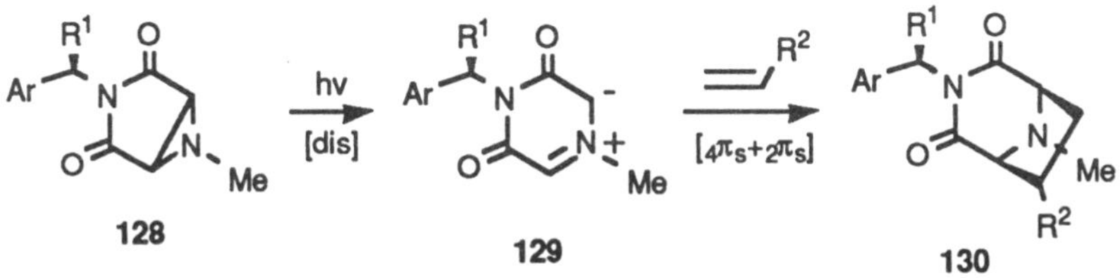

128 129 130

Sisko and Weinreb [91JOC3210] utilized an intramolecular variant of this reaction in a recently reported synthesis of the tricyclic core of the marine alkaloid sarain A. Thus, the thermolysis of aziridine **131** led to an azomethine ylide intermediate which was trapped stereospecifically by a pendant olefin to give the bicyclic lactam **132**.

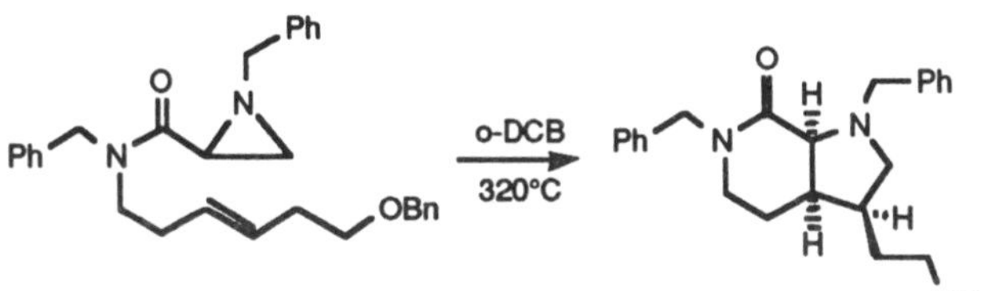

131 132

Certain vinylaziridines **133** have been shown to undergo highly stereoselective conversion to the ring-expanded vinyl azetidinone derivatives **135** in the presence of carbon monoxide and palladium (0) catalyst, *via* ring opening, carbonylation and ring closure [91SL91]. N-Arylsulfonylaziridines 136 undergo ring expansion to the corresponding azetidines **137** when treated with dimethyloxosulfonium methylide. The reaction is stereospecific and appears to be general, although it is not applicable to the synthesis of fused azetidines [91PTI2015].

Schaumann and co-workers [91TL1299] observed that treatment of arylsulfonyl aziridine **138** with n-butyllithium induced an unusual intramolecular nucleophilic aromatic addition reaction to give the tricyclic anion **139**, which was then quenched with an alkyl iodide producing **140**.

Aziridines exhibit interesting chemistry when exposed to nitrosating agents. Thus, the nitrosoaziridinium ion **142** undergoes ring expansion, ultimately forming azoxyalkene **145**. The overall process represents a new synthetic route to this growing class of antibiotic agents [91JA2308].

3.6 References

91AC403 C. Bolm, *Ang. Chem. Int. Ed. Engl.* **1991**, *30*, 403.

91BSJ1810 Y. Fukuda, S. Matsubara, C. Lambert, H. Shiragami, T. Nanko, K. Utimoto, and H. Nozaki, *Bull. Chem. Soc. Japan* **1991**, *64*, 1810.

91CB227 W. Adam, L. Hadjiarapoglou, and A. Smerz, *Chem. Ber.* **1991**, *124*, 227.

91CC297 P. Jankowski, S. Marczak, M. Masnyk, and J. Wicha, *J. Chem. Soc., Chem. Commun.* **1991**, 297.

91CC1064 U. Goergens and M.P. Schneider, *J. Chem Soc., Chem. Commun.* **1991**, 1064.

91HET879 V. Ratovelomanana, L. Vidal, J. Royer, and H.-P. Husson, *Heterocycles* **1991**, *32*, 879.

91JA106 S.S. Woodward, M.G. Finn, and K.B. Sharpless, *J. Am. Chem. Soc.* **1991**, *113*, 106.

91JA870 L.M. Atagi, D.E. Over, D.R. McAlister, and J.M. Mayer, *J. Am. Chem. Soc.* **1991**, *113*, 870.

91 JA2308 R.N. Loeppky, Q. Feng, A. Srinivasan, R. Glaser, C.L. Barnes, and P.R. Sharp, *J. Am. Chem. Soc.* **1991**, *113*, 2308.

91JA5106 S. Kim, S. Lee, and J.S. Koh, *J. Am. Chem. Soc.* **1991**, *113*, 5106.

91JA8005 W. Adam, L. Hadjiarapoglou, T. Mosandl, C.R. Saha-Möller, and D. Wild, *J. Am. Chem. Soc.* **1991**, *113*, 8005.

91JOC499 J. Aubé, M. Hammond, E. Gherardini, and F. Takusagawa, *J. Org. Chem.* **1991**, *56*, 499.

91JOC717 E. Schaumann, A. Kirschning, and F. Narges, *J. Org. Chem.* **1991**, *56*, 717.

91JOC809 F.A. Davis, R. ThimmaReddy, J.P. McCauley, Jr., R.M. Przeslawski, M.E. Harakal, and P.J. Carroll, *J. Org. Chem.* **1991**, *56*, 809.

91JOC1143 F.A. Davis, A. Kumar, and B.-C. Chen, *J. Org. Chem.* **1991**, *56*, 1143.

91JOC1153 J.K. Crandall, D.J. Batal, D.P. Sebesta, and F. Lin, *J. Org. Chem.* **1991**, *56*, 1153.

91JOC1685 J.A. Marshall and W.J. DuBay, *J. Org. Chem.* **1991**, *56*, 1685.

91JOC1976 W.H. Pearson and S.C. Bermeier, *J. Org. Chem.* **1991**, *56*, 1976.

91JOC2225 J.A. Marshall and B.E. Blough, *J. Org. Chem.* **1991**, *56*, 2225.

91JOC2296 W. Zhang and E.N. Jacobsen, *J. Org. Chem.* **1991**, *56*, 2296.

91JOC3210 J. Sisko and S.M. Weinreb, *J. Org. Chem.* **1991**, *56*, 3210.

91JOC4141 C.H. Fotsch and R.A. Chamberlin, *J. Org. Chem.* **1991**, *56*, 4141.

91JOC4513 V. Ceré, C. Paolucci, S. Pollicino, E. Sandri, and A. Fava, *J. Org. Chem.* **1991**, *56*, 4513.

91JOC4598 T. Hudlicky and G. Barbieri, *J. Org. Chem.* **1991**, *56*, 4598.

91JOC5480 G.D. Maynard and L.A. Paquette, *J. Org. Chem.* **1991**, *56*, 5480.

91JOC5893 P. Garner, W.B. Ho, S.K. Grandhee, W.J. Youngs, and V.O. Kennedy, *J. Org. Chem.* **1991**, *56*, 5893.

91JOC6206 C. Bonini, G. Righi, and G. Sotgiu, *J. Org. Chem.* **1991**, *56*, 6206.

91JOC6483 M. Miyashita, M. Hoshino, and A. Yoshikoshi, *J. Org. Chem.* **1991**, *56*, 6483.

91JOC6497 H. Fu, G.C. Look, W. Zhang, E.N. Jacobsen, and C.-H. Wong, *J. Org. Chem.* **1991**, *56*, 6497.

91JOC6744 D.A. Evans, M.M. Faul, and M.T. Bilodeau, *J. Org. Chem.* **1991**, *56*, 6744.

91PTI897 M. Ashwell, W. Clegg, and R.F.W. Jackson, *J. Chem. Soc., Perkin Trans. I* **1991**, 897.

91PTI2015 U.K. Nadir, R.L. Sharma, and V.K. Koul, *J. Chem. Soc., Perkin Trans. I*

	1991, 2015.
91PTI2277	D. Hobbs-Mallyon and D.A. Whiting, J. Chem. Soc., *Perkin Trans. I* 1991, 2277.
91SL91	G.W. Spears, K. Nakanishi, and Y. Ohfune, *Synlett* 1991, 91.
91SL97	Y. Niibo, T. Nakata, J. Otera, and H. Nozaki, *Synlett* 1991, 97.
91SL491	K. Maruoka, R. Bureau, T. Ooi, and H. Yamamoto, *Synlett* 1991, 491.
91SL691	N. Hosoya, R. Irie, Y. Ito, and T. Katsuki, *Synlett* 1991, 691.
91SL774	M. Hayashi, K. Kohmura, and N. Oguni, *Synlett* 1991, 774.
91SL787	A. Abad, C. Agulló, M. Arnó, M. Cuñat, and R.J. Zaregozá, *Synlett* 1991, 787.
91TET1291	A. Messeguer, F. Sánchez-Baeza, and J. Casas, *Tetrahedron* 1991, *47*, 1291.
91TET2991	I. Shimizu, K. Hayashi, N. Ide, and M. Oshima, *Tetrahedron* 1991, *47*, 2991.
91TET3075	A. Nuhrich and J. Moulines, *Tetrahedron* 1991, *47*, 3075.
91TL121	G. Cainelli, M. Panunzio, and D. Giacomini, *Tetrahedron Lett.* 1991, 121.
91TL533	B.A. Marples, J.P. Muxworthy, and K.H. Baggaley, *Tetrahedron Lett.* 1991, 533.
91TL615	P. Lohse, H. Loner, P. Acklin, F. Sternfeld, and A. Pfaltz, *Tetrahedron Lett.* 1991, 615.
91TL667.	S. Saito, N. Takahashi, T. Ishikawa, and T. Moriwake, *Tetrahedron Lett.* 1991, 667.
91TL715	K.R. Guertin and T.-H. Chan, *Tetrahedron Lett.* 1991, 715.
91TL867	F.A. Davis, A. Kumar, and B.-C. Chen, *Tetrahedron Lett.* 1991, 867.
91TL1041	W. Adam, D. Golsch, and L. Hadjiarapoglou, *Tetrahedron Lett.* 1991, 1041.
91TL1295	W. Adam, L. Hadjiarapoglou, and X. Wang, *Tetrahedron Lett.* 1991, 1295.
91TL1299	H.-J. Breternitz, E. Schaumann, and G. Adiwidjaja, *Tetrahedron Lett.* 1991, 1299.
91TL2637	L. Lambs, N.P. Singh, and J.-F. Biellmann, *Tetrahedron Lett.* 1991, 2637.
91TL2783	M. Taniguchi, K. Oshima, and K. Utimoto, *Tetrahedron Lett.* 1991, 2783.
91TL3021	J.S. Bajwa and R.C. Anderson, *Tetrahedron Lett.* 1991, 3021.
91TL3457	B. Santiago, C. Lopez, and J.A. Sonderquist, *Tetrahedron Lett.* 1991, 3457.
91TL5055	N.H. Lee, A.R. Muci, and E.N. Jacobsen, *Tetrahedron Lett.* 1991, 5055.
91TL6533	N.H. Lee and E.N. Jacobsen, *Tetrahedron Lett.* 1991, 6533.
91TL6575	S. Kim and S. Lee, *Tetrahedron Lett.* 1991, 6575.
91TL6617	M. Chini, P. Crotti, L. Favero, and F. Macchio, *Tetrahedron Lett.* 1991, 6617.
91TL6931	M. Canas, M. Poch, X. Verdaguer, A. Moyano, M.A. Pericás, and A. Riera, *Tetrahedron Lett.* 1991, 6931.
91TL6935	M. Poch, X. Verdaguer, A. Moyano, M.A. Pericás, and A. Riera, *Tetrahedron Lett.* 1991, 6935.

CHAPTER 4

Four-Membered Ring Systems

J. PARRICK AND L.K. MEHTA
Brunel University, Uxbridge, UK

4.1 INTRODUCTION

This review of recent work is selective and publications are chosen to illustrate the developments of special significance. Of necessity, many publications of merit are omitted. A review, noteworthy at this point, describes the synthesis of periannelated systems including (1) (90AHC1).

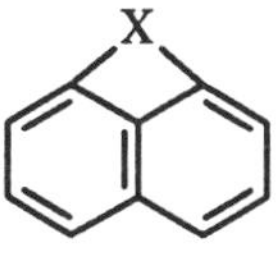

X = NH, O, S,
(1)

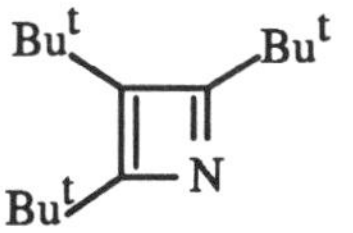

(2)

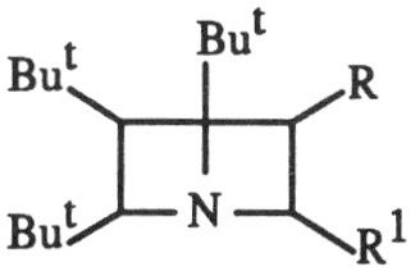

(3)

4.2 AZETES AND AZETIDINES

The preparation and properties of the kinetically stabilized tri-*tert*-butylazete (2) have been reviewed (91MI9). Studies of the addition of organic cyanides (90SL401) and moderately electron-rich alkynes (90CC1456) to the C=N bond of the azete to give pyrimidines and 1-Dewar-pyridine are available. In the case of the terminal alkynes, e.g. $HC{\equiv}COCH_2Ph$, a mixture of two regioisomers is formed, i.e. (3, R=H, R^1=OCH_2Ph and R=OCH_2Ph, R^1=H), but internal alkynes such as $MeC{\equiv}COCH_2Ph$ give only 3-oxy-Dewar-pyridine (3, R=OCH_2Ph, R^1=Me).

Photochemical methods have been used to cause dimerization of 1-azadiene (4) in the solid state to give mainly (5) as a result of topochemical control of the reaction process (91JOC6840), and to produce cyclization of the *N*-phenacylpyrrolidones (6) to yield the separable 1-azabicyclo[3.2.0]heptanes (7) and (8). These are readily converted to the azetidinols, e.g. (9) from (8), or their more stable *N*-acetyl derivatives (90T7751).

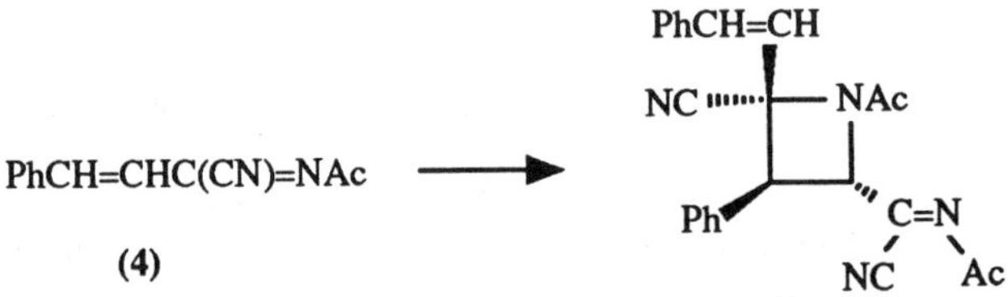

PhCH=CHC(CN)=NAc → (5)

(4) (5)

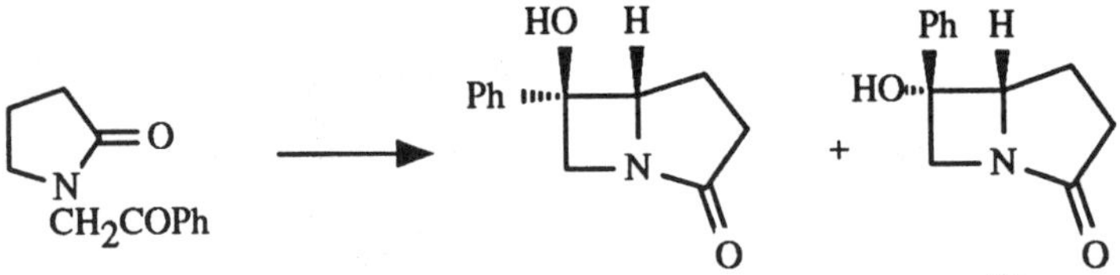

(6) → (7) + (8)

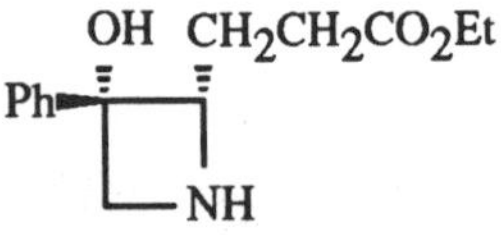

(9) 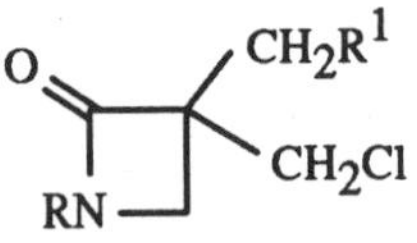 (10) 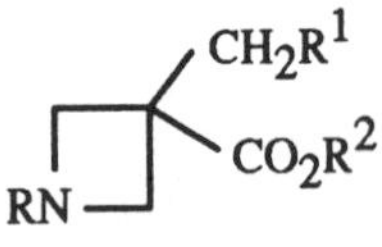 (11)

Other routes to azetidines include the highly selective reduction of β-lactams with diisobutylaluminium hydride (91JOC5263) and the base catalysed rearrangement of 3-(chloromethyl)azetidin-2-ones (10) to azetidine-3-carboxylic acid esters (11) (91TL4795).

A review of the synthesis and reactions of four-membered cyclic nitrones has appeared (90BSF704).

4.3 OXETANES, THIETANES AND DIHYDROPHOSPHETES,

Monocyclic oxetanes are obtained by base catalysed rearrangement of 2-(benzyloxymethyl)-3-methyloxirane (12) to yield mainly the oxetane (13) (90SC1777) or by treatment of oxiranes with oxosulfonium methylide (90MI83). Studies of photochemical addition of benzophenone to substituted ethenes bearing a 1-methylthio group show high regioselectivity to give (14) (91JCS(P1)865) and similar addition of allenic esters (e.g. $R_2C{=}C{=}CHCO_2Et$) gives spirooxetanes (15) (91T2211). A novel addition reaction, catalysed by crown ethers or cationic phase transfer catalysts, to yield (17) from the oxetane (16) and phenyl thioacetate is described (91CL697).

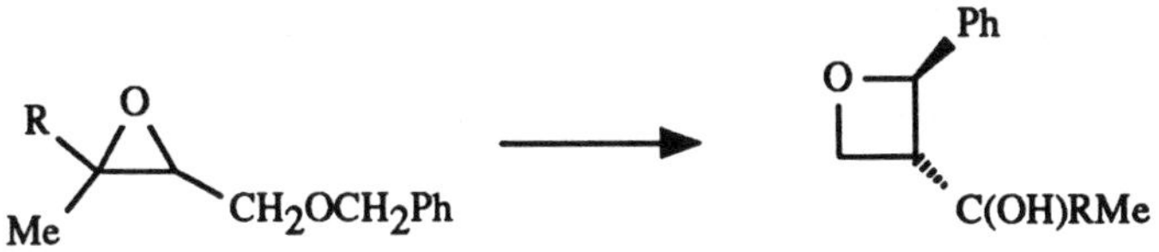

(12) (13)

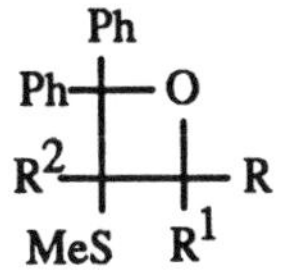

(14) 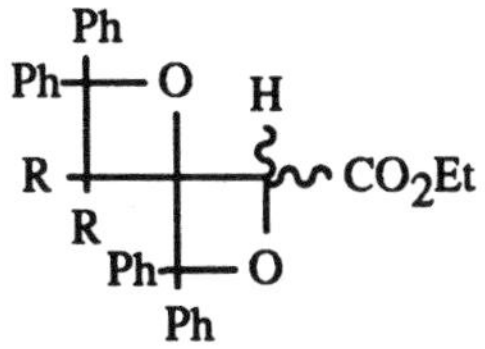 (15)

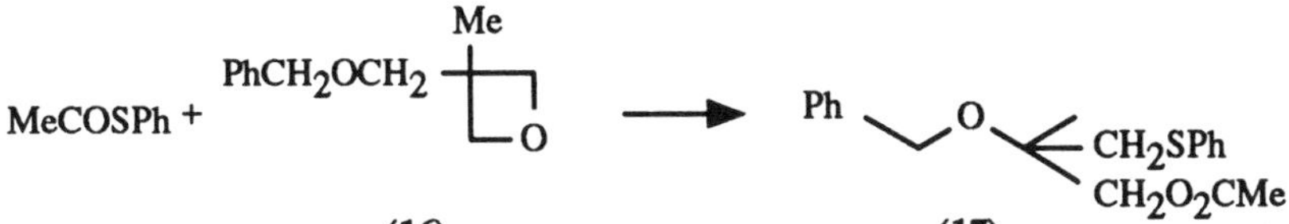

(16) (17)

The 3-methylene-2-oxetanone (β-lactone) (18, R=Ph) has been prepared (90S855) as have some 4-substituted derivatives of (18, R=H) (91JOC5778). The latter have been shown to serve as useful precursors of allenes by thermolysis (91JOC5782). The β-lactone antibiotic (19) has been synthesised (91LA1057) and a series of oxetane and azetidinone analogues have been prepared (91TL3337). The β-lactone, ebelactone A, (20) has been synthesised (90TL7513).

Treatment of 3,4-di-*tert*-butylthiophene 1,1-dioxide with peracid gave (22) apparently from the initially formed epoxide (21) by protonation, opening of the epoxide and rearrangement of the carbocation (91JOC4001). Thietan-2-ones have been identified as degradation products of the sodium salt of the antibiotics naficillin and oxacillin (90JCS(P2)1559).

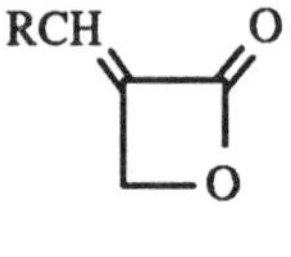

(18) 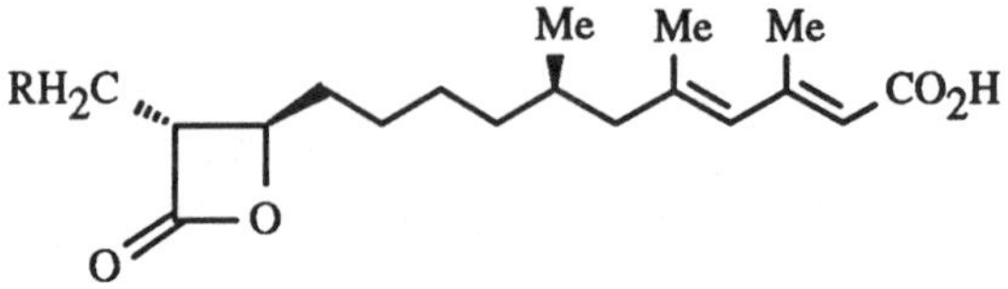 (19)

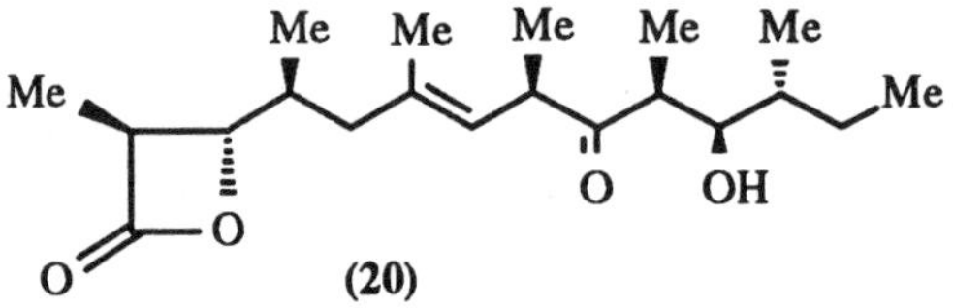

(20)

(21) (22)

4.4 DIOXETANES, DITHIETANES AND DIAZETE

1,2-Dioxetanes have been prepared from olefins by photooxygenation in polar solvents (90MI182) and, in good yields, by photooxygenation of benzofurans in the presence of tetraphenylporphyrin (91LA33). The benzofuran dioxetanes (e.g. 23) are strongly mutagenic and on photodeoxygenation in the presence of dimethylsulphide, yield benzofuran epoxides, e.g. (24) (91JA8005) (91AG187). Both 1,2- and 1,3-dioxetanes are formed on radiolysis of adamantanone (90MI2183).

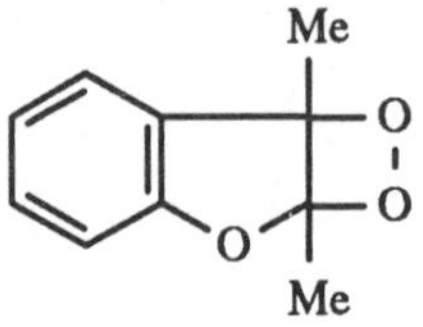

(23) 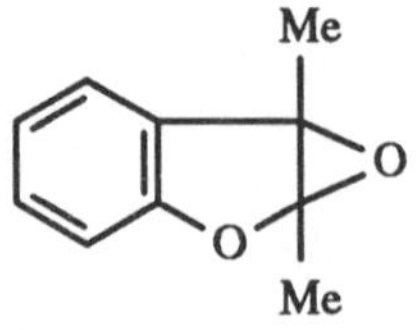 (24)

1,3-Dithietanes are formed on reaction of 1,1-ethylenedithiolates with either BrC≡CCOPh or Cl_2C=CHCOPh in polar solvents (90ZOR2097). Flash vacuum thermolysis of 1,3-dithietane (25) yields dichlorothioketene (26) which is trapped with cyclopentadiene to give the adduct (27) (91CB1485).

The 1,2-diazete *N*-oxide (29) has been obtained (59% yield) by ring expansion of the nitrosoaziridinium ion formed by nitrosation of (28) (91JA2308).

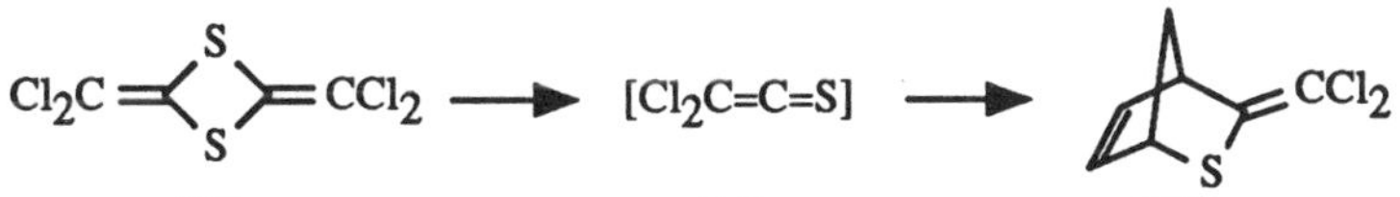

$Cl_2C{=}C(S)_2C{=}CCl_2$ → $[Cl_2C{=}C{=}S]$ → (27) CCl_2, S

(25) (26) (27)

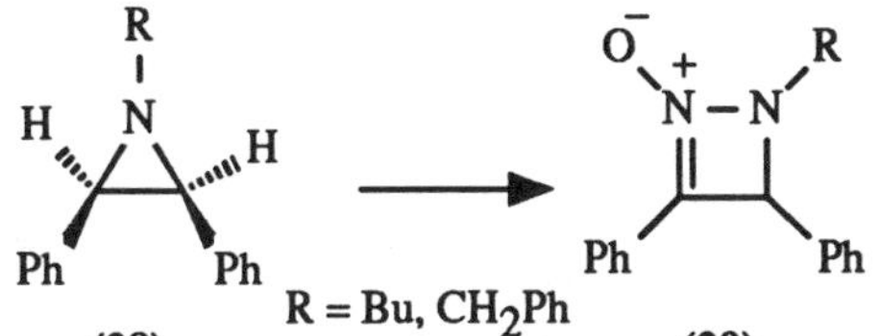

(28) R = Bu, CH_2Ph → (29)

4.5 OXAZETIDINES, OXATHIETIDINES AND THIAZETIDINES

N-Substituted α-hydroxyhydroxamic acids on treatment with 1,1′-carbonyldiimidazole yield cyclic carbonates which decompose with loss of CO_2 to give 1,2-oxazetidin-3-ones (90AP967).

The preparation and reaction of 1,2-oxathietidin-2,2-dioxides (β-sultones) and 1,2-thiazetidine-1,1-dioxides (β-sultams) have been reviewed (91MI789). Fluorinated β-sultones have been prepared from fluorinated vinyl ethers (90MI349). Unstable 2,3-dialkyl-β-sultams were obtained by treatment of β-alkylethene-sulphonic acid fluoride with aliphatic primary amines (91MI455). 4-Acylation and 4,4-dimethylation of *N*-silyl protected β-sultams, reduction of the ketone to the corresponding alcohol, and desilylation of the ring substituted products have been described (91LA171) (91AP15). *N*-Silylated β-sultams (30) on treatment with aldehydes or ketones (R^1COR^2) yielded (31). Similarly *N*-amino-alkylated 4-trimethylsilyl β-sultams reacted with R^1COR^2 to give (32) (91AP213). An effective method for the deprotection of *N*-(3,4-dimethoxybenzyl)-β-sultams has been described (91TL751).

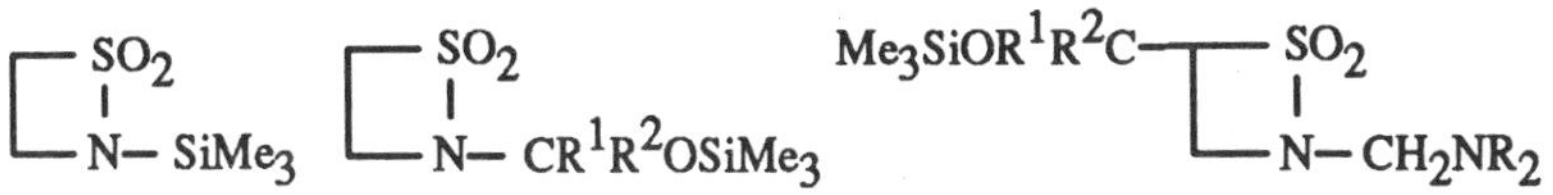

SO_2, N– $SiMe_3$ (30) SO_2, N– $CR^1R^2OSiMe_3$ (31) $Me_3SiOR^1R^2C$–, SO_2, N–CH_2NR_2 (32)

Novel *N*-alkylsulphonylamidines on treatment with LDA yielded 4*H*-thiazete 1,1-dioxide (33) among other products (91T1937). *N*-Benzoyl-*N*′-arylthiourea and diiodomethane gave 2-imino-1,3-thietanes (34), which, on treatment with *S*-methyl-thiourea gave substituted 1,3,5-triazines (91JHC177).

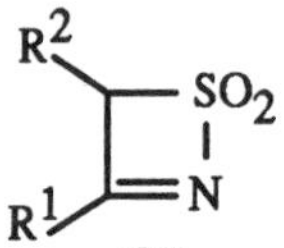

(33) 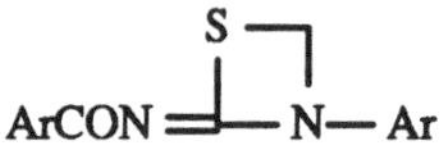 (34)

4.6 RINGS CONTAINING BORON, SILICON OR PHOSPHORUS

Cyclization of (35) with Na-K alloy gave the diboretane (36) in high yield (90ZN(B)1019). The heterocycles (37, X=O, S) were obtained by [2+2]cycloaddition of isocyanate or isothiocyanate (RNCX) to (dimethylamino)bis(trifluoromethyl)boranes. In some cases these heterocycles rearranged at 60°C to the iminiumazaborataoxacyclobutane (38) (91JOM283).

$(Pr^i_2NBClCH_2)_2$ ⟶ (36)

(35) (36)

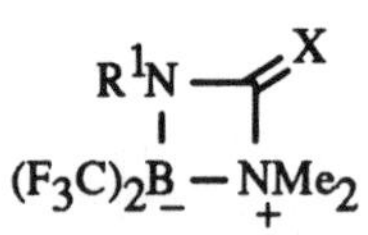

(37) 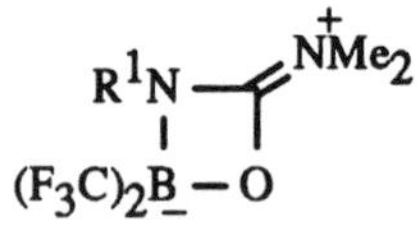 (38)

Trichlorosilylethene is used in the efficient and simple synthesis of silacyclobutenes (39) (91AG1172) and intramolecular bis-silylation of a carbon-carbon double bond yields the silacyclobutane (40) (91JA3987). A 1-chloro substituent on a silacyclobutane (41, R=Cl) can be utilized to introduce an ethyne substituent (41, R = $C{\equiv}CSiMe_3$) (90MI1423) while treatment with a base of a 1-oxiranylsilacyclobutane causes a rearrangement to a silacyclopentane (91TL4545). Stereoselective reaction of lithium carbenoids with silacyclobutanes also yields silacyclopentanes (90TL6055). The relative reactivity of the two C-Si bonds in (42, R=Me, X=CH) in reactions with methanol has been studied (91MI57) and the reactions of 3,4-benzo-1,1-dimethyldisilacyclobutene (42, R=Me, X=$SiEt_2$) in the presence of dienophiles has been shown to proceed through [4+2]cycloaddition reactions of *o*-quinodisilanes (91MI3173).

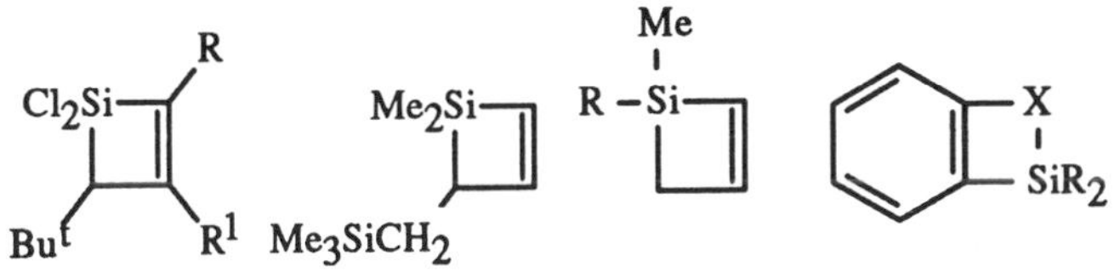

(39) (40) (41) (42)

Calculations are reported for the structure and inversion barrier of 1,2-dihydrophosphete (43, R=H) (91JOC2205). The addition of 1,3,4-triphenyl-1,2-dihydrophosphete (43, R=Ph) to dimethyl maleate occurs by an ionic mechanism, in contrast to the concerted process for the aza analogues, and gives (44) and (45) as the kinetic and thermodynamic products, respectively (90CC1649).

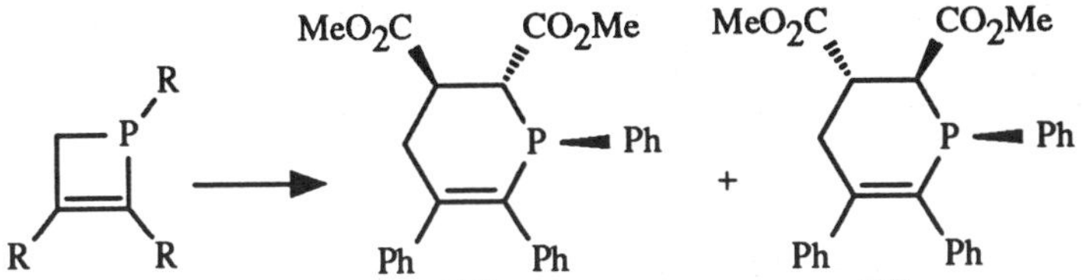

(43) (44) (45)

4.7 AZETIDIN-2-ONES (β-LACTAMS)

Reviews have appeared of routes to enantiomerically pure, monocyclic azetidinones (90MI533), intermediates of use in the synthesis of β-lactam antibiotics (91MI294), and of the synthesis of hydroxyethylazetidinones useful in the preparation of penem-type antibiotics (90MI616) (90MI620).

The [2+2]cycloaddition reactions between imines and ketenes is described and the constraints necessary to obtain stereoselective synthesis of *cis*-3,4-disubstituted azetidinones, where the 4-substituent is formyl (91TL803), 2-phenylethenyl (90TL6707) (91JOC4418), alkylthio (91T6759) or aryl (91TL581) (91JOC6118), are discussed. Chiral ketenes derived from carbohydrates provide *cis*-3-hydroxy-4-(2-phenylethenyl)azetidinones (91TL1039) and chiral imines from *N*-protected α-aminoaldehydes yield azetidinones with complete diastereoselection (91TL3109). The diastereoselective synthesis of *bis*-β-lactams is reported (91T5379).

A novel enantioselective synthesis utilises optically pure *o*-substituted benzylideneanilinetricarbonylchromium complexes in addition reactions with ester enolates to afford β-lactams (ee>98%) (91CC982). The reaction of *N,N*-diprotected glycine enolates with a range of substituted imines (91JOC5868) and glycoaldehyde imines

(91JOC1933) is reported. The factors affecting the stereoselectivity of zinc enolate and imine addition are discussed in some detail (91JOC5147) and the merits of organocopper enolates in similar reactions are described (90CC1390).

Recent contributions to the formation of β-lactams by cyclization reactions include the ring closure of β-aminoacids with methanesulfonyl chloride and bicarbonate (91TL2299), phenyl phosphodichloridate (91JOC2244), other chloro derivatives of phosphorus acids (90TL2905) and other derivatives of phosphorus (91MI456) (91MI457). L-Glutamic acid derivatives (obtained from *O*-protected 5-(hydroxymethyl)pyrrolidin-2-one) are used in a diastereoselective synthesis of (46) (91TL283). Cyclization of substituted ornithine using the Ohno procedure (di-2-pyridyl-disulphide and triphenylphosphine) yields proclayaminic acid which is expected to be (47) from the synthetic route (90JCS(P1)1521).

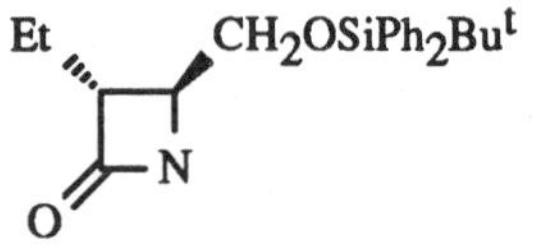

(46) 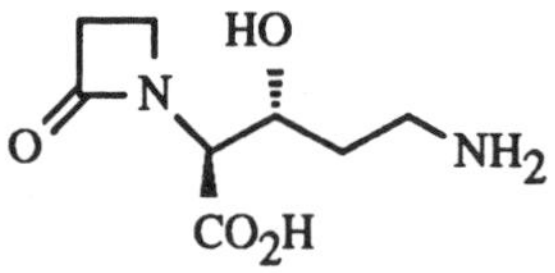 **(47)** 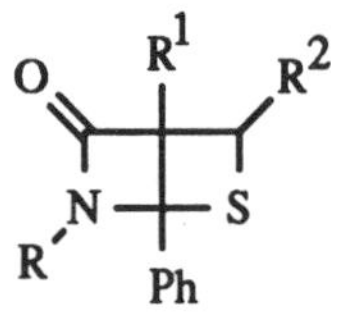 **(48)**

The readily available thioamides ($PhCSNRCOCR^1{=}CHR^2$) and ($Me_2C{=}CPhCSNHCH_2Ph$) undergo photochemical cyclization to yield the bicyclic β-lactam (48) (91JCS(P1)403) and the thione (49) (90CC1214) respectively. Suitably substituted 3-hydroxy-hydroxamic acid *O*-ethers undergo Mitsunobu cyclization to yield (50) (90TL4317) and (51) (91T1137).

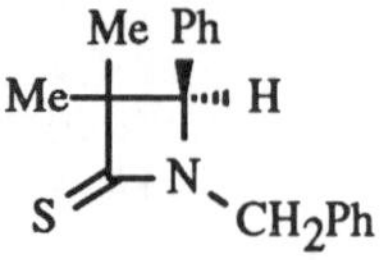

(49) 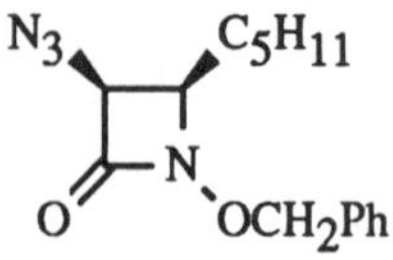 **(50)** 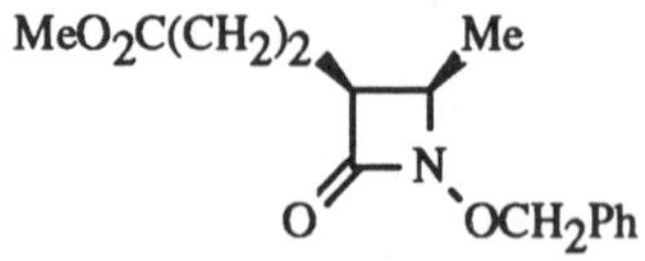 **(51)**

Further development of the free radical approach to the formation of the 2,3-bond in azetidinones is reported (91TL259) in a new synthesis of thienamycin using a thermal 4-*exo*-trig cyclization of the hexenylcarbamoylcobalt salophen (52) to give exclusively the *trans* product (53) in 40%yield. Tributyltin hydride may be used for the efficient cyclization of α-bromoacid enamides (91TL2335).

[Co] = Cobalt(III) salophen

(52) **(53)**

The introduction of a 4-acetoxy substituent on to a β-lactam by the action of peracetic acid is catalysed by osmium trichloride (91TL2145) or ruthenium (90JA7820). Such an acetoxy substituent may be converted to a trimethylsilyl (54, R=$SiMe_3$) or tributylstannyl (54, R=$SnBu_3$) group, and the latter substituent may be subsquently replaced by an acyl group (90TL2637). Other reactions of 4-substituents include the substitution of sulphinyl by heteronucleophiles in the presence of zinc iodide to give trans products in high yield (91TL2375); oxidative decarboxylation to yield a benzoyloxy substituent (90S691); and the oxidation of 4-hydroxymethyl to formyl by bis(trichloromethyl)carbonate in DMSO (91JOC5948). Interestingly, the tricyclic azetidinone (55) undergoes solvolysis by opening of the 1,4-azetidinone bond (91MI204).

Azetidin-2,3-diones provide useful starting materials for the synthesis of 3-alkylidene β-lactams (90TL6425). Certain (±)-*cis*-3-amino-4-substituted β-lactams are enantioselectively *N*-acetylated by methyl phenylacetate or methyl phenoxyacetate in the presence of Penicillin G amidase to give, for instance (56, R=Ph or PhO) (91TL1621).

(54) **(55)** **(56)**

REFERENCES

90AHC1 V. V. Mezheritskii and V. V. Tkachenko; *Adv. Heterocycl. Chem.*, 1990, **51**, 1.

90AP967 A. Burchardt and D. Geffken; *Arch. Pharm. (Weinheim, Ger.).*, 1990, **323**, 967.

90BSF704	W. Verboom and D. N. Reinhoudt; *Bull. Soc. Chim. Fr.*, 1990, 704.
90CC1214	M. Sakamoto, M. Kimura, T. Shimoto, T. Fujita and S. Watanabe; *J. Chem. Soc., Chem. Commun.*, 1990, 1214.
90CC1390	C. Palomo, J. M. Aizpurua and R. Urchegui; *J. Chem. Soc., Chem. Commun.*, 1990, 1390.
90CC1456	G. Mass, M. Regitz, R. Rahm, J. Schneider, P. J. Stang and C. M. Crittel; *J. Chem. Soc., Chem. Commun.*, 1990, 1456.
90CC1649	K. M. Doxsee, G. S. Shen and C. B. Knobler; *J. Chem. Soc., Chem. Commun.*, 1990, 1649.
90JA7820	S. Murahashi, T. Naota, T. Kuwabara, T. Saito, H. Kumobayashi and S. Akutagawa; *J. Am. Chem. Soc.*, 1990, **112**, 7820.
90JCS(P1)1521	K. H. Baggaley, S. W. Elson, N. H. Nicholson and J. T. Sime; *J. Chem. Soc., Perkin Trans. 1*, 1990, 1521.
90JCS(P2)1559	K. A. Ashline, R. P. Attrill, E. K. Chess, J. P. Clayton, E. A. Cutmore, J. R. Everett, J. H. C. Nayler, D. E. Pereira, W. J. Smith III, J. W. Tyler, M. L. Vieira and M. Sabat; *J. Chem. Soc., Perkin Trans. 2*, 1990, 1559.
90MI83	K. Okuma, Y. Tanaka, K. Nakamura and H.Ohta; *Fukuoka Daigaku Rigaku Shuho,* 1990, **20**, 83.
90MI182	Z. Huang, X. Liang and Y. Chan; *Chin. J. Chem.*, 1990, 182.
90MI349	J. Mohtasham, F. E. Behr and G. L. Gard; *J. Fluorine Chem.*, 1990, **49**, 349.
90MI533	R. C. Thomas in *"Recent Progress in the Chemistry and the Synthesis of Antibiotics", eds. G. G. Lukacs and M. Ohno, Springer, Berlin,* 1990, 533.
90MI616	G. Sedelmeir; *Nachr. Chem. Tech. Lab.*, 1990, **38**, 616.
90MI620	G. Sedelmeir; *Nachr. Chem. Tech. Lab.*, 1990, **38**, 620.
90MI1423	M. G. Voronkov, O. G. Yarosh, V. K. Roman and A. I. Albanov; *Metalloorg. Khim.*, 1990, **3**, 1423.
90MI2183	G. L. Sharipov, A. I. Voloshin, L. M. Khalilov, V. P. Kazakov and G. A. Tolstikov; *Izv. Akad. Nauk SSSR, Ser. Khim.*, 1990, 2183.
90S691	M. Shiozaki; *Synthesis*, 1990, 691.
90S855	E. M. Campi, K. Dyall, G. Fallon, W. R. Jackson, P. Perlmutter and A. J. Smallridge; *Synthesis*, 1990, 855.
90SC1777	C. W. Bird and N. Hormozi; *Synth. Commun.*, 1990, **20**, 1777.
90SL401	U. Hees, M. Ledermann and M. Regitz; *Synlett*, 1990, 401.
90T7751	L. Quazzani-Chadi, J. C. Quirion, Y. Troin and J. C. Gramain; *Tetrahedron*, 1990, **46**, 7751.
90TL2637	C. Nativi, A. Ricci and M. Taddei; *Tetrahedron Lett.*, 1990, **31**, 2637.

90TL2905	C. W. Kim, B. Y. Chung, J. Y. Namkung, J. M. Lee and S. Kim; *Tetrahedron Lett.*, 1990, **31**, 2905.
90TL4317	B. M. Kim and K. B. Sharpless; *Tetrahedron Lett.*, 1990, **31**, 4317.
90TL6055	K. Matsumoto, K. Oshima and K. Utimoto; *Tetrahedron Lett.*, 1990, **31**, 6055.
90TL6425	C. Palomo, J. M. Aizpurua, M. C. Lopez, N. Aurrekoetxea and M. Oiarbide; *Tetrahedron Lett.*, 1990, **31**, 6425.
90TL6707	T. E. Gunda, S. Vieth, K. E. Kover and F. Sztaricskai; *Tetrahedron Lett.*, 1990, **31**, 6707.
90TL7513	I. Paterson and A. N. Hulme; *Tetrahedron Lett.*, 1990, **31**, 7513.
90ZN(B)1019	A. Kraemer, J. K. Uhm, S. E. Garner, H. Pritzkow and W. Siebert; *Z. Naturforsch., B: Chem. Sci.*, 1990, **45**, 1019.
90ZOR2097	E. N. Komarova, L. B. Dmitriev and V. N. Drozd; *Zh. Org. Khim.*, 1990, **26**, 2097.
91AG187	W. Adam, L. Hadjiarapoglou, T. Mosandl, C. R. Saha-Moeller and D. Wild; *Angew. Chem.*, 1991, **103**, 187.
91AG1172	N. Auner, C. Seindenscwarz and E. Herdtweck; *Angew. Chem.*, 1991, **103**, 1172.
91AP15	M. Mueller and H. H. Otto; *Arch. Pharm. (Weinheim, Ger.)*, 1991, **324**, 15.
91AP213	M. Mueller and H. H. Hartwig; *Arch. Pharm. (Weinheim, Ger.)*, 1991, **324**, 213.
91CB1485	G. Adiwidjaja, C. Kirsch, F. Pedersen, E. Schaumann, G. C. Schmerse and A. Senning; *Chem. Ber.*, 1991, **124**, 1485.
91CC982	C. Baldoli and P. Del Buttero; *J. Chem. Soc., Chem. Commun.*, 1991, 982.
91CL697	T. Nishikubo and K. Sato; *Chem. Lett.*, 1991, 697.
91JA2308	R. N. Loeppky, Q. Feng, A. Srinivasan, R. Glaser, C. L. Barnes and P. R. Sharp; *J. Am. Chem. Soc.*, 1991, **113**, 2308.
91JA3987	M. Murakami, P. G. Andersson, M. Suginome and Y. Ito; *J. Am. Chem. Soc.*, 1991, **113**, 3987.
91JA8005	W. Adam. L. Hadjiarapoglou, T. Mosandl, C. R. Saha-Moeller and D. Wild; *J. Am. Chem. Soc.*, 1991, **113**, 8005.
91JCS(P1)403	M. Sakamoto, T. Yanase, T. Fujita, S. Watanabe, H. Aoyama and Y. Omote; *J. Chem. Soc., Perkin Trans. 1*, 1991, 403.
91JCS(P1)865	N. Khan, T. H. Morris, E. H. Smith and R. Walsh; *J. Chem. Soc., Perkin Trans. 1*, 1991, 865.
91JHC177	N. Okajima and Y.Okada; *J. Heterocycl. Chem.*, 1991, **28**, 177.
91JOC1933	M. J. Brown and L. E. Overman; *J. Org. Chem.*, 1991, **56**, 1933.
91JOC2205	S. M. Bachrach; *J. Org. Chem.*, 1991, **56**, 2205.

91JOC2244	C. Palomo, J. M. Aizpurua, R. Urchegui, M. Iturburu, A. Ochoa de Retana and C. Cuevas; *J. Org. Chem.*, 1991, **56**, 2244.
91JOC4001	J. Nakayama and Y. Sugihara; *J. Org. Chem.*, 1991, **56**, 4001.
91JOC4418	C. Palomo, F. P. Cossio, J. M. Odiozola, M. Oiarbide and J. M. Ontoria; *J. Org. Chem.*, 1991, **56**, 4418.
91JOC5147	F. H. Van der Steen, H. Kleijn, J. T. B. H. Jastrzebski and G. Van Koten; *J. Org. Chem.*, 1991, **56**, 5147.
91JOC5263	I. Ojima, M. Zhao, T. Yamato, K. Nakahashi, M. Yamashita and R. Abe; *J. Org. Chem.*, 1991, **56**, 5263.
91JOC5778	W. Adams, R. Albert, N. Dachs Grau, L. Hasemann, B. Nestler, E. M. Peters, K. Peters, F. Prechtl and H. G. Von-Schnering; *J. Org. Chem.*, 1991, **56**, 5778.
91JOC5782	W. Adam, R. Albert, L. Hasemann, V. O. Nava Salgado, B. Nestler, E. M. Peters, K. Peters, F. Prechtl and H. G. Von-Schnering; *J. Org. Chem.*, 1991, **56**, 5782.
91JOC5868	F. H. Van der Steen, H. Kleijn and A. L. Spek; *J. Org. Chem.*, 1991, **56**, 5868.
91JOC5948	C. Palomo, F. P. Cossio, J. M. Ontoria and J. M. Odriozola; *J. Org. Chem.*, 1991, **56**, 5948.
91JOC6118	W. T. Brady and M. M. Dad; *J. Org. Chem.*, 1991, **56**, 6118.
91JOC6840	M. Teng, J. W. Lauher and F. W. Fowler; *J. Org. Chem.*, 1991, **56**, 6840.
91JOM283	A. Ansorge, D. J. Brauer, H. Buerger, F. Doerrenbach, T. Hagen, G. Pawelke and W. Weuter; *J. Organomet. Chem.*, 1991, **407**, 283.
91LA33	W. Adam, O. Albrecht, E. Feineis, I. Reuther, C. R. Saha-Moeller, P. Seufert-Baumbach and D. Wild; *Leibigs Ann. Chem.*, 1991, 33.
91LA171	M. Mueller and H. H. Otto; *Liebigs Ann. Chem.*, 1991, 171.
91LA1057	K. Mori and Y. Takahashi; *Liebigs Ann. Chem.*, 1991, 1057.
91MI9	M. Regitz; *Nachr. Chem. Tech. Lab.*, 1991, **39**, 9.
91MI57	K. T. Kang, U. C. Yoon, H. C. Seo, K. N. Kim, H. Y. Song and J. C. Lee; *Bull. Korean Chem. Soc.*, 1991, **12**, 57.
91MI204	C. Nisole, P. Uriac, J. Huet and J. Toupet; *J. Chem. Res., Synop.*, 1991, 204.
91MI294	C. H. Frydrych; *Amino Acids Rept.*, 1991, **22**, 294.
91MI455	J. Chanet-Ray, N. Lahbabi and R. Vessiere; *C. R. l'Academie Sci., Ser. II Univers.*, 1991, **312**, 455.
91MI456	B. Y. Chung, K. C. Paik and C. S. Nah; *Bull. Korean Chem. Soc.*, 1991, **12**, 456.
91MI457	B. Y. Chung, W. Goh and C. S. Nah; *Bull. Korean Chem. Soc.*, 1991, **12**, 457.
91MI789	A. J. Buglass and J. G. Tillett in *"The Chemistry of Sulphonic Acids, Esters and Their Derivatives", eds. S. Patai and Z.*

Rappoport, Wiley, Chichester and New York, 1991, 789.

91MI3173 M. Ishikawa, H. Sakamoto and T. Tabuchi; *Organometallics,* 1991, **10,** 3173.

91T1137 M. Kahn and K. Fujta; *Tetrahedron,* 1991, **47,** 1137.

91T1937 F. Clerici, D. Pocar and A. Rozzi; *Tetrahedron,* 1991, **47,** 1937.

91T2211 M. P. S. Ishar and R. P. Gandhi; *Tetrahedron,* 1991, **47,** 2211.

91T5379 A. K. Bose, J. F. Womelsdorf, L. Krishnan, Z. Urbanczyk-Lipkowska, D. C. Shelly and M. S. Manhas; *Tetrahedron,* 1991, **47,** 5379.

91T6759 E. Grochowski and K. Pupek; *Tetrahedron,* 1991, **47,** 6759.

91TL259 G. Pattenden and S. J. Reynolds; *Tetrahedron Lett.,* 1991, **32,** 259.

91TL283 P. Somfai, H. M. He and D. Tanner; *Tetrahedron Lett.,* 1991, **32,** 283.

91TL581 G. I. Georg, P. M. Mashava and X. Guan; *Tetrahedron Lett.,* 1991, **32,** 581.

91TL751 E. Grunder-Klotz and J. D. Ehrhardt; *Tetrahedron Lett.,* 1991, **32,** 751.

91TL803 B. Alcaide, Y. Martin-Cantalejo, J. Plumet, J. Rodriguez-Lopez and M. A. Sierra; *Tetrahedron Lett.,* 1991, **32,** 803.

91TL1039 B. C. Borer and D. W. Balogh; *Tetrahedron Lett.,* 1991, **32,** 1039.

91TL1621 M. J. Zmijewski Jr., B. S. Briggs, A. R. Thompson and I. G. Wright; *Tetrahedron Lett.,* 1991, **32,** 1621.

91TL2145 S. Murahashi, T. Saito, T. Naota, H. Kumobayashi and S. Akutagawa; *Tetrahedron Lett.,* 1991, **32,** 2145.

91TL2299 M. F. Loewe, R. J. Cvetovich and G. G. Hazen; *Tetrahedron Lett.,* 1991, **32,** 2299.

91TL2335 S. L. Fremont, J. L. Belletire and D. M. Ho; *Tetrahedron Lett.,* 1991, **32,** 2335.

91TL2375 Y. Kita, N. Shibata, N. Yoshida and T. Tohjo; *Tetrahedron Lett.,* 1991, **32,** 2375.

91TL3109 C. Palomo, F. P. Cossio and C. Cuevas; *Tetrahedron Lett.,* 1991, **32,** 3109.

91TL3337 K. L. Thompson, M. N. Chang, Y. P. Chiang, S. S. Yang, J. C. Chabala, B. Y. Arison, M. D. Greenspan, D. P. Hanf and J. Yudkovitz; *Tetrahedron Lett.,* 1991, **32,** 3337.

91TL4545 K. Matsumoto, Y. Takeyama and K. Oshima; *Tetrahedron Lett.,* 1991, **32,** 4545.

91TL4795 D. Bartholomew and M. J. Stocks; *Tetrahedron Lett.,* 1991, **32,** 4795.

CHAPTER 5.1

Five-Membered Ring Systems: Thiophenes & Se & Te Analogs

JEFFERY B. PRESS & RONALD K. RUSSELL
The R. W. Johnson Pharmaceutical Research Institute, Spring House, PA and Raritan, NJ, USA

5.1.1 INTRODUCTION

Utilization of thiophenes increased in intensity in 1991 as evidenced by nearly twice as many references uncovered in our literature search this year as compared to last. New synthesis strategies as well as greater understanding of the electronic properties of thiophene have been the prime reasons for this increase in publication activity. As usual, there are very few reports of selenophenes and tellurophenes.

The organization of this review follows that of previous years. The aromaticity and electronic character of thiophene are discussed first which leads to ring substitution reactions. Thiophene ring formation, annelation of rings onto thiophene and the use of thiophenes as intermediates to non-thiophene molecules follow. Thiophene oligomers have unique electroconductive properties and many interesting molecules have been synthesized for this purpose, especially in the search for new superconductors. Selenophene and tellurophene syntheses are then discussed. Instead of the biologically active derivatives highlighted last year, we have added some interesting thiophene derivatives that are difficult to categorize in any of the other areas. There is overlap among these topics and we have chosen an organization that best illustrates the utility of thiophene for each report.

The nature of the deadlines for these reviews necessitates performing the literature searches prior to the complete indexing of all of the year's publications. Consequently, as in the past, we include not only 1991 references but also those from the latter part of 1990.

5.1.2 AROMATICITY, CONFORMATION AND STABILIZATION

Theoretical studies continue to examine the thiophene nucleus as the most aromatic of the 5-membered heteroaromatic rings. Calculated thermodynamic properties of benzo[*b*]thiophenes are comparable to literature values <91JCT759>. Computer-assisted methods to predict boiling points of furans and thiophenes also have good correlation to measured properties <91MI301>. Some interpretations of the vibrational overtone spectra of 5-membered heterocycles have shown thiophene to be similar to benzene <91JPC7659>. Semi-empirical MO calculations using SINDO1 suggest that the d orbitals of sulfur influence the photoisomerization mechanism of substituted thiophenes <91JOC129>. The surface chemistry of thiophenes and other 5-membered heterocycles on Pt(III) electrodes shows that thiophene forms a vertically oriented mixed layer which oxidizes to CO_2 and SO_4^{2-} <91MI37>. Electrochemical stability of fused thiophene rings varies as shown by reduction of thieno[2,3-*b*]pyrazines which only leads to reduction of the pyrazine ring while the isomeric thieno[3,4-*b*]-system gives a dihydrothiophene derivative <91JOC4840>.

Issues surrounding the theoretically interesting dimethylenethiophenes remain under study. Two precursors to 2,3-bis(methylene)-2,3-dihydrothiophene (**1**) include cyclic sulfone **2** which loses SO_2 at 130°C <91CC1287> and trimethylsilyl derivative **3** which produces **1** upon treatment with fluoride <90S915>. The unique spiro dimer **4** forms through the intermediacy of the 5-carboxylate derivative of **1** <91JOC6948>. Studies of transient signals in ^{13}C NMR spectra show that the isomeric biradical 3,4-dimethylenethiophene forms in rigid matrices at 77°K and exhibits an intense purple color which disappears upon warming <91JA2318>.

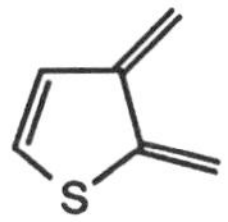

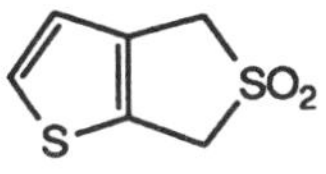

Very detailed ^1H and ^{13}C NMR studies show that the structure of tris(2-thienyl)methyl cation differs dramatically from the 3-thiophene isomer or phenyl isostere. Possible through-space charge delocalization by the three proximate sulfur atoms may account for these differences <91JOC3224>. A stable dication of 1,3,4,6-tetrakis(isopropylthio)-2-λ^4,δ^2-thieno[3,4-*c*]thiophene (**5**), which forms by nitrosyl fluorborate oxidation, reacts with a variety of nucleophiles to produce interesting mono- and bicyclic products <91CC520>. Reactions of the *t*-butyl analogue of **5** also lead to a variety of interesting products <91JCS(P1)909>.

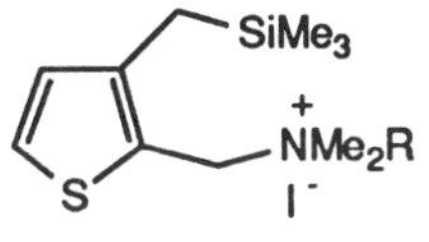

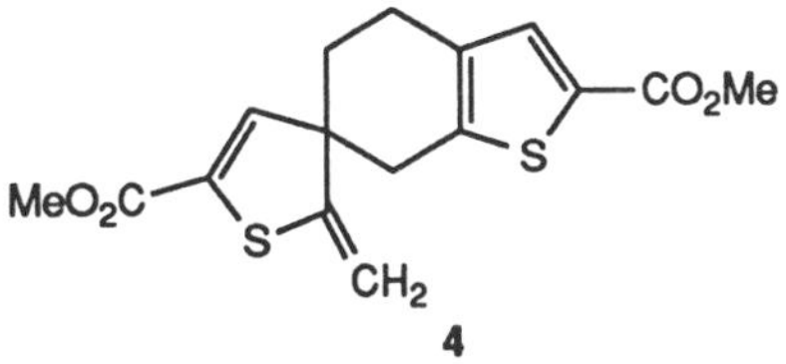

Some very detailed conformational investigations of 2-methylsulfinyl thiophene show that the S,O-*cis* conformer is preferred in solution <91MI81>. Studies using lanthanide-induced shifts also show these conformational preferences <91JCS(P2)269>. Halogenated derivatives of 2- and 3-methylsulfinylthiophene have varying conformational preferences as determined by x-ray diffraction <91MI99>. MO calculations of ground states and transition states for 2- and 3-substituted thiophenes and furans predict that solvent polarity affects these conformers <91MI71>. In a somewhat related study, ESR spectra of radical anions of furan- and thiophene-2-thioaldehydes show a conformational preference similar to that of the neutral parent compounds <91JOC6337>.

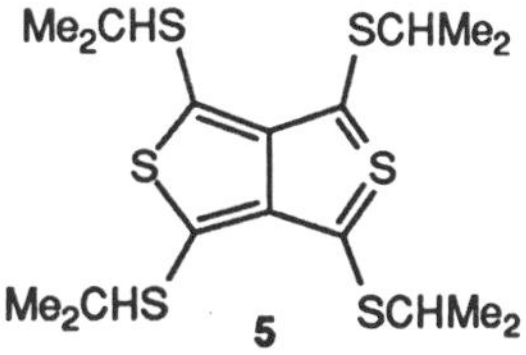

The mechanism of hydrosulfurization continues to have theoretical and commercial import. Reaction of (pentamethylcyclopentadienyl)(2,5-dimethylthiophene)iridium with iron carbonyls gives **6** which desulfurizes upon treament with carbon monoxide <91JA2544>. Reaction of $(C_5Me_5)Rh(PMe_3)(Ph)H$ with thiophene leads to the elimination of benzene and insertion of Rh across the thiophene C-S bond to give **7** which may then undergo Diels-Alder reactions <91JA559>. Reactions of 2,5-dihydrothiophene with Mo(110) gives facile and selective desulfurization <91JA820>. *S*-Bound thiophene complexes (**8**) form by reaction of $Cp'(CO)_5Re(THF)$ with thiophene and are activated for further reactions on the thiophene ring <91MI2436>. Similar reaction occurs for benzo[*b*]thiophene <91JA4005>. Desulfurization of benzo[*b*]thiophene may be accomplished using aqueous platinum metal species <91CJC590> or nickel boride <90CC819>.

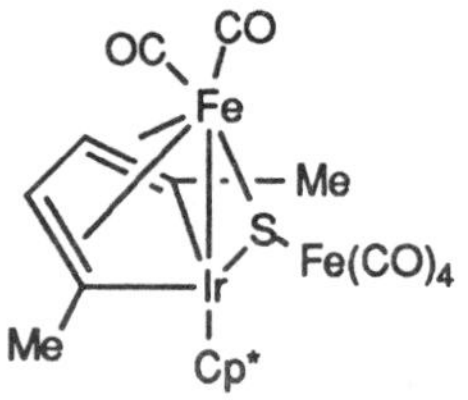

6 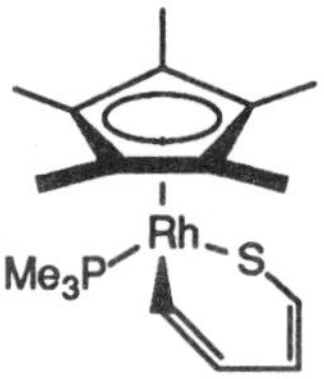 **7** 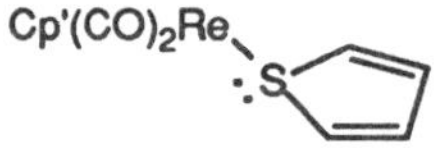 **8**

5.1.3 RING SUBSTITUTION

Electrophilic substitution reactions provide one of the most direct and well-utilized methods to functionalize electron-rich aromatic systems such as thiophene rings. Studies of acid-catalyzed hydrogen exchange of thiophene and other aromatic derivatives show that electrophilic rate constants are described by a Hammett-type linear free energy relationship <91MI233>. Protonation of some 5-substituted-3-nitro-2-pyrrolidin-1-yl thiophenes in aqueous perchloric acid correlates with the modified Hammett, Bunnett-Olsen and Marziano-Cox-Yates methods to calculate pK_{BH} <91JCS(P2)1477>. Substitution of thiophene with phenylium ion in the gas and liquid phase gives predominantly the α-substituted product <91CJC732>. Similarly, CH_3CO^+ reacts to form the α-substituted product in the gas phase <91CJC740>. Core-electron spectroscopy provides new insights into the rate-determining step for electrophilic reaction on thiophene and the origin of substituent effects <91JOC3935>. Orientation effects for the nitration of dithieno[3,4-*b*:3',4'-*d*]pyridine <91JOC1590> as well as the [3,4-*b*:3',2'-*d*]- and [2,3-*b*:3',2'-*d*] isomers <91JHC351> may be explained by the electron distribution of the transition state structures. In a preparatively useful study, zinc chloride-catalyzed reaction of 3,6-dimethoxy-3,6-dimethylcyclohexa-1,4-diene with electron-rich heterocycles such as thiophene gives 2-arylated thiophene derivatives in a Scholl-type reaction <91T313>. Tri(ethoxycarbonyl)methylation of electron-rich aromatic systems including thiophene is catalyzed by manganese (III) acetate and occurs in good to excellent yields <91S567>.

Halogenation of thiophene provides an excellent precursor for further transformations. Phenylsulfinyl chloride and aluminum halide is an efficient, regioselective reagent to give α-substituted thiophenes and furans although it is ineffective for pyrroles <90PS29>. Alternatively, benzyltrimethylammonium polyhalides and zinc chloride also give excellent yields of chloro-, bromo- or iodo-thiophene derivatives <91BCJ2566>. ^{18}F-Labelled thiophene derivatives form using $[^{18}F]$-F_2 under controlled conditions <90MI1109>.

Nucleophilic addition reactions may also occur on the thiophene ring. Meisenheimer-type adducts form nitrothiophenes and lead to some interesting insights into the reactivity of both thiophene and benzene <91JCS(P2)1631>. A kinetic study of the benzenethiolate debromination of some thiophenes shows a good Hammett correlation with *o*-substituents and is unaffected by the geometry of the substituent <91JCR(S)270>. Other ways of activating thiophene for nucleophilic attack include oxidation to a sulfoxide; this may mimic P450 metabolic activation in biological systems <91JA7825>.

α-Anions generated directly by deprotonation of thiophene or by transmetalation of bromothiophenes also provide a powerful means to substitution. Butyl lithium deprotonates 3-methoxythiophene regiospecifically and subsequent reaction with ethylene oxide and an acylating agent produces **9** which is an intermediate for ketal formation <91M185>. 3-(Silyloxycarbonyl)thiophenes are deprotonated by LDA at -78°C and the resulting α-anion undergoes a 1,4-*O,C* migration to provide 2-silylated 3-carboxythiophene derivatives <91SL33>. Other α-lithiated thiophenes are used for the preparation of 2,5-disubstituted thiophene derivatives <90PS75>. A detailed study of the chemistry of the dianions of *N-tert*-butylthiophene-2-sulfonamide shows that the 5-lithiated species forms under equilibrium conditions and has synthetic utility <91JOC4260>. LDA deprotonates 2-iodothiophene to allow reaction with aldehydes to give 5-substituted-2-iodothiophene (**10**). Sonogashira alkyne coupling of **10** promoted by palladium(II) chloride/triphenylphosphine gives **11** <90SL755>. Pd(0)-Cu(I)-catalyzed coupling reactions also produce alkynyl substituted thiophenes <91SC1875>.

OMe; $(CH_2)_2OCOEt$; S — **9**

HO; C_8H_{17}; S; I — **10**

CuI, $PdCl_2(Ph_3P)_2$, $(i\text{-}Pr)_2NH$, H—≡—R →

HO; C_8H_{17}; S; R — **11**

R = $CHOH(CH_2)_3CO_2Me$

Pd(0) as well as other transition metal-catalyzed coupling provides access to a variety of useful thiophene derivatives. 2-Thienylzinc chloride reacts with aryl iodides to produce novel *p*-terphenoquinone analogues containing a central thiophene diylidene <91JA4576>. Such quinones have interesting amphoteric redox properties which will be discussed further in the polymer section **5.1.7**. Similar couplings produce PAF antagonists <91JMC1209>. Palladium-catalyzed reaction of 5-acetoxymercurinucleosides with 2-iodothiophenes produce cytosine nucleosides <91CCC1295>. Similar nucleoside structures are formed by Pd(0)-catalyzed coupling of 2-stannylated thiophene with 5-iodo-2'-deoxyuridines

<91JMC2383, 91JHC529>. 3-Bromothiophene reacts with Grignard reagents as catalyzed by nickel chloride complexes to give 3-substituted thiophenes <90NJC359>. Two stable arylnickel(II) complexes promote the reaction of 3,4-dibromothiophene to produce the cyclic tetramer cyclotetrathiophene <91MI119>.

Radical reactions also give access to substituted thiophene derivatives. 3-Bromothiophene reacts with allyl bromide in the presence of AIBN and butyllithium to provide a precursor for self-doped conducting polymers <90CC1694>. 2-Perfluoro-oxa-alkyl substituted thiophenes are formed by thermal decomposition of peroxide precursors <91JFC117>. Photolysis of 5-iodo-2'-deoxycytidine in the presence of thiophene gives the 2-thienyl derivative <91NN1277>. Alternatively, photolysis of 2-iodothiophene derivatives allows coupling with benzimidazoles <91H1059>. Nickel-phosphine catalyzed reaction of 4-aryl-2-chlorothiophene gives symmetrical bithienyls <91BCJ864>. In a more theoretical study, thiophene reacts over pyridine (1.3:1) in competitive free radical 4-nitrophenylation experiments <91JHC1153>.

Oxidation of 1,3,4,6-tetrakis(alkylthio)thieno[3,4-*c*]thiophenes with iodine in the presence of aniline gives the corresponding thiophenimine derivative in high yields through the intermediacy of radical cations<91JHC1643>.

5.1.4 THIOPHENE RING FORMATION

Preparation of the thiophene ring system is at the heart of the most interesting thiophene chemistry. Among the more remarkable examples of thiophene formation reported last year, ylide **12** reacts with esters and subsequently rearranges to tetracycle **13** <91TL4359>. Further mechanistic studies of the enediyne antibiotics have focused on the kinetics of the trisulfide bond cleavage of calicheamicin to form **14** <91TL4635>.

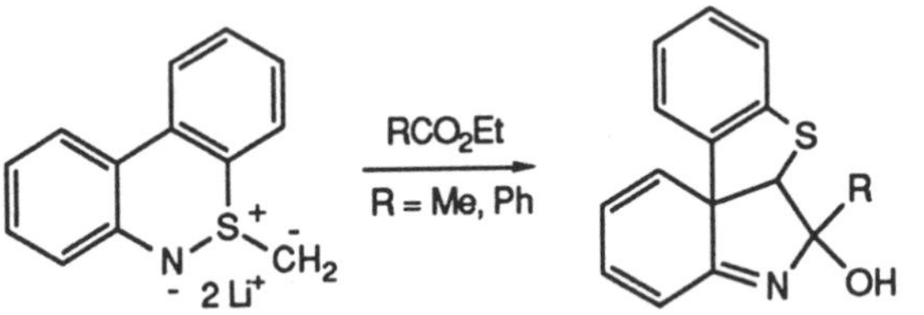

RCO_2Et, R = Me, Ph

12 **13** 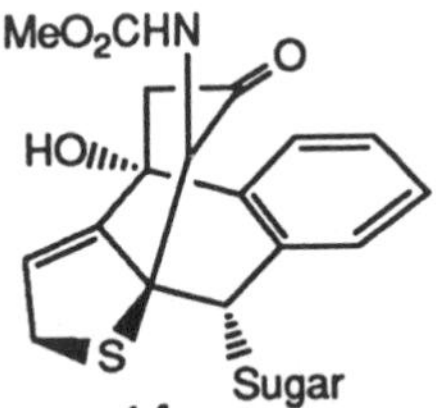 **14**

Electrocyclic reactions provide a powerful, non-classical approach to the construction of thiophene rings. Intramolecular approaches include the cycloaddition of sulfonylallene **15** to produce mixtures of [4+2] and [2+2] adducts <91TL1351>. α,β-Unsaturated sulfines are generated by a Peterson alkylidenation of sulfur dioxide to form **16** which then closes to form 2-toluenesulfonylthiophenes by an unprecedented cycloaromatization <91TL3867>. Bridged heteroannulenes thermally isomerize to cycloheptathiophenes <90JCS(P1)2035>. Alkylation of β-ketothiones with propargyl bromide derivatives leads to **17** which subsequently ring-closes to α-substituted thienothiopyran derivatives <91SL595>. Photochemically driven intramolecular ylide-alkene cycloadditions give rise to some novel tetracyclic thiophene derivatives <90TL3821>.

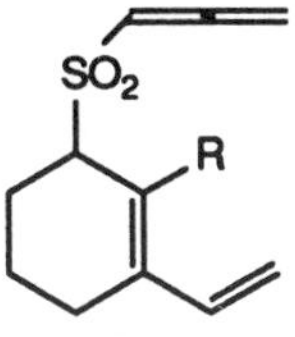

Intermolecular cycloaddition reactions also provide thiophenes. Thioaldehydes react with dienes to give mixtures of thiopyrans and thiabicyclohept-5-enes <91S785>. Allenic episulfides catalyzed by Pd(0) react with dimethylacetylene dicarboxylate to give thiophene **18** as well as thiepin derivatives <91TL4573>. Regioselective cyclotrimerization of (η^2-thiophosphinito)manganese with methylpropynoate gives **19** which may be oxidized with cerric ion to a 2,4-disubstituted thiophene <91CB1985>. Cobalt-catalyzed reaction of heterocyclic diynes produces [2,2](2,5)thiophenophane derivatives <91CB357>. 1,3-Dimetallated acetylenes react with non-enolizable thiocarbonyl compounds to produce 2,3-disubstituted thiophenes <90SC3427>. Pentadiyne reacts with potassium *t*-butoxide and CS_2 to produce **20** which subsequently ring closes to form thieno[2,3-*b*]thiophene **21** <91SC145>.

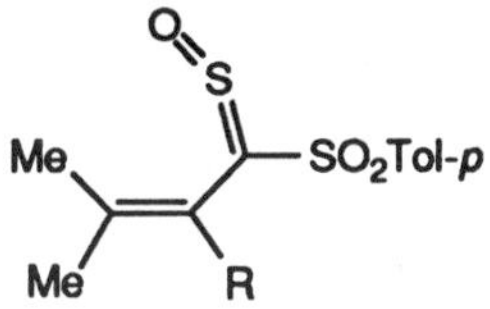

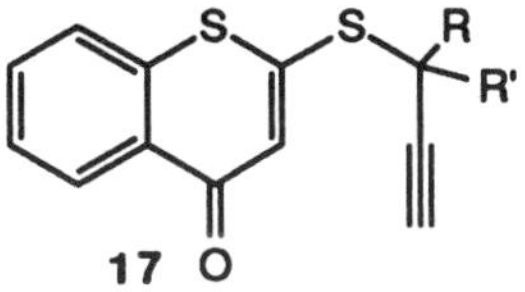

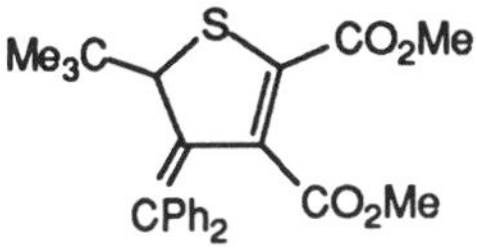

Thiophene rings may be formed by addition of a sulfur "atom" to a four carbon framework. *o*-Phthaloyl chloride reacts with sodium sulfide under phase-transfer catalysis to provide an anhydride which reacts with PCl_5 to give **22** <91JOC6024>. Reaction of *m*-terphenyl with chlorosulfonic acid results in the formation of a dibenzothiophene 5,5-dioxide derivative <91JHC187>. Reaction of 2-vinylbenzothiophenes with thionyl chloride produces thieno[3,2-*b*][1]benzothiophene **23** <91JHC109>. Elemental sulfur may also serve as the atom source in a Gewald synthesis. Alkyl-substituted heterocyclic carbonitriles react with sulfur to form 2-aminothiophene derivatives <90LA1215> as do β-keto esters with cyanoacetates to form allosteric phosphofructokinase ligands <91JOC7179>. Lawesson's reagent is a frequent source of sulfur in thiophene ring formation. Reaction of diketones such as **24** with Lawesson's reagent give **25** as an example of this approach <91JOC78>. Other diketones give similar results to produce *c*-annelated thiophenes <90PS209>, quaterthienyls <91JOC5095>, thiophenes from acyl chloride-allyl chloride adducts <91MI109> and dithiophthalides <91H1559>.

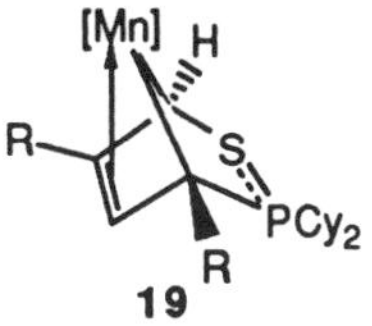

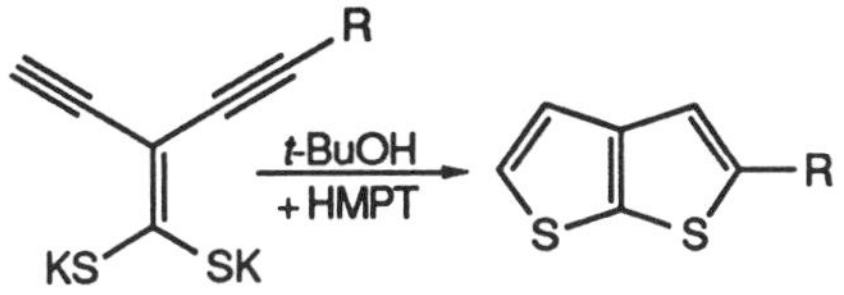

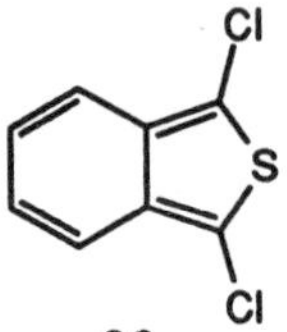

The thiophene ring may also form by combining 2- and 3-atom fragments. In a unique synthesis of 5- and 6-ring heterocycles, 2-diazo-1,3-diketones condense with readily available phosphane derivatives such as **26** to give **27** <91LA331>. Condensation of mercaptoacetates with α,β-unsaturated carbonyl derivatives has been one of the most utilized methods of preparing the thiophene ring system. An extremely relevant example of this approach is used in reactions of golfomycin-A to demonstrate the mechanism of the DNA-cleaving properties of the enediyne antibiotics (**28** -> **29**) <90AG(E)1064>. Using examples of this technology, β-aminoenones produce 2-functionalized thiophenes regioselectively <90SC2537> and α-chloroketoesters give thiophene-2-carboxylate derivatives <91JMC2186>. Thienopyrimidines <91P26, 91PPI413> and thienopyridines <91JCR(S)178, 91CCC1749, 91MI27> may also be formed. Use of β-chloro-α,β-unsaturated aldehydes leads to thiophenes in classic syntheses <91JHC999, 90SC2749> while condensations with β-chloro-α,β-unsaturated nitriles give 3-aminothiophene derivatives <91M413>.

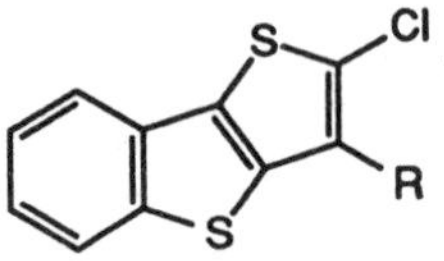

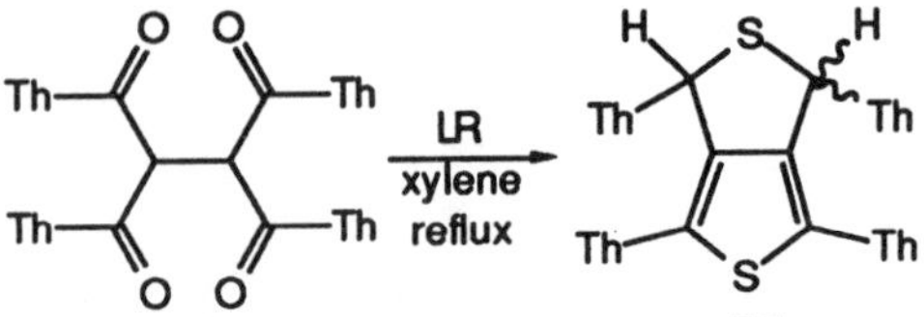

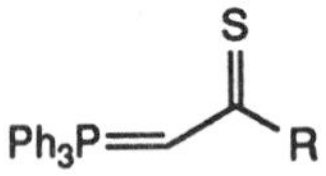

Some other interesting methodology also allows the formation of novel thiophene derivatives. Thus, **30** is formed by reaction of pyridinium methylketone derivatives and arylmethylenecyanothioacetamides which then collapses to the 2-amino-3-cyanodihydothiophene **31** <91S277>. Base-catalyzed condensation of benzylthioacetic acid with diethyloxalate gives 3,4-dihydroxythiophene derivatives <91JHC1449>. Lithiation of *S*-(2-methylphenyl) *N,N,N',N'*-tetramethylphosphodiamidothioate and reaction with esters gives **32** which produces benzo[*b*]thiophene **33** in refluxing formic acid <91JHC173>. Symmetrical benzo[*c*]thiophenes are produced by cyclization of *o*-dibromoxylenes <91CB645>.

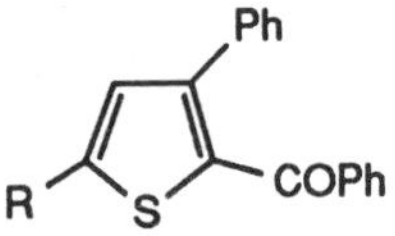

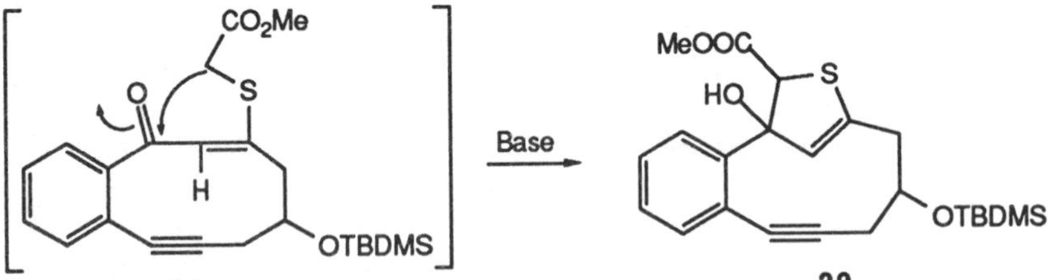

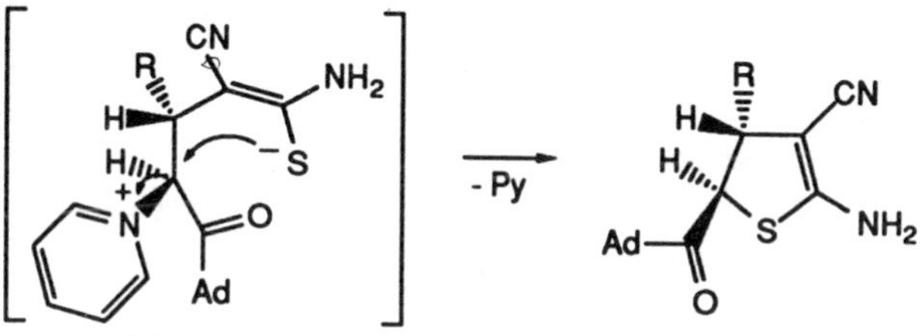

5.1.5 RING ANNELATION ON THIOPHENE

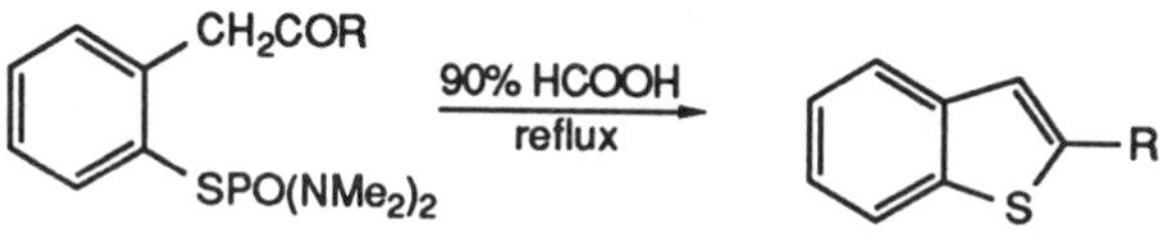

In contrast to thiophene ring-formation, preparation of thieno-fused compounds frequently is based on extensions of the thiophene substitution chemistry discussed in section **5.1.3**. Derivatives produced in these reactions are frequently of enormous interest as pharmaceutical and agricultural agents. This section is organized as to increasing fused-ring size (*i.e.* [5.5], [5.6]...[5.x] bicyclic systems).

The formation of the [5.5] system 2,3-dihydrothieno[2,3-*b*]thiophene **34** results from a 5-*exo*-trig anionic Michael ring closure on the precursor *trans*-nitroolefin. The reaction does not work for the non-oxidized sulfide analogue <91JHC13>. Another synthesis of this ring system utilizes a lithium-bromine exchange of 3-bromothiophene derivatives to promote intramolecular acylation <91TL721>. Alternatively, use of the Pummerer reaction on **35** promotes a 3,3-sigmatropic shift giving 2-mecapto-3-allyl-substituted thiophene intermediates which close to **36** <91MI19>. A more classical approach to thieno[2,3-*b*]thiophene derivatives utilizes base-induced intramolecular thiophene ring formation from a 2-thioacetate 3-aldehyde derivative <91JMC1805>. Other [5.5] thiophenes prepared using routine chemistry include thieno[3,2-*c*]pyrazoles <91AP469>, thieno[2,3-*b*]pyrroles <91AP219>, thieno[3,2-*c*]pyrazoles <91JCR(S)235> and isoxazolino[4,5-*b*]benzo[*b*]thiophenes <91TL3699>.

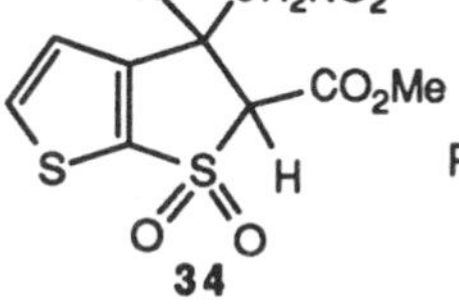

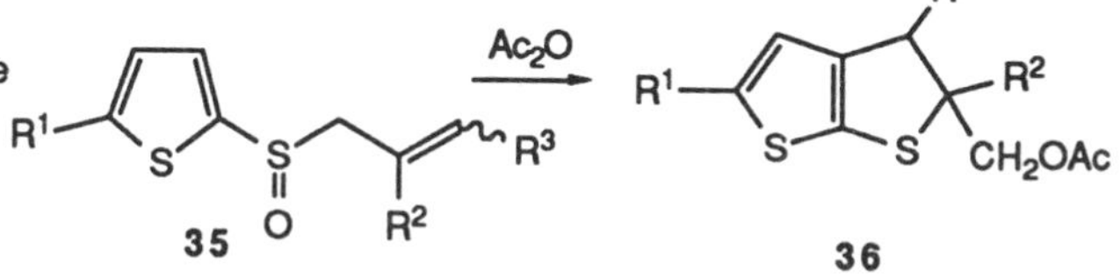

The use of cycloaddition reactions to form [5.6] fused ring systems provides a powerful route into novel thieno-compounds. Reaction of 2-vinylthiophenes with tetrabromocyclopropene results in [4+2] addition to give **37** <90TL4581>. Diene **38** reacts with quinones to produce novel thiophene-fused derivatives with interesting psychotropic activity <90KFZ27>. Diradical cycloaddition of 2,5-dimethyl-3,4-dimethylenethiophene with styrenes or benzaldehydes gives cycloadducts resembling those of normal electron-demand Diels-Alder reactions <91TL5305>. An analogous 2,3-dimethylenethiophene may be generated by zinc reduction of 2,3-bis(bromomethyl)thiophene and reacts with 4-cyclopentene-1,3-dione to give the isostere of ninhydrin <91SC1055>. Photochemically-driven cyclization of **39** produces **40** which is reversible by use of longer wavelength light <91BCJ202>. Electron-transfer reaction pathways are presumed to result in the 1,6-cyclodimerization of 1,1-di-2-thienylethylene to form a trithienylbenzo[*b*]thiophene <91CB1203>. Similar reaction with TCNE or DDQ quantitatively produces [4+2] cycloadducts through radical ion pairs <90H1873>. Irradiation of a number of 3-halothiophene derivatives produce a variety of novel thiophene-fused compounds via radical intermediates <91JHC203, 91JHC737, 90JHC2047>.

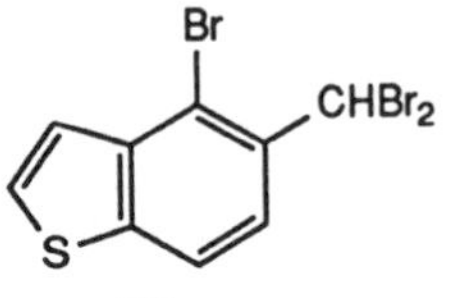

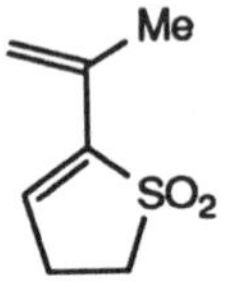

2-Aminothiophene-3-carboxylates, and analogously their isomers, react with dimethylacetylene dicarboxylate to form **41.** Cyclization in the presence of base via a ketene produces thienopyridine **42** which undergoes further reaction <91JHC205>. Other thienopyridine <91PS305> as well as thienopyrimidine <91HCA579, 91IJC(B)618, 91JPR229, 91EJM323> derivatives are prepared by conventional methodology. Retro-malonate addition reactions lead to intermediates that react with hydrazine to produce thienopyridazines <91MI183>. Intramolecular electrophilic substitution of the acid-induced carbonium ion of **43** gives the thienomorphan derivatives **44** <91H107>. An interesting application of the Pummerer reaction allows intramolecular electrophilic substitution on thiophene to form tricyclic intermediates of ergoline sulfur analogues <91CJC1011>. Palladium-catalyzed cyclocarbonylation of 3-(thienyl)allyl acetates provides a synthetically useful route to substituted 7-acetoxybenzo[*b*]thiophenes <91JOC1922>. A novel thienotriptycene derivative forms from the reaction of 1,1,1-tris(2-lithio-5-methyl-3-thienyl)ethane with diethylcarbonate <91CC751>.

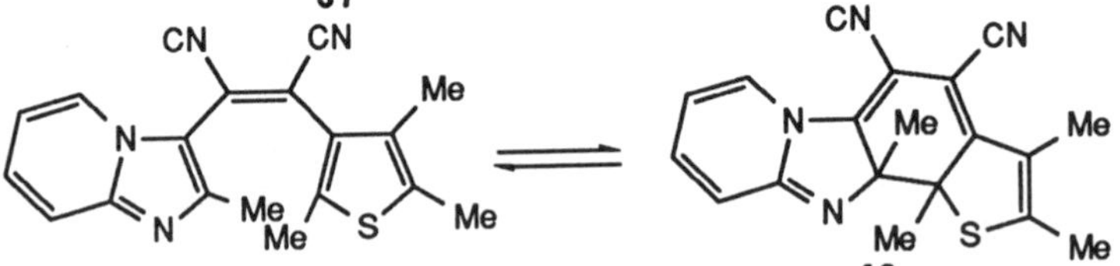

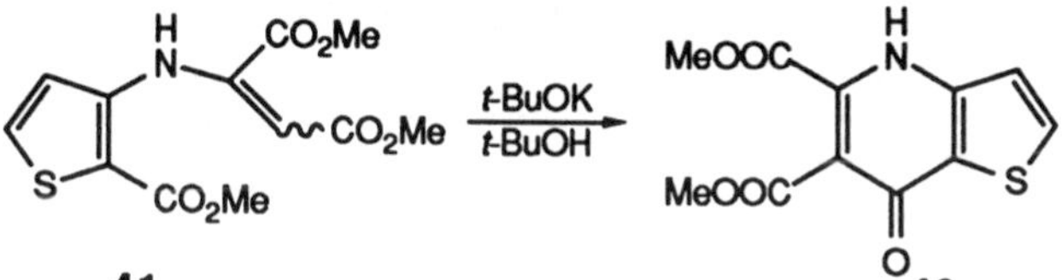

Thiophene-fused 7- and larger rings are represented by a diverse group of compounds. The thieno[2,3-*b*]thiepine system is constructed by reaction of sodium sulfide with 2-(4-chlorobutyl)-3-chlorotetrahydrothiophene <91JCS(P1)285>. Thieno[3,2-*b*]azepine-5,8-diones form by acylation of a 3-aminothiophene-2-carboxylate and subsequent Dieckmann condensation <91AP579>. Thienodiazepine **45** is prepared by intramolecular *N*-alkylation of the benzimidazole <91JHC1121>. Alternatively, the thienodiazepine ring system may be formed by 1-carbon delivery (via an aldehyde) to amine **46** to give **47** <91JHC945>. Similar Mannich-type reaction on an enaminothiolactone produces thieno[4,3-*b*][1,5]benzodiazepin-1-ones <91MI495>. Imine formation by aldehyde trapping of an amine arising from a Curtius rearrangement also produces a thienodiazepine derivative <91JHC81>. Trithienocyclotriyne **48** forms by a Stephens-Castro coupling of 3-ethynyl-2-iodothiophene <90MI427>.

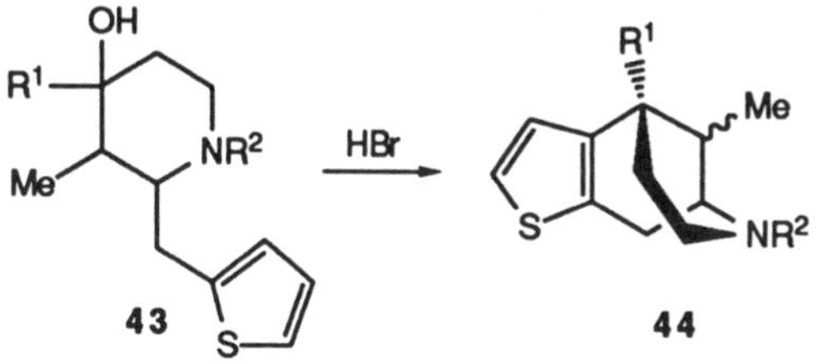

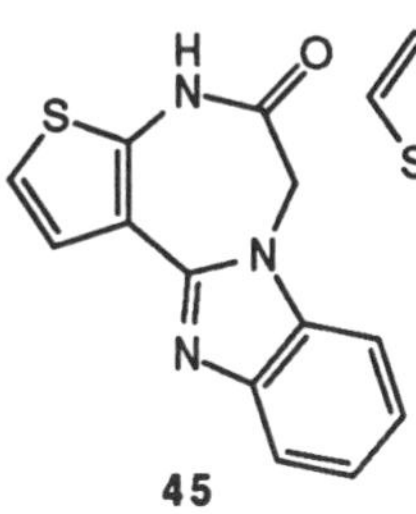

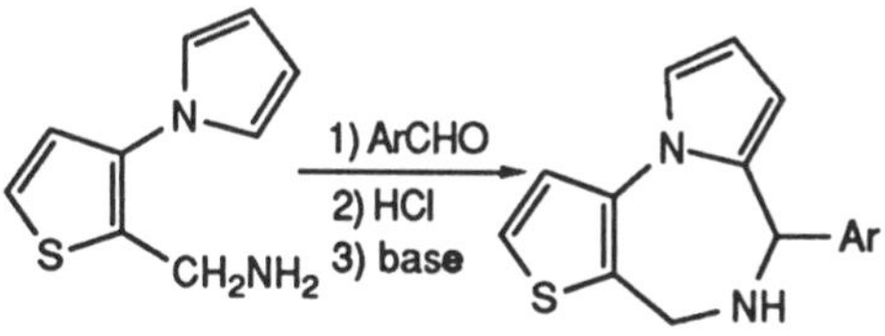

5.1.6 THIOPHENES AS INTERMEDIATES

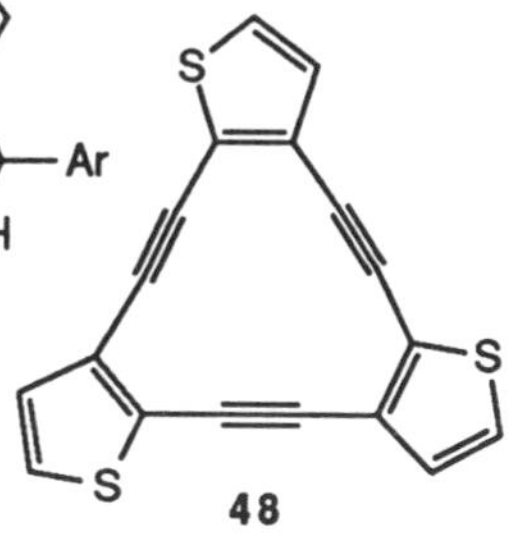

The use of thiophene as a precursor to non-thiophene molecules utilizes the unique electronic properties of sulfur as well as the steric constraints of a 5-ring to produce some very interesting results. The diene system of thiophene can be induced to undergo Diels-Alder reactions by the activation of 1,1-dioxide derivatives. In an extremely interesting application of this approach, a tandem dimerization of 3-bromo-2-alkylthiophenes and subsequent ring-opening provides a synthetically useful route to 1,2,3-trisubstituted benzenes (Eq. 1) <91ACS636>. Treatment of the same starting bromothiophene with Grignard reagents leads to the remarkable formation of a heterocycloheptane derivative <91JOC4064>.

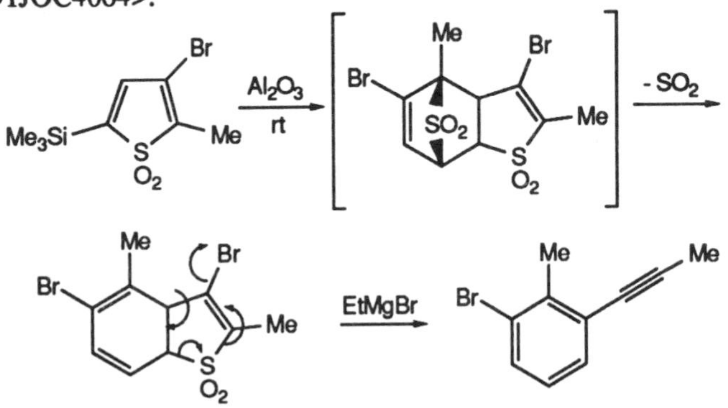

(Eq. 1)

3-Sulfolenes are precursors to *cis*-1,3-butadienes. Regiospecific substitution of 3-sulfolenyl dianions leads to synthetically useful di- and trisubstituted 3-sulfolenes <90JOC5410>. 3-Vinylsulfolene undergoes thermal cheleotropic extrusion of SO_2 to produce [3]dendralene (**49**) which functions as a tandem annelating reagent <91CC114>. Compound **50** provides an intramolecular example of a latent *cis*-1,3-butadiene derivative which subsequently undergoes azidocarbonyl Diels-Alder reaction <91SL31>. Less activated thiophenes may also react in a Diels-Alder fashion; **51** reacts with dimethylacetylene dicarboxylate or benzyne to give naphthalene or anthracene derivatives <90JCS(P1)1919>. Enantiomerically pure alkanols form by derivatization (and resolution) of alkylthienylcarbinols with chiral acetals and, ultimately, desulfurization <91M705>.

49 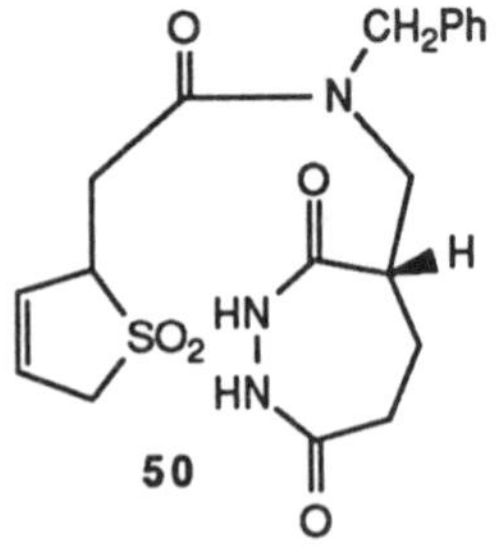 **50** 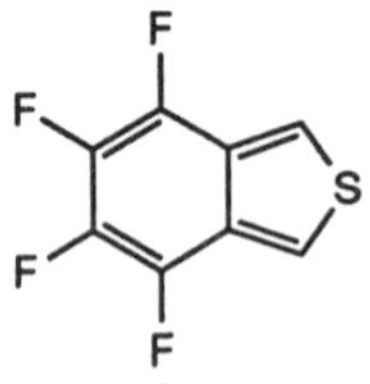 **51**

Thiophenes may also undergo ring contraction or expansion to produce novel ring systems. Oxidation of 3,4-di-*tert*-butylthiophene-1,1-dioxide affords thiete 1,1-dioxide and sulfone derivatives <91JOC4001>. Reductive rearrangement of dimethylamino(thienyl)fulvene with lithium naphthalide produces the thiophene ring-expanded cyclopentathienylidenethiopyran by the intermediacy of an allene <91TL3499>. Photooxygenation of 2,3-dihydrothiophenes forms the first examples of stable sulfur-substituted 1,2-dioxetanes. These compounds may be cleaved to linear dicarbonyl compounds at low temperatures <91JOC4027>. Thiophene-2,3-dione **52** undergoes [2+2] cycloaddition with diphenylketene to give spirothiete **53** which loses CO_2 thermally to reform a thiophene ring (Eq. 2) <91T3045>. Extreme conditions such as flash pyrolysis <91CB2613> or flash photolysis <91MI697> of thiophene derivatives produce linear annulenes or free radicals.

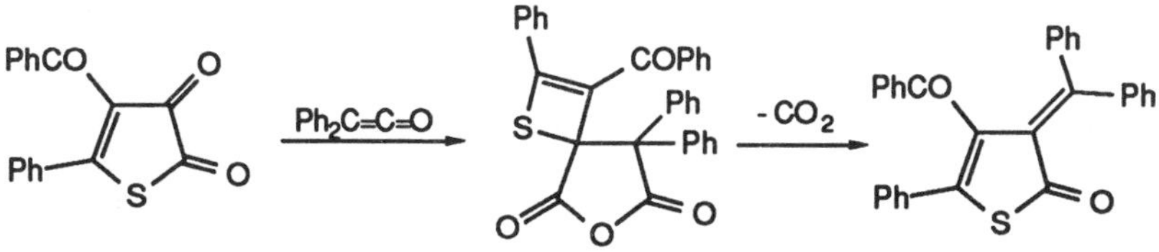

Eq. 2

52 **53**

5.1.7 THIOPHENE CYCLOPHANES AND POLYMERS

There have been a few examples of thiophene-containing cyclophanes reported in 1991. Synthesis of [2]paracyclo[2](2,5)- and -(2,4)thiophenophanes allows the study of desulfurization as well as ring inversion reactions <91CB1403>. [7]- and [8]Metacyclophanes result from reductive ring-opening of the precursor thiophenes <91CB411>. Dithia[3]metacyclo[3]thiophenophanes and [2]metacyclo[2]thiophenophanes form by dithiol bis-alkylation and their conformational preferences are the subject of study <91JOC2837>. None of these reports investigate metal-ion coordination.

Molecules with extended conjugation play an important role in the development of organic redox chemistry as well as of improving the properties of conducting organic polymers. Quinones **54**, for example, form by reaction of 2-thienylzinc chloride with iodoanisoles and exhibit half-wave oxidation and reduction potentials indicative of a new type of amphoteric redox molecule <91JA4576>. Related radialenes which incorporate an additional quinoid into conjugation using a cyclopropane link also have amphoteric redox properties <91TL3507>. Other quinone derivatives of thieno[3,2-*b*]thiophenes are electron-acceptors under redox conditions <91CL1117, 91Z353, 91TL4367>. Alternatively, extended conjugation is achieved by multisulfur fulvalenic systems <91MI2205>. Tailoring of monomers by fusion with thiophene is successful in controlling the redox potentials of conducting polymers <91MI1841>.

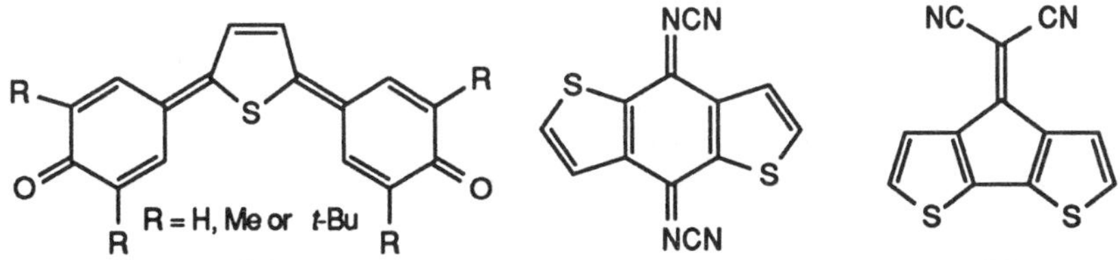

R = H, Me or *t*-Bu

54 **55** **56**

Metal complexes of molecules with extended π-frameworks are also of interest because of their metallic behavior with high conductivities at low temperatures. The new acceptor DTQI (**55**) forms a Cu(I) complex with extremely interesting semi-conducting properties <91MI1847, 91CL1033>.

Developing polymers with narrow band gaps (E_{gap}) to improve conductivity has been successful by using the unique properties of thiophene. Cyclopenta[2,1-*b*;3,4-*b'*]dithiophen-4-one <91CC752> and the derived dicyanomethylene analogue **56** <91CC1268> are both monomers that produce electroactive polymers. Monocyclic thienoquinoids are also precursors of low-gap polymers <91CB1597>.

Thiophene oligomers continue to receive intense study. Polarons and bipolarons of thiophene oligomers in zeolites lead to radical ion-generated conduction in polymers <91JA600, 91MI463>. Electrically-conductive polythiophene forms from thiophene in the presence of a small amount of bithiophene <91MI888> while the longest characterized oligothiophene is prepared by thiophene construction from 1,4-diketones <91JA5887>. α-Terthienyl may be prepared similarly <91S462>. Alternatively, palladium-catalyzed coupling of 2-thienylzinc chloride with 2-iodothiophenes produce terthiophene derivatives <90GCI793>. Native thiophene oligomerizes in the presence of $PdCl_4^{2-}$ by novel thiophene-palladium(II) complexes <91MIC29>. Other novel functionalized polythiophenes form by electropolymerization <91MI3037>. Linear poly(2,5-thienylene) forms as thin layer films arranged along the direction of surface-rubbed polyimides <91CL1483>. Viologen-functionalized poly(3-alkylthienylenes) form by electrooxidative polymerization <90MI185>. Tetrathienylsilane is a precursor of highly conducting electrogenerated polythiophene <91MI277>. Conducting polyheteroarenediylvinylenes are prepared by palladium-catalyzed coupling of stannanes <91CC364> or by Wittig reaction of 2- or 3-thienyl carboxaldehydes <91JCS(P1)799, 91H991>.

Structural and electronic properties of n- and p-doped poly(thiophene) on Au electrodes were studied with an electrochemical quartz crystal microbalance <91MI872>. Photochemical reactions of terthienylmethanol derivatives are inhibited by singlet oxygen quenchers <91JCR(S)166>. Alkyl substituents on thiophene oligomers have significant effects on the conformation and solubility properties of these materials <91MI579>. Solid-state ^1H NMR studies of substituted polythiophene show two distinct types of side-chain motion <91JA8243>. Orthogonally-fused conducting polythiophenes form by using a spiro silicon tether. These polymers have two reversible cyclic voltametric waves <91JA7064>. Progress has been made in the preparation of the phosphorus analogues of polythiophenes <90AG(E)655>. Quaterthiophenediphosphonic acid (QDP) was prepared in three steps from 2,2'-bithiophene and provides the basis for the first example of a zirconium phosphonate multilayer <91MI699>. Electrochemical properties of poly[2,5-dialkoxy-1,4-bis(2-thienyl)benzenes] are examined by cyclic voltammetry <91MI182>.

5.1.8 SELENOPHENES AND TELLUROPHENES

As was the case in previous years, very few examples of selenophenes and only one case of tellurophene are reported. In the most unusual formation of selenophene derivatives, selenoselenophene **57** forms by tandem [3,3]sigmatropic rearrrangement and double Michael-type addition to an allene diselenide <90T5759>. Cycloaddition of selones with dienes gives bicyclic selenophene derivatives <91CL1053>. Reaction of 2,2'-dilithio-1,1'-binaphthyl with selenium or tellurium metal gives mixtures of products including the selenophene or tellurophene **58** <91JHC433>. The thiophene ring may be activated by oxidation to a sulfone which then reacts with selenium metal at elevated temperatures to form 3,4-disubstituted selenophenes <90TL4473>. The selenophene **59** as well as the furan and *N*-methylpyrrole analogues of the amphoteric 2,5-dimethylene-2,5-dihydrothiophene may provide interesting contrast to the electrochemical properties of the thiophene standard <91TL4313>.

X = Se or Te

57 **58** **59**

5.1.9 INTERESTING THIOPHENE DERIVATIVES

Squaraine **60** may be prepared from 2-aminothiophenes with squaric acid and are a novel class of intensely colored panchromatic dyes <91DP19>. Hydrotris[3-(2'-thienyl)pyrazol-1-yl]borate has the second-lowest steric hindrance among known poly(pyrazolyl)borates <91IC2795>. Platinum complex **61** has strong luminescence with a long excited-state half-life; self-annihilation occurs in a diffusion controlled process <91IC2476>. The thiophene isostere of ninhydrin (**62**) may be prepared by either of two routes and has reactivity similar to that of ninhydrin toward aminoacids with similar color generation <91BSF260>.

60 **61** **62**

5.1.10 REFERENCES

We greatly appreciate the enormous assistance provided by Ms. Carol Whitham and Ms. Jean Ellis-Fleming of the Information Services Department in gathering these references.

90AG(E)655 M. O. Bevierre, F. Mercier, L. Ricard and F. Mathey; *Angew. Chem., Int. Ed. Engl.*, 1990, **29**, 655.

90AG(E)1064 K. C. Nicolaou, G. Skokotas, S. Furuya, H. Suemune and D. C. Nicolaou; *Angew. Chem., Int. Ed. Engl.*, 1990, **29**, 1064.

90CC819 T. G. Back and K. Yang; *J. Chem. Soc., Chem. Commun.*, 1990, 819.

90CC1694 Y. Ikenoue, Y. Saida, M. Kira, H. Tomozawa, H. Yashima and M. Kobayashi; *J. Chem. Soc., Chem. Commun.*, 1990, 1694.

90GCI793 R. Rossi, A. Carpita, M. Ciofalo and J. L. Houben; *Gazz. Chim. Ital.*, 1990, **120**, 793.

90H1873 T. Varea, B. Abarca, R. Ballesteros and G. Asensio; *Heterocycles*, 1990, **31**, 1873.

90JCS(P1)1919 G. M. Brooke and S. D. Mawson; *J. Chem. Soc., Perkin Trans. 1*, 1990, 1919.

90JCS(P1)2035 H. Kato, S. Toda, Y. Arikawa, M. Masuzawa, M. Hashimoto, K. Ikoma, S. Z.

	Wang and A. Miyasaka; *J. Chem. Soc., Perkin Trans. 1*, 1990, 2035.
90JHC2047	J. K. Luo, L. Steven and R. N. Castle; *J. Heterocycl. Chem.*, 1990, **27**, 2047.
90JOC5410	T. Chou, C.-Y. Tsai and L.-J. Huang; *J. Org. Chem.*, 1990, **55**, 5410.
90KFZ27	T. G. Tolstikova, V. A. Davydova, E. E. Shul'ts, G. F. Vafina, G. M. Safarova, F. A. Zarudii, D. N. Lazareva and G. A. Tolstikov; *Khim. -Farm. Zh.*, 1990, **24**, 27; [*Chem. Abstr.* 1990, **113**, 224124h].
90LA1215	M. H. Elnagdi and A., W. Erian; *Liebigs Ann. Chem*, 1990, 1215.
90MI185	P. Bäuerle and K. Gaudl; *Adv. Mater. (Weinheim, Fed. Repub. Ger.)*, 1990, **2**, 185.
90MI427	D. Solooki, V. O. Kennedy, C. A. Tessier and W. J. Youngs; *Synlett*, 1990, 427.
90MI1109	M. E. Crestoni; *J. Labelled Compd. Radiopharm.*, 1990, **28**, 1109.
90NJC359	M. Lemaire, R. Garreau, D. Delabouglise, J. Roncali, H. K. Youssoufi and F. Garnier; *New J. Chem.*, 1990, **14**, 359.
90PS29	N. Kamigata, T. Suzuki and M. Yoshida; *Phosphorus, Sulfur Silicon Relat. Elem.*, 1990, **53**, 29.
90PS75	B. Garrigues; *Phosphorus, Sulfur Silicon Relat. Elem.*, 1990, **53**, 75.
90PS209	W. Volz and J. Voss; *Phosphorus, Sulfur Silicon Relat. Elem.*, 1990, **54**, 209.
90S915	A. Plant and D. J. Chadwick; *Synthesis*, 1990, 915.
90SC2537	A. Alberola, J. M. Andres, A. Gonzalez , R. Pedrosa and P. Pradonos; *Synth. Commun.*, 1990, **20**, 2537.
90SC2749	N. Vasumathi, D. V. Ramana and S. R. Ramadas; *Synth. Commun.*, 1990, **20**, 2749.
90SC3427	R. L. P. de Jong and L. Brandsma; *Synth. Commun.*, 1990, **20**, 3427.
90SL755	P. T. DeSousa and R. J. K. Taylor; *Synlett*, 1990, 755.
90T5759	S. Braverman and M. Freand; *Tetrahedron*, 1990, **46**, 5759.
90TL3821	J. P. Dittami, X. Y. Nie, C. J. Buntel and S. Rigatti; *Tetrahedron Lett*, 1990, **31**, 3821.
90TL4473	J. Nakayama, Y. Sugihara, K. Terada and E. L. Clennan; *Tetrahedron Lett.*, 1990, **31**, 4473.
90TL4581	J. M. Keil, T. Kaempchen and G. Seitz; *Tetrahedron Lett*, 1990, **31**, 4581.
91ACS636	S. Gronowitz, G. Nikitids, A. Hallberg and C. Staelhandske; *Acta Chem. Scand.*, 1991, **45**, 636.
91AP219	D. Binder, H. Schnait, F. Rovensky, R. Enzenhofer and H. Stroissnig; *Arch. Pharm.* (*Weinheim, Ger.*), 1991, **324**, 219.
91AP469	R. M. Mohareb and S. M. Sherif; *Arch. Pharm.* (*Weinheim, Ger.*), 1991, **324**, 469.
91AP579	C. Kunick; *Arch. Pharm.* (*Weinheim, Ger.*), 1991, **324**, 579.
91BCJ202	Y. Nakayama, K. Hayashi and M. Irie; *Bull. Chem. Soc. Jpn.*, 1991, **64**, 202.
91BCJ864	T. Sone, Y. Umetsu and K. Sato; *Bull. Chem. Soc. Jpn.*, 1991, **64**, 864.
91BCJ2566	T. Okamoto, T. Kakinami, H. Fujimoto and S. Kajigaeshi; *Bull. Chem. Soc. Jpn.*, 1991, **64**, 2566.
91BSF260	P. Dallemagne, S. Rault and M. Robba; *Bull. Soc. Chim. Fr.*, 1991, 260.
91CB357	R. Gleiter, S. Rittinger and H. Langer; *Chem. Ber.*, 1991, **124**, 357.
91CB411	M. Takeshita, A. Tsuge and M. Tashiro; *Chem. Ber.*, 1991, **124**, 411.
91CB645	R. P. Kreher and J. Kalischko; *Chem. Ber.*, 1991, **124**, 645.
91CB1203	T. Varea, B. Abarca, R. Ballesteros and G. Asensio; *Chem. Ber.*, 1991, **124**, 1203.
91CB1403	M. Takeshita, M. Tashiro and A. Tsuge; *Chem. Ber.*, 1991, **124**, 1403.
91CB1597	G. Hieber, M. Hanack, K. Wurst and J. Strähle; *Chem. Ber.*, 1991, **124**, 1597.
91CB1985	E. Lindner, C. Haase and H. A. Mayer; *Chem. Ber.*, 1991, **124**, 1985.
91CB2613	G. Maier, J. Schrot and H. P. Reisenauer; *Chem. Ber.*, 1991, **124**, 2613.
91CC114	J. I. G. Cadogan, S. Cradock, S. Gillam and I. Gosney; *J. Chem. Soc., Chem. Commun.*, 1991, 114.
91CC364	R. Galarini, A. Musco, R. Pontellini, A. Bolognesi, S. Destri, M. Catellani, M. Mascherpa and G. Zhuo; *J. Chem. Soc., Chem. Commun.*, 1991, 364.
91CC520	A. Tsubouchi, N. Matsumura and H. Inoue; *J. Chem. Soc., Chem. Commun.*,

1991, 520.

91CC751 A. Ishii, M. Kodachi, J. Nakayama and M. Hoshino; *J. Chem. Soc., Chem. Commun.*, 1991, 751.

91CC752 T. L. Lambert and J. P. Ferraris; *J. Chem. Soc., Chem. Commun.*, 1991, 752.

91CC1268 J. P. Ferraris and T. L. Lambert; *J. Chem. Soc., Chem. Commun.*, 1991, 1268.

91CC1287 T. S. Chou and C. Y. Tsai; *J. Chem. Soc., Chem. Commun.*, 1991, 1287.

91CCC1295 M. E. Hassan; *Collect. Czech. Chem. Commun.*, 1991, **56**, 1295.

91CCC1749 I. M. A. Awad, R. A. E. Abdel and E. A. Bakhite; *Collect. Czech. Chem. Commun.*, 1991, **56**, 1749.

91CJC590 P. D. Clark, J. F. Fait, C. G. Jones and M. J. Kirk; *Can. J. Chem.*, 1991, **69**, 590.

91CJC732 A. Filippi, G. Occhiucci and M. Speranza; *Can. J. Chem.*, 1991, **69**, 732.

91CJC740 A. Filippi, G. Occhiucci, C. Sparapani and M. Speranza; *Can. J. Chem.*, 1991, **69**, 740.

91CJC1011 P. D. Clark, N. M. Irvine and P. Sarkar; *Can. J. Chem.*, 1991, **69**, 1011.

91CL1033 K. Kobayashi, Y. Mazaki, K. Kikuchi, K. Saito, I. Ikemoto and S. Hino; *Chem. Lett*, 1991, 1033.

91CL1053 K. Okuma, K. Kojima, I. Kaneko and H. Ohta; *Chem. Lett*, 1991, 1053.

91CL1117 Y. Mazaki, N. Takiguchi and K. Kobayashi; *Chem. Lett*, 1991, 1117.

91CL1483 T. Yamamoto, C. Mori, H. Wakayama, Z. H. Zhou, T. Maruyama, R. Ohki and T. Kanbara; *Chem. Lett*, 1991, 1483.

91DP19 D. Keil, H. Hartmann and T. Moschny; *Dyes Pigm*, 1991, **17**, 19.

91EJM323 S. Vega, J. Alonso, J. A. Diaz, F. Junquera, C. Perez, V. Darias, L. Bravo and S. Abdallah; *Eur. J. Med. Chem.*, 1991, **26**, 323.

91H107 M. Ban, Y. Baba, K. Miura, Y. Kondo and M. Hori; *Heterocycles*, 1991, **32**, 107.

91H991 J. Nakayama and T. Fujimori; *Heterocycles*, 1991, **32**, 991.

91H1059 M. D'Auria; *Heterocycles*, 1991, **32**, 1059.

91H1559 P. N. Nugara, N. Z. Huang, M. V. Lakshmikantham and M. P. Cava; *Heterocycles*, 1991, **32**, 1559.

91HCA579 R. Neidlein and Z. Sui; *Helv. Chim. Acta*, 1991, **74**, 579.

91IC2476 M. Maestri, D. Sandrini, A. von Zelewsky and C. Deuschel-Cornioley; *Inorg. Chem.*, 1991, **30**, 2476.

91IC2795 J. C. Calabrese, P. J. Domaille, S. Trofimenko and G. J. Long; *Inorg. Chem.*, 1991, **30**, 2795.

91IJC(B)618 U. S. Pathak, S. Singh and J. Padh; *Indian J. Chem., Sect. B*, 1991, 618.

91JA559 W. D. Jones and L. Dong; *J. Am. Chem. Soc.*, 1991, **113**, 559.

91JA600 J. V. Caspar, V. Ramamurthy and D. R. Corbin; *J. Am. Chem. Soc.*, 1991, **113**, 600.

91JA820 A. C. Liu and C. M. Friend; *J. Am. Chem. Soc.*, 1991, **113**, 820.

91JA2318 M. M. Greenberg, S. C. Blackstock, J. A. Berson, R. A. Merrill, J. C. Duchamp and K. W. Zilm; *J. Am. Chem. Soc.*, 1991, **113**, 2318.

91JA2544 J. Chen, L. M. Daniels and R. J. Angelici; *J. Am. Chem. Soc.*, 1991, **113**, 2544.

91JA4005 M. G. Choi, M. J. Robertson and R. J. Angelici; *J. Am. Chem. Soc.*, 1991, **113**, 4005.

91JA4576 K. Takahashi, T. Suzuki, K. Akiyama, Y. Ikegami and Y. Fukazawa; *J. Am. Chem. Soc.*, 1991, **113**, 4576.

91JA5887 H. W. Ten, H. Wynberg, E. E. Havinga and E. W. Meijer; *J. Am. Chem. Soc.*, 1991, **113**, 5887.

91JA7064 J. M. Tour, R. Wu and J. S. Schumm; *J. Am. Chem. Soc.*, 1991, **113**, 7064.

91JA7825 D. Mansuy, P. Valadon, I. Erdelmeier, G. P. Lopez, C. Amar, J. P. Girault and P. M. Dansette; *J. Am. Chem. Soc.*, 1991, **113**, 7825.

91JA8243 A. C. Kolbert, N. S. Sariciftci, K. U. Gaudl, P. Bäuerle and M. Mehring; *J. Am. Chem. Soc.*, 1991, **113**, 8243.

91JCR(S)166 N. R. Krishnaswamy, C. S. S. R. Kumar and S. R. Prasanna; *J. Chem. Res., Synop.*, 1991, 166.

91JCR(S)178	D. Briel, S. Dumke, G. Wagner and B. Olk; *J. Chem. Res., Synop.*, 1991, 178.
91JCR(S)235	J. M. Barker, P. R. Huddleston, G. D. Khandelwal and M. L. Wood; *J. Chem. Res., Synop.*, 1991, 235.
91JCR(S)270	R. Noto, V. Frenna, G. Consiglio and D. Spinelli; *J. Chem. Res., Synop.*, 1991, 270.
91JCS(P1)285	P. A. Delaney, R. A. W. Johnstone, P. A. Leonard and P. Regan; *J. Chem. Soc., Perkin Trans. 1*, 1991, 285.
91JCS(P1)799	C. W. Spangler, P. K. Liu, A. A. Dembek and K. O. Havelka; *J. Chem. Soc., Perkin Trans. 1*, 1991, 799.
91JCS(P1)909	A. Tsubouchi, N. Matsumura, H. Inoue and K. Yanagi; *J. Chem. Soc., Perkin Trans. 1*, 1991, 909.
91JCS(P2)269	U. Folli, D. Iarossi, A. Mucci, L. Schenetti and F. Taddei; *J. Chem. Soc., Perkin Trans. 2*, 1991, 269.
91JCS(P2)1477	R. Noto, M. Gruttadauria, D. Dattolo, C. Arnone, G. Consiglio and D. Spinelli; *J. Chem. Soc., Perkin Trans. 2*, 1991, 1477.
91JCS(P2)1631	C. Dell'Erba, F. Sancassan, M. Novi, D. Spinelli and G. Consiglio; *J. Chem. Soc., Perkin Trans. 2*, 1991, 1631.
91JCT759	R. D. Chirico, S. E. Knipmeyer, A. Nguyen and W. V. Steele; *J. Chem. Thermodyn.*, 1991, **23**, 759.
91JFC117	H. Sawada and M. Nakayama; *J. Fluorine Chem*, 1991, **51**, 117.
91JHC13	T. M. Williams, R. J. Hudcosky, C. A. Hunt and K. L. Shepard; *J. Heterocycl. Chem.*, 1991, **28**, 13.
91JHC81	B. Decroix and J. Morel; *J. Heterocycl. Chem.*, 1991, **28**, 81.
91JHC109	J. L. Jackson; *J. Heterocycl. Chem.*, 1991, **28**, 109.
91JHC173	M. Watanabe, M. Date, K. Kawanishi, R. Akiyoshi and S. Furukawa; *J. Heterocycl. Chem.*, 1991, **28**, 173.
91JHC187	L. H. Klemm, W. Lane and E. Hall; *J. Heterocycl. Chem.*, 1991, **28**, 187.
91JHC203	M. J. Musmar and R. N. Castle; *J. Heterocycl. Chem.*, 1991, **28**, 203.
91JHC205	J. K. Luo and R. N. Castle; *J. Heterocycl. Chem.*, 1991, **28**, 205.
91JHC351	S. Gronowitz, K. J. Szabo and J. O. Oluwadiva; *J. Heterocycl. Chem.*, 1991, **28**, 351.
91JHC433	S. Murata, T. Suzuki, A. Yanagisawa and S. Suga; *J. Heterocycl. Chem.*, 1991, **28**, 433.
91JHC529	D. Peters, A. B. Hoernfeldt, S. Gronowitz and N. G. Johansson; *J. Heterocycl. Chem.*, 1991, **28**, 529.
91JHC737	J. K. Luo, A. S. Zektzer and R. N. Castle; *J. Heterocycl. Chem.*, 1991, **28**, 737.
91JHC945	S. Vega and M. S. Gil; *J. Heterocycl. Chem.*, 1991, **28**, 945.
91JHC999	G. K. Kar, A. C. Karmakar and J. K. Ray; *J. Heterocycl. Chem.*, 1991, **28**, 999.
91JHC1121	A. Walser, T. Flynn and C. Mason; *J. Heterocycl. Chem.*, 1991, **28**, 1121.
91JHC1153	L. H. Klemm and J. Dorsey; *J. Heterocycl. Chem.*, 1991, **28**, 1153.
91JHC1449	D. W. Rangnekar and S. V. Mavlankar; *J. Heterocycl. Chem.*, 1991, **28**, 1449.
91JHC1643	A. Tsubouchi, C. Kitamura, N. Matsumura and H. Inoue; *J. Heterocycl. Chem.*, 1991, **28**, 1643.
91JMC1209	A. Walser, T. Flynn, C. Mason, H. Crowley, C. Maresca, B. Yaremko and M. O'Donnell; *J. Med. Chem.*, 1991, **34**, 1209.
91JMC1805	J. D. Prugh, G. D. Hartman, P. J. Mallorga, B. M. McKeever, S. R. Michelson, M. A. Murcko, H. Schwam, R. L. Smith, J. M. Sondey, J. P. Springer and M. F. Sugrue; *J. Med. Chem.*, 1991, **34**, 1805.
91JMC2186	M. D. Mullican, R. J. Sorenson, D. T. Connor, D. O. Thueson, J. A. Kennedy and M. C. Conroy; *J. Med. Chem.*, 1991, **34**, 2186.
91JMC2383	P. Wigerinck, C. Pannecouque, R. Snoeck, P. Claes, C. E. De and P. Herdewijn; *J. Med. Chem.*, 1991, **34**, 2383.
91JOC78	A. Ishii, J. Nakayama, J. Kazami, Y. Ida, T. Nakamura and M. Hoshino; *J. Org. Chem.*, 1991, **56**, 78.

91JOC129 K. Jug and H. P. Schluff; *J. Org. Chem.*, 1991, **56**, 129.
91JOC1590 K. J. Szabo, A. B. Hörnfeldt and S. Gronowitz; *J. Org. Chem.*, 1991, **56**, 1590.
91JOC1922 M. Iwasaki, Y. Kobayashi, J. P. Li, H. Matsuzaka, Y. Ishii and M. Hidai; *J. Org. Chem.*, 1991, **56**, 1922.
91JOC2837 M. Takeshita and M. Tashiro; *J. Org. Chem.*, 1991, **56**, 2837.
91JOC3224 B. Abarca, G. Asensio, R. Ballesteros and T. Varea; *J. Org. Chem.*, 1991, **56**, 3224.
91JOC3935 L. J. Saethre and T. D. Thomas; *J. Org. Chem.*, 1991, **56**, 3935.
91JOC4001 J. Nakayama and Y. Sugihara; *J. Org. Chem.*, 1991, **56**, 4001.
91JOC4027 K. Gollnick and K. Knutzen-Mies; *J. Org. Chem.*, 1991, **56**, 4027.
91JOC4064 G. Nikitidis, S. Gronowitz, A. Hallberg and C. Staalhandske; *J. Org. Chem.*, 1991, **56**, 4064.
91JOC4260 S. L. Graham and T. H. Scholz; *J. Org. Chem.*, 1991, **56**, 4260.
91JOC4840 J. Armand, C. Bellec, L. Boulares, P. Chaquin, D. Masure and J. Pinson; *J. Org. Chem.*, 1991, **56**, 4840.
91JOC5095 D. M. Perrine, D. M. Bush, E. P. Kornak, M. Zhang, Y. H. Cho and J. Kagan; *J. Org. Chem.*, 1991, **56**, 5095.
91JOC6024 Y. Okuda, M. V. Lakshmikantham and M. P. Cava; *J. Org. Chem.*, 1991, **56**, 6024.
91JOC6337 R. Borghi, M. A. Cremonini, L. Lunazzi, G. Placucci and D. Macciantelli; *J. Org. Chem.*, 1991, **56**, 6337.
91JOC6948 M. Takeshita, M. Koike, S. Mataka and M. Tashiro; *J. Org. Chem.*, 1991, **56**, 6948.
91JOC7179 N. S. Sampson and P. A. Bartlett; *J. Org. Chem.*, 1991, **56**, 7179.
91JPC7659 M. G. Sowa, B. R. Henry and Y. Mizugai; *J. Phys. Chem.*, 1991, **95**, 7659.
91JPR229 K. Gewald, T. Jeschke and M. Gruner; *J. Prakt. Chem.*, 1991, **333**, 229.
91LA331 L. Capuano, S. Drescher and V. Huch; *Liebigs Ann. Chem*, 1991, 331.
91M185 C. R. Noe, M. Knollmüller, K. Dungler and P. Gärtner; *Monatsh. Chem.*, 1991, **122**, 185.
91M413 K. Czech, N. Haider and G. Heinisch; *Monatsh. Chem.*, 1991, **122**, 413.
91M705 C. R. Noe, M. Knollmuller, K. Dungler, C. Miculka and P. Gartner; *Monatsh. Chem.*, 1991, **122**, 705.
91MI19 V. V. Litvinova, A. V. Anisimov and E. A. Viktorova; *Sulfur Lett*, 1991, **13**, 19.
91MI27 S. A. S. Ghozlan; *Sulfur Lett*, 1991, **13**, 27.
91MI37 J. Y. Gui, D. A. Stern, F. Lu and A. T. Hubbard; *J. Electroanal. Chem. Interfacial Electrochem*, 1991, **305**, 37.
91MI71 R. Benassi, U. Folli, A. Mucci, L. Schenetti and F. Taddei; *THEOCHEM*, 1991, **74**, 71.
91MI81 R. Benassi, U. Folli, D. Iarossi, A. Mucci, L. Schenetti and F. Taddei; *J. Mol. Struct.*, 1991, **246**, 81.
91MI99 U. Folli, D. Iarossi, A. Mucci, A. Musatti, M. Nardelli, L. Schenetti and F. Taddei; *J. Mol. Struct.*, 1991, **246**, 99.
91MI109 E. I. Mamedov, A. G. Ismailov, N. V. Zyk, A. G. Kutateladze and N. S. Zefirov; *Sulfur Lett*, 1991, **12**, 109.
91MI119 Z. H. Zhou and T. Yamamoto; *J. Organomet. Chem.*, 1991, **414**, 119.
91MI182 A. D. Child, J. P. Ruiz, Y. J. Qiu and J. R. Reynolds; *Polym. Mater. Sci. Eng.*, 1991, **64**, 182.
91MI183 C. K. Sha and C. P. Tsou; *J. Chin. Chem. Soc. (Taipei)*, 1991, **38**, 183.
91MI233 R. A. Cox; *J. Phys. Org. Chem.*, 1991, **4**, 233.
91MI277 J. Roncali, A. Guy, M. Lemaire, R. Garreau and H. A. Hoa; *J. Electroanal. Chem. Interfacial Electrochem*, 1991, **312**, 277.
91MI301 D. T. Stanton, P. C. Jurs and M. G. Hicks; *J. Chem. Inf. Comput. Sci.*, 1991, **31**, 301.
91MI463 D. Fichou, B. Xu, G. Horowitz and F. Garnier; *Synth. Met.*, 1991, **41**, 463.
91MI495 K. Matsuo, A. Takeuchi and J. Kawanishi; *Chem. Express*, 1991, **6**, 495.

91MI579	D. A. dos Santos, D. S. Galvão, B. Laks and M. C. dos Santos; *Chem. Phys. Lett.*, 1991, **184**, 579.
91MI697	R. Venkatasubramanian and S. L. N. G. Krishnamachari; *Indian J. Pure Appl. Phys.*, 1991, **29**, 697.
91MI699	H. E. Katz, M. L. Schilling, C. E. D. Chidsey, T. M. Putvinski and R. S. Hutton; *Chem. Mater.*, 1991, **3**, 699.
91MI872	R. Borjas and D. A. Buttry; *Chem. Mater.*, 1991, **3**, 872.
91MI888	Y. Wei, C. C. Chan, J. Tian, G. W. Jang and K. F. Hsueh; *Chem. Mater.*, 1991, **3**, 888.
91MI1841	A. Berlin, G. A. Pagani, F. Sannicolo, G. Schiavon and G. Zotti; *Polymer*, 1991, **32**, 1841.
91MI1847	K. Kobayashi, Y. Mazaki, K. Kikuchi, K. Saito, I. Ikemoto and S. Hino; *Synth. Met.*, 1991, **42**, 1847.
91MI2205	H. Santalo, J. Veciana, C. Rovira, E. Molins, C. Miravitlles and J. Claret; *Synth. Met.*, 1991, **42**, 2205.
91MI2436	M. G. Choi and R. J. Angelici; *Organometallics*, 1991, **10**, 2436.
91MI3037	P. Bäuerle and K. U. Gaudl; *Synth. Met.*, 1991, **43**, 3037.
91MIC29	D. L. Wang and W. S. Hwang; *J. Organomet. Chem.*, 1991, **406**, C29.
91NN1277	M. E. Hassan; *Nucleosides Nucleotides*, 1991, **10**, 1277.
91P26	S. El-Bahaie, A. El-Deeb and M. C. Assy; *Pharmazie*, 1991, **46**, 26.
91PPI413	S. Tumkevicius and J. Mickiene; *Org. Prep. Proced. Int.*, 1991, **23**, 413.
91PS305	I. M. A. Awad, A. E. Abdelrahman and E. A. Bakhite; *Phosphorus, Sulfur Silicon Relat. Elem.*, 1991, **61**, 305.
91S277	A. M. Shestopalov, O. P. Bogomolova and V. P. Litvinov; *Synthesis*, 1991, 277.
91S462	A. Merz and F. Ellinger; *Synthesis*, 1991, 462.
91S567	I. S. Cho and J. M. Muchowski, *Synthesis*, 1991, 567.
91S785	R. Sato and S. Satoh; *Synthesis*, 1991, 785.
91SC145	R. L. P. de Jong and L. Brandsma; *Synth. Commun.*, 1991, **21**, 145.
91SC1055	R. J. Heffner and M. Joullié; *Synth. Commun.*, 1991, **21**, 1055.
91SC1875	R. Rossii, A. Carpita and T. Messeri; *Synth. Commun.*, 1991, **21**, 1875.
91SL31	M. Kahn, R. A. Chrusciel, S. Ting and X. Tong; *Synlett*, 1991, 31.
91SL33	G. Beese and B. A. Keay; *Synlett*, 1991, 33.
91SL595	K. C. Majumdar, A. T. Khan and S. Saha; *Synlett*, 1991, 595.
91T313	F. Alonso and M. Yus; *Tetrahedron*, 1991, **47**, 313.
91T3045	E. Terpetschnig, G. Penn, G. Kollenz, K. Peters, E. M. Peters and H. G. von Schnering; *Tetrahedron*, 1991, **47**, 3045.
91TL721	H. G. Selnick, E. M. Radzilowski and G. S. Ponticello; *Tetrahedron Lett*, 1991, **32**, 721.
91TL1351	K. Kanematsu, N. Sugimoto, M. Kawaoka, S. Yeo and M. Shiro; *Tetrahedron Lett*, 1991, **32**, 1351.
91TL3499	T. Kawase, S. Fujino and M. Oda; *Tetrahedron Lett*, 1991, **32**, 3499.
91TL3507	K. Takahashi and M. Ogiyama; *Tetrahedron Lett*, 1991, **32**, 3507.
91TL3699	M. Soufiaoui, B. Syassi, B. Daou and N. Baba; *Tetrahedron Lett*, 1991, **32**, 3699.
91TL3867	S. Braverman, A. P. F. T. M. Van, d. L. J. B. Van and B. Zwanenburg; *Tetrahedron Lett*, 1991, **32**, 3867.
91TL4313	A. Ishii, Y. Horikawa, I. Takaki, J. Shibata, J. Kakayama and M. Hoshino; *Tetrahedron Lett.*, 1991, **32**, 4313.
91TL4359	H. Shimizu, K. Hamada, M. Ozawa, T. Kataoka and M. Hori; *Tetrahedron Lett*, 1991, **32**, 4359.
91TL4367	Y. Mazaki, S. Murata and K. Kobayashi; *Tetrahedron Lett*, 1991, **32**, 4367.
91TL4573	N. Choi, Y. Kabe and W. Ando; *Tetrahedron Lett*, 1991, **32**, 4573.
91TL4635	K. D. Cramer and C. A. Townsend; *Tetrahedron Lett*, 1991, **32**, 4635.
91TL5305	K. W. Haider, J. A. Clites and J. A. Berson; *Tetrahedron Lett*, 1991, **32**, 5305.
91Z353	J. Salbeck and E. Guenther; *Z. Naturforsch., B: Chem. Sci.*, 1991, **46**, 353.

CHAPTER 5.2

Five-Membered Ring Systems: Pyrroles and Benzo Derivatives

RICHARD J. SUNDBERG
University of Virginia, Charlottesville, VA, USA

Introduction

Two useful reviews appeared early in 1991. A review of the synthesis of pyrroles from ketoximes and acetylenes was published.<90AHC171> The review discusses reaction variables and tabulates yields and physical properties for a variety of alkyl, aryl and heteroaryl derivatives prepared by this method. A review on 1-hydroxypyrroles, 1-hydroxyindoles and 9-hydroxycarbazoles appeared. <90AHC105> This review discusses structure, spectroscopic properties, synthesis and natural occurrence of these *N*-OH heterocycles.

5.2.1 Methods for Ring Synthesis

A new [4 + 1] route to 3-acylpyrroles involves addition of an amine to 2-acyl-3-(phenylsulfonyl)1,3-butadiene.<91S171> Aromatization occurs by elimination of benzenesulfinate and air oxidation. The substituted dienes are prepared in three steps from 3-phenylthiosulfolene and are somewhat unstable.

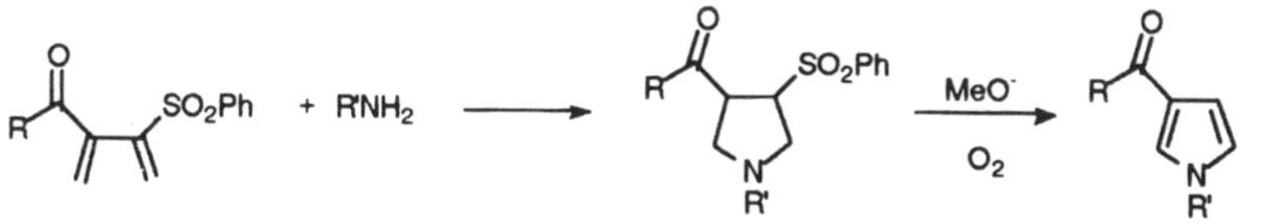

The synthesis of 3-alkylpyrrole-2-carboxylate esters from *N*-*p*-toluenesulfonylglycine and α,β-unsaturated ketones has been improved by substituting DBU for potassium *t*-butoxide as the base in the addition and aromatization steps <91JHC1671>.

The overall yield for benzyl 2-methylpyrrole-2-carboxylate was improved from 30-35% to 53%. The modified conditions also gave much improved results for α,β-unsaturated aldehydes. Acrolein, crotonaldehyde and methacrolein gave the corresponding pyrroles in 20-50% yield.

Several new examples of pyrrole synthesis from isocyanides have been reported. Conjugate addition of nitromethane to 1-isocyano-1-tosyl-1-alkenes gives 4-substituted-3-nitropyrroles.<91T4639>

$$RCH{=}C(NC)SO_2Ar + CH_3NO_2 \xrightarrow{t\text{-BuOK}} \text{3-R-4-}NO_2\text{-pyrrole (N-H)}$$

The isocyanides are made by condensation of an aldehyde with tosylmethyl isocyanide. This route appears to be superior to condensation of tosylmethyl isocyanide with nitroalkenes, at least in some cases <90T7587>. The latter method was used to prepare a series of 3-aryl-4-nitropyrroles in 40-60% yield by reaction of the nitroalkene with the sodium salt of tosylmethyl isocyanide. <91JHC2053>

$$ArCH{=}CHNO_2 + ArSO_2\bar{C}HNC \longrightarrow \text{3-Ar-4-}NO_2\text{-pyrrole (N-H)}$$

α-Isocyanatoesters react with nitroolefins to give pyrrole-2-carboxylates by elimination of nitrite ion.<90T7587>

$$RO_2CCH_2NC + R^3CH{=}CR^4NO_2 \longrightarrow \text{3-}R^3\text{-4-}R^4\text{-pyrrole-2-COOR (N-H)}$$

The tosylmethyl isocyanide synthesis was also applied to 3-trifluoromethyl-pyrroles by using trifluoromethylacrylates in the cycloaddition.<91JFC61> The ester substituent is removed by decarboxylation.

The formation of the pyrrole ring by [2 + 3] cycloaddition has been observed using adducts prepared from an aromatic aldehyde and benzotriazole in the presence of ammonia. The adducts are deprotonated to 2-azaallyl anions by *n*-butyllithium. Cycloaddition occurs with electron-deficient dipolarophiles. Dimethyl acetylene-dicarboxylate leads directly to a pyrrole, whereas methyl acrylate, acrylonitrile and similar dipolarophiles give 3,4-dihydro-2*H*-pyrroles.<91S863>

$$ArCH(Bt)N{=}CHAr \xrightarrow{BuLi} Ar\bar{C}(Bt)N{=}CHAr \xrightarrow[-\,BtH]{CH_2{=}CHZ} \text{2,5-diAr-3-Z-3,4-dihydro-2}H\text{-pyrrole}$$

Bt = benzotriazolyl

A cobalt-mediated cyclization of alkynes with trimethylsilyl cyanide gives tris-TMS derivatives of 5-aminopyrrole-2-carbonitriles.<91JO2166> The reaction is not very regioselective, although terminal alkynes tend to give mainly the 3-substituted isomer.

$$RC{\equiv}CR + NCSiMe_3 \xrightarrow[100\text{-}130^\circ C]{CpCo(CO)_2} \text{NC-pyrrole(R,R)-}N(SiMe_3)_2\ (N\text{-}SiMe_3)$$

Propargyl amines are converted to pyrroles under hydroformylation conditions. <91TL1093>

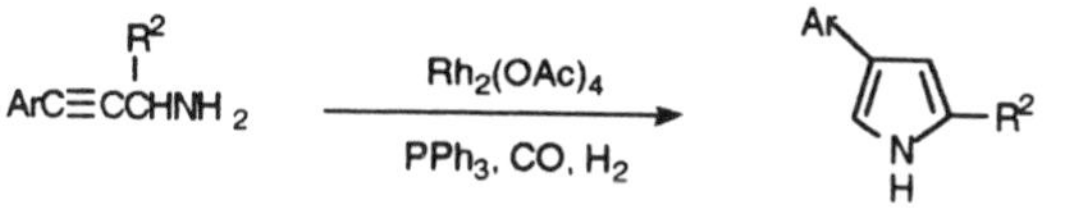

Most of the examples studied gave arylpyrroles but use of 2-heptyneamine gave a fair yield of 3-butylpyrrole.

Some new mechanistic insight into the Paal-Knorr reaction was obtained in a study of the cyclization of meso and dl stereoisomers of 3,4-dialkyl-2,5-hexadienes. <91JO6924> Among the key points which were established was a difference in the rates of cyclization of the diastereomers and the absence of hydrogen exchange via enolization of the starting material. These observations rule out enolization prior to the rate-determining step, as well as intermediates such as B which are common to both diastereomers.

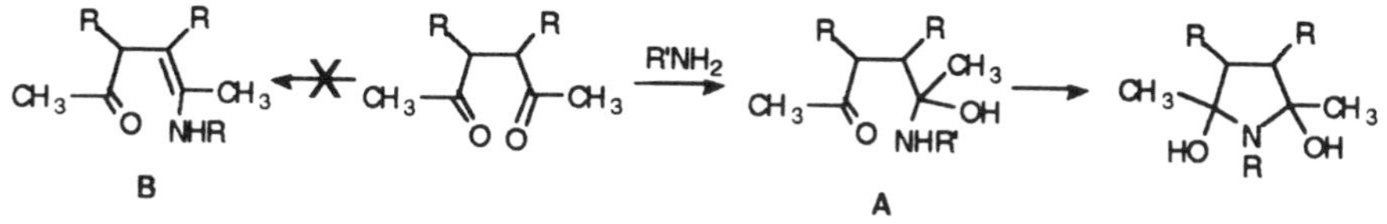

The authors suggest that the carbinolamine A undergoes cyclization in the rate-determining step. Differences in rates between the meso and dl isomers increased with the size of both the 3,4 and amine substituents, indicating that steric interactions are a major factor in determining the rate of cyclization.

New information on the long-standing issue of regioselectivity in the Fischer indole cyclization has been developed. The generalization that use of a strongly acidic medium tends to favor cyclization of methyl ketones to 2-alkylindoles was pursued by using P_2O_5/CH_3SO_3H as a cyclization medium. Selective cyclization to 3-alkylindoles was achieved with product ratios of >50:1 being achieved in some cases. <91JO3001> With the prototype series 1a-c, the ratios were not as high, but did achieve a preference for the 2-alkylindole.

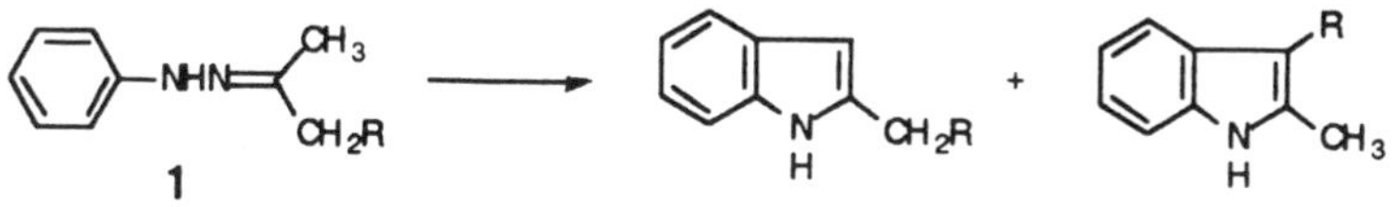

	R	Ratio
a	CH_3	78:22
b	$(CH_3)_2CH$	85:15
c	$(CH_3)_3C$	90:10

Several factors were identified which influence the product ratio. Dilution of the acid with an inert solvent generally increased the ratio of the 2,3-disubstituted product. Increasing steric bulk, either on the phenylhydrazine nitrogen or in the ketone substituent, favored the 3-unsubstituted product.

A new sigmatropic rearrangement which yields *N*-tosylindole from aniline has been developed. The procedure is based on selective ortho alkylation by a Sommelet-Hauser rearrangement of an anilinosulfonium ion. On heating in propionic acid cyclization to the indole occurs, presumably through an episulfonium ion intermediate.<91H1377>

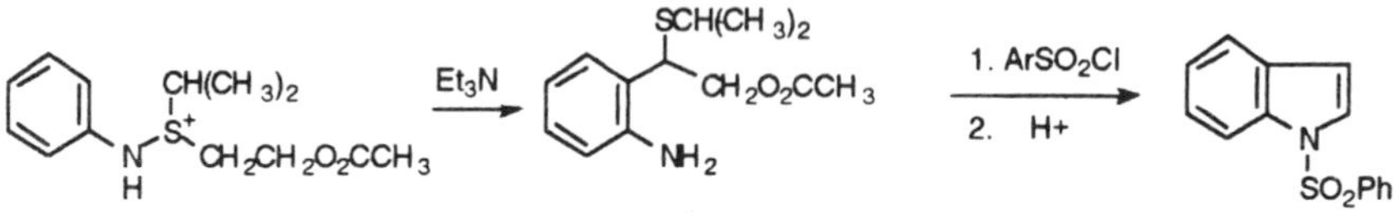

The dilithio derivative of *N-t*-butoxycarbonyl-2-alkylanilines can be cyclized to either indoles or oxindoles by using DMF or CO_2, respectively.<91S871>

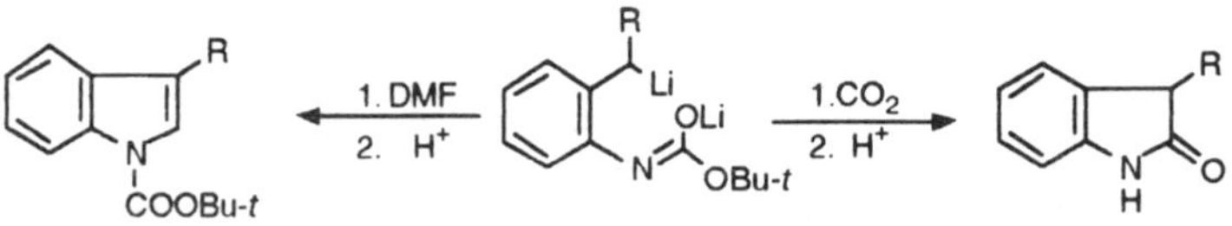

Yields of 60-90% were obtained with both indoles and oxindoles. The method was extended to 2-substituted indoles by using *N*-methyl-*N*-methoxycarboxamide in place of DMF. The reaction is compatible with ring substituents such as Cl, F and OMe.

A new oxindole synthesis based on sigmatropic rearrangement of trimethylsilyl enol ethers of *N*-acyloxyanilines has been developed.<91TL2671> The TMS ethers are prepared at -90 °C from the ester enolate in situ and the rearrangement occurs in the course of warming to ambient temperature. The cyclization is then done with dicyclohexylcarbodiimide.

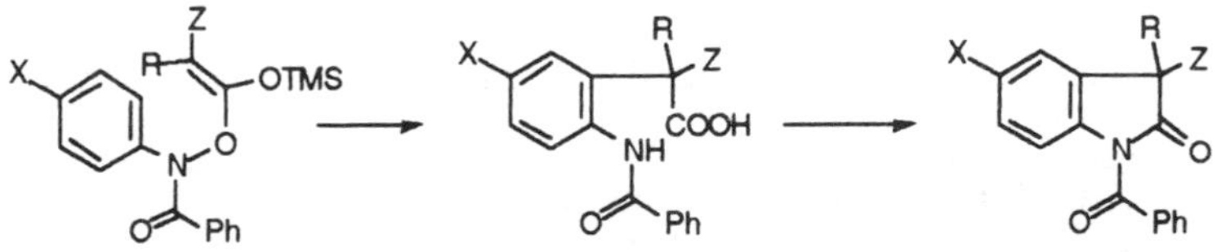

At least one of the substituents must be carbanion-stabilizing, (e.g. Ph, SPh) for good results. A rather similar sequence which apparently extends the scope of the reaction was reported by Endo and co-workers.<91TL2803> Their method involves rearrangement of the potassium enolate at -78 °C. These workers also compared yields with *N-t*-butyl, *N*-benzoyl and *N*-isobutyroyl derivatives as reactants. Using LDA for rearrangement of the amides, comparable yields were obtained.

New methods for conversion of o-haloanilines to indoles continue to be developed. A high-yielding oxindole ring closure applicable to oxindole-3-acetic acids was developed and used as the starting point for synthesis of physostigmine and other calabar alkaloids.<91JCS(P1)3047>

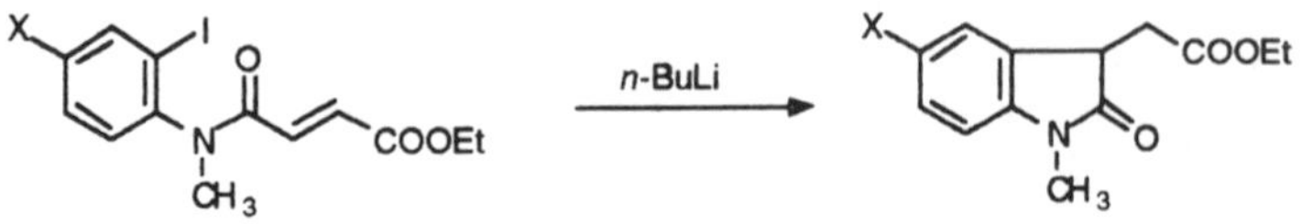

Samarium diodide has been shown to effect a reductive cyclization of *N*-allyl-*o*-bromoanilines. With an *N*-propargyl substituent a fair yield of the corresponding indole was obtained.<91TL1737>

A procedure for converting o-iodoanilines to indoles by cyclization with acetylenes has been developed.<91JA6689> The regioselectivity is such as to place the smaller alkyne substituent at the 3-position of the indole ring.

An interesting extension of the *N*-allyl-o-haloaniline cyclization was achieved by use of a zirconocene intermediate. The cyclization occurs with concomitant zirconation at C-4 and this intermediate can be utilized to form 4-substituted derivatives. The iodomethyl products can be dehydrohalogenated to 3-methylene-indolines which are reactive towards electrophiles so as to permit a tandem chain extension.<91JA4685>

X=Y : $EtO_2CC{\equiv}CCO_2Et$; NCCH=CHCN ; $O{=}C(CO_2Et)_2$; $CH{=}N^+R_2$

Oxidative cyclization of derivatives of *o*-aminophenylethanol, which has previously been applied to simple indoles <90JO580> was used with more elaborate structures in the preparation of teleocidin analogs.<91JO4706>

A new sequence for synthesis of indoles from nitroaromatics has been introduced. An acetonitrile substituent is introduced following the "vicarious nucleophilic aromatic substitution" route pioneered by Makosza.<88LA203>

Treatment of the acetonitrile under Mitsunobu conditions then generates an anion which can be alkylated by primary, benzylic and allylic alcohols. Reductive cyclization then gives a 3-substituted indole. <91TL7195>

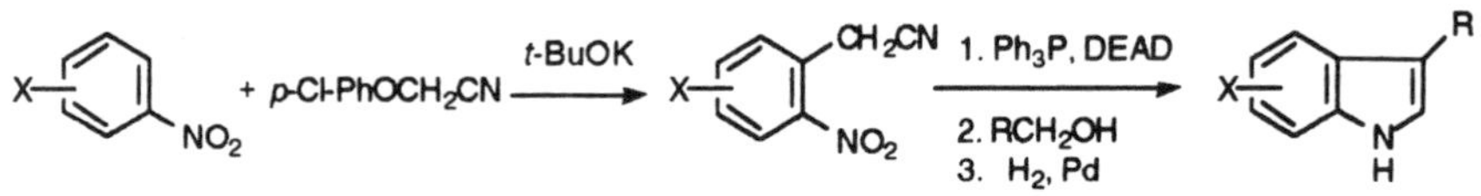

Improvement was reported in the synthesis of 7-alkoxyindoles by reaction of alkoxynitrobenzenes with vinylmagnesium bromide.<91SC611> The best results were achieved with benzhydryl as the oxygen substituent. This reaction presumably begins with addition of the Grignard reagent ortho to the nitro group.

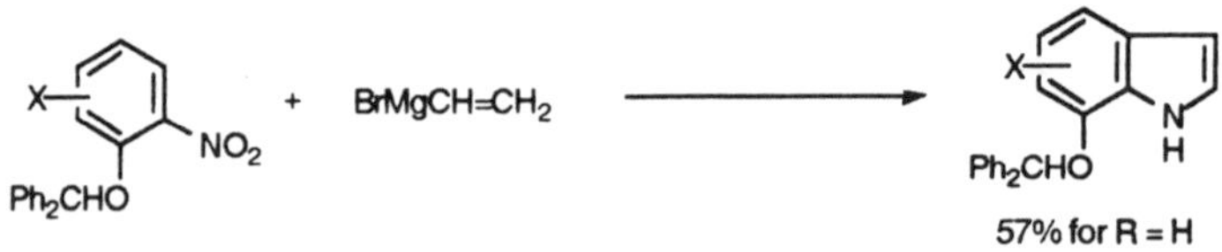

57% for R = H

Addition of 1-trimethylsilylvinylmagnesium bromide to nitrosoarenes generates 2-trimethylsilylindoles in 15-50% yield.<91JCS(P1)2757> Competing reactions include formation of azo and azoxy compounds. The reaction is formulated as occurring through a hydroxylamine derivative which undergoes sigmatropic rearrangement followed by cyclization.

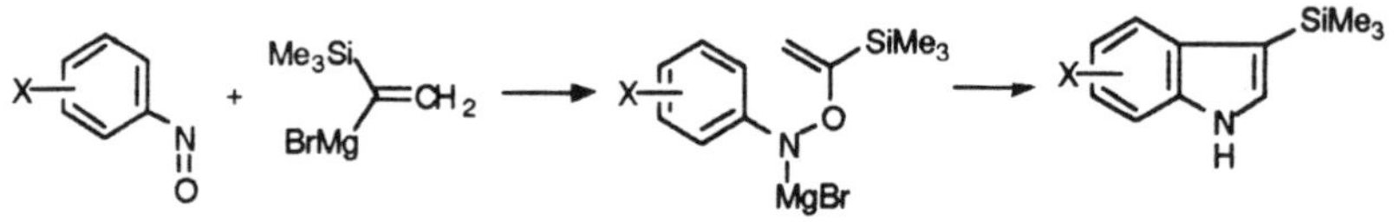

Reduced titanium has proven to be able to reductively form double bonds from a variety of dicarbonyl compounds. This methodology has now been applied to formation of indoles from *o*-acylanilides. <91TL6695> The reagent used was titanium on graphite prepared from potassium-graphite and $TiCl_3$.

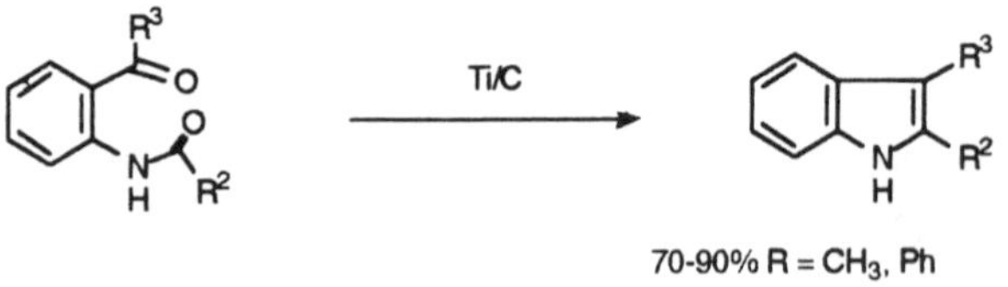

70-90% R = CH_3, Ph

Highly substituted indoles have been constructed by intramolecular inverse electron demand Diels-Alder reactions involving diazines. The diazene, which is itself constructed by a Diels-Alder addition, undergoes nucleophilic substitution by 1-amino-2,3-diene. Thermal cycloaddition-elimination after acetylation gives trikentin, an example of antimicrobial indoles isolated from marine sponges. <91JA4230>

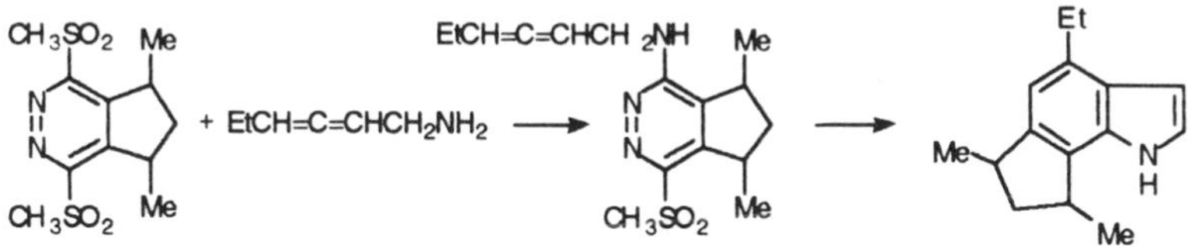

5.2.2 Ring Substitution and Modification Reactions

Both ring and side chain lithiation continue to play a prominent role in new methods for synthetic elaboration of indoles. Metallation of 2,3-dialkyl-indoles by *n*-BuLi/*t*-BuOK takes place at the 2 position after prior nitrogen deprotonation. Successful alkylation of the dianion by methyl bromide has been reported. <91JO2256>

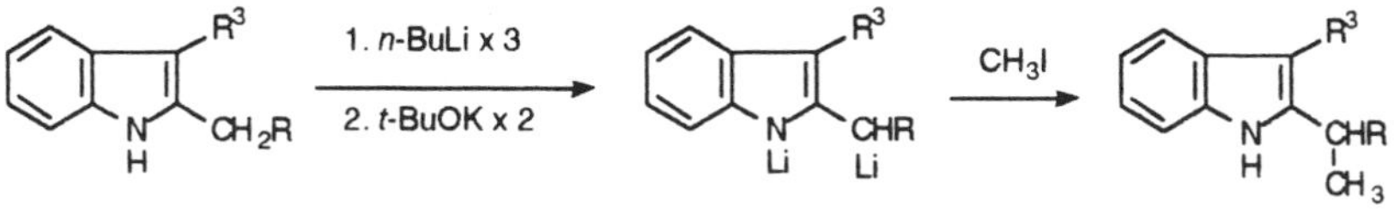

1-Methoxyindole was shown to be cleanly lithiated at C-2 with *n*-BuLi. After reaction with typical carbonyl electrophiles, 2-substituted products were obtained.<91H221> Since the methoxy group is easily removed by hydrogenolysis, it can serve as a protecting group in the preparation of indoles. The preparation of the 1-alkoxyindoles was done by oxidation of indoline with $NaWO_4/H_2O_2$. <91CPB1905>

The use of the *N,N*-diethylcarbamoyl moiety as the protecting group for lithiation of 2-(2-indolyl)dithiane proved to be more effective than arenesulfonyl groups which were found to promote 3-lithiation of the indole ring in competition with dithiane lithiation. Although some competition from this process is still observed with aldehydes as electrophiles, halides give exclusively alkylation of the dithiane. <91T7911>

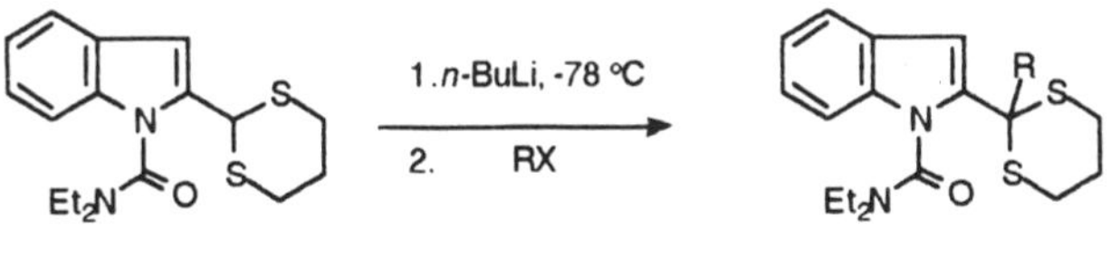

The *N-t*-butylcarbamoyl group has been shown to have excellent properties as a substituent for 2-lithiation of pyrroles and indoles. *t*-Butyllithium (2.2 equiv.) is used for deprotonation and lithiation at -78 °C, followed by warming to room temperature. The lithiated intermediates react with aldehydes, CO_2, TMS-Cl and *N,N*-dimethylbenzamide to give the expected products in 45-95% yield. The decarbamoylation can be carried out with LiOH in methanol/THF.<91S1079>

A reaction sequence which affords 2-monoalkylated 1H-indol-3-ones has been developed. 1-Acetyl-2-methoxy-1,2-dihydroindol-3-one, which is readily available by oxidation of 1-acetylindole <84CPB3945> can be alkylated under phase-transfer conditions. The indolone is then reduced with $NaBH_4$ and a pinacol rearrangement is induced with $SnCl_4$ or TMS triflate.<91JCS(P1)2445>

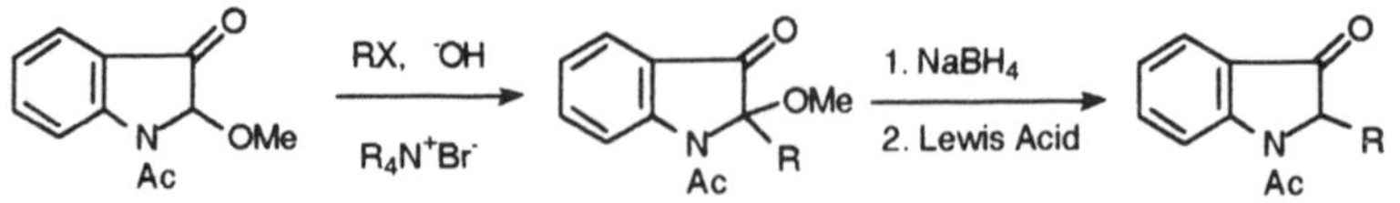

The alkylation of 1*H*-pyrrol-3(2*H*)-ones with alkyl halides typically gives mainly C-alkylated products. The O-alkylation products, which are 3-alkoxypyrroles, can be obtained by using methyl tosylate in a dipolar aprotic solvent such as *N,N*-dimethylimidazolidinone (DMI). <91JCS(P1)3245> These results follow the usual hard-soft reactivity relationships for enolate alkylation.

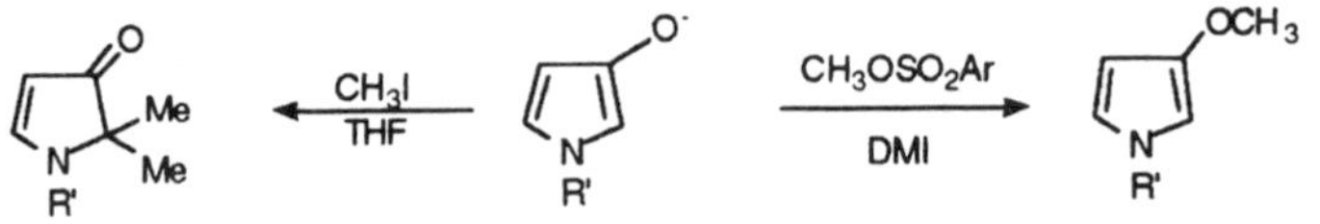

Several indoleacetic acid esters and indolylmethyl ketones were prepared by a Wittig reaction between substituted methylenetriphenylphosphoranes and *N*-acetylindol-3-one. <91S701>

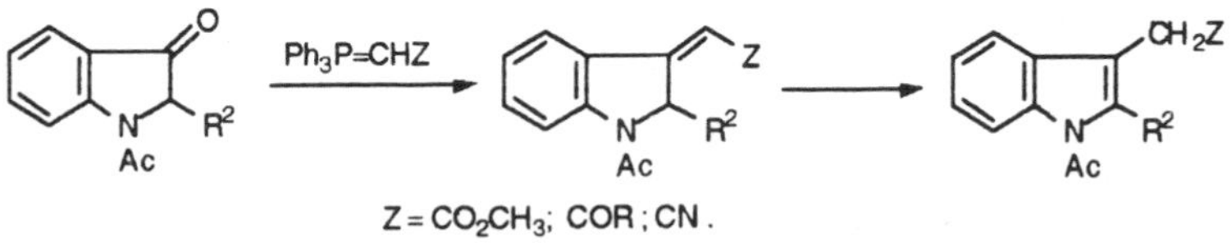

Z = CO_2CH_3; COR ; CN .

Both arylboronic acids and arylstannanes have proven to be useful intermediates in cross-coupling reactions catalyzed by palladium. The preparation of *t*-Boc-protected pyrrole intermediates via lithiation have been described. <91S613>

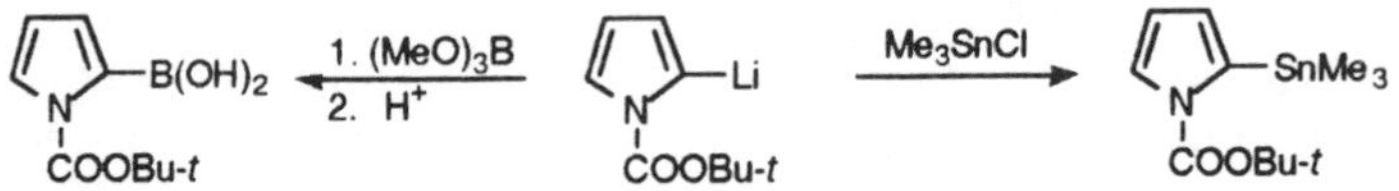

Cross-coupling of *N*-*t*-butoxycarbonyl-2,5-dibromopyrrole with *N*-*t*-butoxycarbonyl-2-trimethylstannylpyrrole gave a tripyrrole in about 30% yield. <91SM403>

The scope of nucleophilic addition to 1-(2-indolyl)but-2-enecarbonitrile has been explored. <91TL7237> Good yields of addition of methyl, vinyl and allyl groups were achieved with organocerium or organoytterbium reagents. Conjugate addition also occurred with lithiated ethyl vinyl ether and the product was cyclized to a pyrrolo[1,2-a]indole.

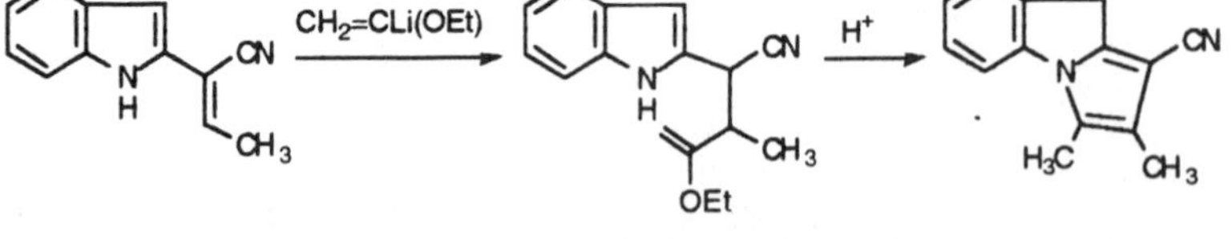

Diethyl malonate also gave a cyclic product derived from conjugate addition.

A reactivity pattern which is well-known for simple *N*-magnesioindoles was extended to fused ring indoles to obtain 3-methyl-3*H*-indole derivatives. <91JHC1293> spiro-Pseudoindoxyls were observed as by-products and could be obtained in good yield (except for n = 3) when the magnesioindole was intentionally oxygenated.

$(CH_2)_n$ MgX CH_3I CH_3 $(CH_2)_n$ + O $(CH_2)_n$ H

n = 3 - 6

Intramolecular radical addition has provided a route to pyrrolo[1,2-a]indole ring systems such as found in the mitomycins. 3-Bromo or 3-iodoallylindole can by cyclized with tri-*n*-butylstannane.<91JO3479> The reaction seems to be quite insensitive to substituents with 2- or 3-carboethoxy and 3-carboethoxymethyl substituents having little effect on yields. Even the *N*-3-iodopropyl derivative cyclized in 50% yield.

Z Y $CH_2CH{=}CHX$ Bu_3SnH AIBN Z Y

X = Br; I
Y = H; COOR
Z = H; COOR; CH2COOR

Indolylcarbenes were generated from indole-2-carboxaldehyes by thermal decomposition of *N*-(diphenylaziridino)imines or by base-catalyzed decomposition of tosylhydrazones.<91JCS(P1)1721> The carbene derived from **2** added to toluene to give a cycloheptatriene and gave other reactions indicative of carbene formation but intramolecular addition to the 1-alkynylmethyl group was not observed.

Ph CH=NN Ph CH_2-C≡CCH$_2$OCH$_3$ **2** Toluene 105 °C CH_3 CH_2-C≡CCH$_2$OCH$_3$

On the other hand, addition did occur to an *N*-methoxyimine bond.

CH=NNSO$_2$Ar CH_2-C=NOCH$_3$ NaH NOCH$_3$

These are interesting results in that they suggest indolylcarbenes may not undergo ring expansion reactions as easily as many other aryl carbenes.

The ketene derived from indole-3-carboxylic acid was generated both from the acid chloride and the enol ester **3**. The product isolated is the tetramer **4**, whose structure was determined by X-ray crystallography and NMR. <91JHC1569>

Et_3N

3 **4**

5.2.3. Annulation Reactions

3-Thioacetylpyrroles can be converted to α-benzylthio-3-vinylpyrroles by S-alkylation with benzyl chloride. The 3-vinylpyrroles undergo cycloaddition reactions with typical electron-deficient dienophiles. The products, after oxidative aromatization, are 4-benzylthioindoles.<91CPB489>

SCH_2Ph + XCH=CHY → SCH_2Ph

X = COOMe; CN; CH=O
Y = COOMe;

X, Y = –C(=O)–O–C(=O)–

Similarly, a 2-(1-trimethylsiloxyvinyl)pyrrole prepared from a 2-acetylpyrrole was used for the preparation of a 7-hydroxyindole.<91H1199>

+ HC≡CCOOCH$_3$ →

Studies of the [2 + 2] photocyclization of 1-benzoylindole and 1-carboethoxyindole with alkenes have yielded some mechanistic and stereochemical insight. With cyclopentene, cyclohexene or cycloheptene, two stereoisomeric adducts were isolated. For cyclopentene these could be unambiguously assigned as **5** and **6**. For cycloheptene a trans ring fusion is observed as the major product.

5 + **6** **7**

With acyclic alkenes mixtures of all four possible stereoisomers are formed, regardless of alkene stereochemistry. The results are compatible with formation of a

1,4-diradical intermediate having a structure-dependent ratio of cyclization and reversion to reactant.<91CJC1171>

Indole has been shown to undergo electron transfer catalyzed cycloaddition with cyclohexadiene. A photoactivated pyrylium salt is used as the electron transfer agent.<91JO1405> The adducts must be trapped in situ by acylation to prevent their deleterious photo-oxidation. Neither 2- nor 3-substituted indoles react as well as indole. The cyclohexadiene can carry an alkyl or acetoxy substituent. The adducts are generally mixtures of endo and exo isomers with the endo predominating.

hv, R'COCl

endo exo

The regioselectivity is dominated by a donor substituent on the diene. The more electron-rich site in the diene becomes bound to the 3-position of the indole ring. This reactivity pattern is consistent with the general mechanistic designation involving initial bond formation to the indole radical cation at the 3-position.

The scope of dienophiles which can successfully trap the benzoylindole quinodimethane intermediate has been explored. With highly electrophilic dienophiles such as maleimide yields are excellent. Good yields are also reported for dimethyl acetylenedicarboxylate and benzoquinone.<91T1925> Reactions with mono-substituted dienophiles such as phenyl vinyl sulfone or methyl vinyl ketone give mixtures of regioisomers with the 3-substituted tetrahydrocarbazole derivative predominating.

+ CH_2=CHZ

Z = $COCH_3$; SO_2Ph

Several 7-substituted indoles were prepared by intramolecular palladium-mediated vinylation.<91CPB2830> In each case at least some of the initial product proceed directly to the aromatic structure. The starting materials were prepared beginning with a Friedel-Crafts acylation of 1-phenylsulfonylpyrrole.

$PdCl_2$

X, Y = O; H, OAc; H, H; Z = OH; OAc; H

Both indoles and pyrroles were annulated by a Pd-catalyzed carbonylation reaction of reactants with 3-acetoxy-1-propenyl substituents. <91JO1922>

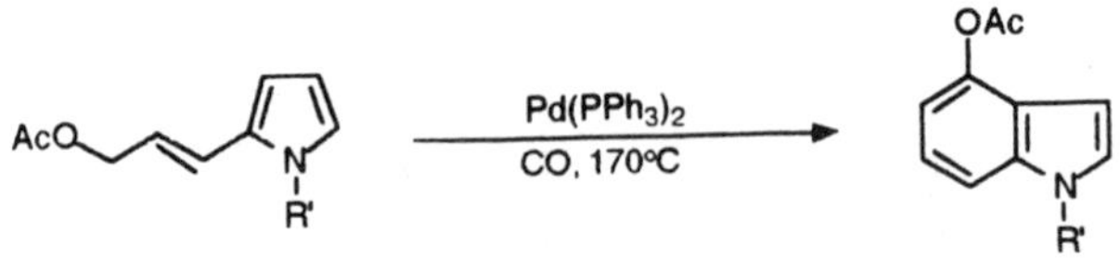

AcO
N
R'
Pd(PPh3)2
CO, 170°C
OAc
N
R'

Intramolecular arylation of indoles with a variety of 2-bromophenyl groups attached via the 1-position has been achieved using $Pd(PPh_3)_4$. <91TL3317>

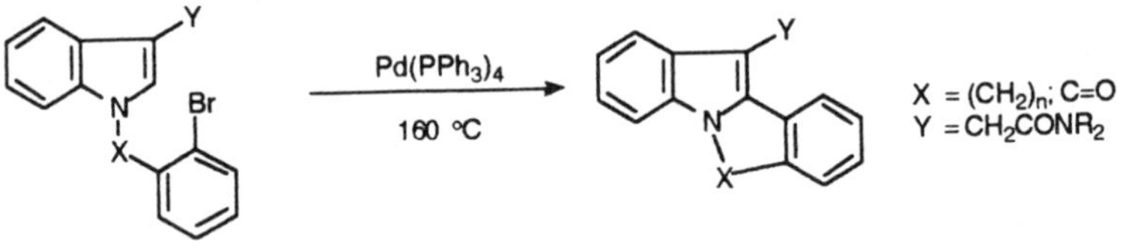

Y
N
Br
X
Pd(PPh3)4
160 °C
X = (CH2)n; C=O
Y = CH2CONR2

A tandem strategy for indole annulation based on a borate intermediate has been illustrated.<91CC1219> In the first step a terminally unsaturated allylic acetate is hydroborated with 9-BBN and then allowed to react with 2-lithio-1-methylindole, generating a borate intermediate.

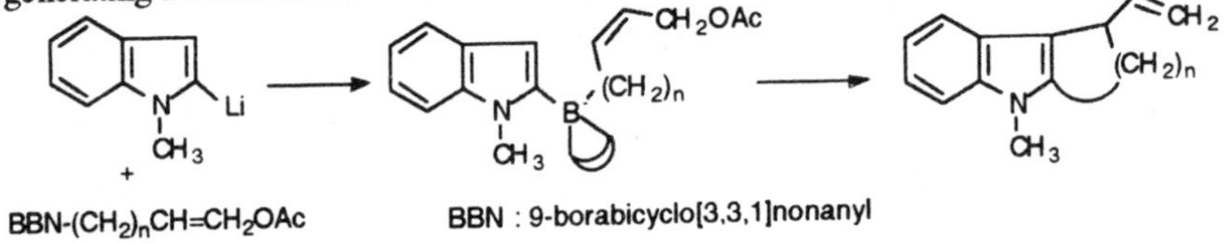

N
Li
CH3
+
CH2OAc
(CH2)n
B
CH2
BBN-(CH2)nCH=CH2OAc
BBN : 9-borabicyclo[3,3,1]nonanyl

The cyclization is then induced with Pd(0). The allylic Pd intermediate effects electrophilic substitution at the 3-position of the indole ring which, in turn, triggers B-C migration and aromatization.

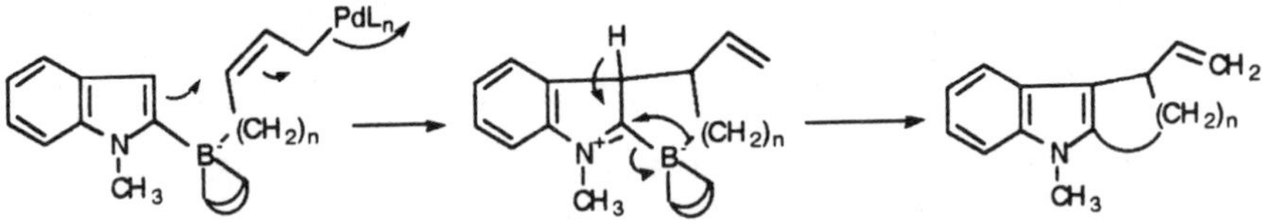

PdLn
(CH2)n
B
N
CH3
H
N+
CH2

Operationally, the procedure can be done in "one-pot" as illustrated by the synthesis of **8**.

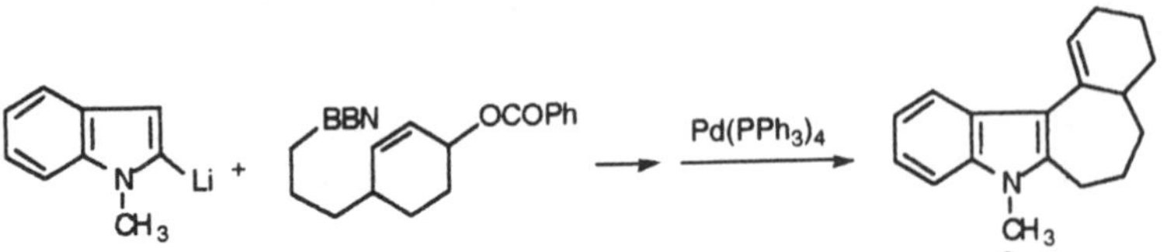

N
Li
CH3
+
BBN
OCOPh
Pd(PPh3)4

8

References

84CPB3945 C.-S. Chien, T. Suzuki, T. Kawasaki and M. Sakamoto, *Chem. Pharm. Bull.*, **32**, 3945-3951, 1984.

88LA203 M. Makosza, W. Danikiewicz and K. Wojciechowski, *Liebigs Ann. Chem.*, 203-214 (1988).

90AHC105 R. M. Acheson, *Adv. Heterocycl. Chem.*, 51, 105-175 (1990).

90AHC177 B. A. Trofimov, *Adv. Heterocycl. Chem.*, 51, 177-301 (1990).

90JO580 Y. Tsuji, S. Kotachi, K.-T. Huh and Y. Watanabe, *J. Org. Chem.*, **55**, 580-584 (1990).

90T7587 D. H. R. Barton, J. Kervagoret an S. Z. Zard, *Tetrahedron*, **46**, 7587-7598 (1990).

91CC1219 M. Ishikura, M. Terashima, K. Okamura and T. Date, *J. Chem. Soc., Chem. Commun.*, 1219-1221 (1991).

91CJC1171 D. J. Hastings and A. C. Weedon, *Can. J. Chem.*, 69, 1171-1181 (1991).

91CPB489 M. Murase, S. Yoshida, T. Hosaka and S. Tobinaga, *Chem. Pharm. Bull.*, **39**, 489-492 (1991).

91CPB1905 M. Somei, T. Kawasaki, K. Shimizu, Y. Fukui and T. Ohta, *Chem. Pharm. Bull.*, **39**, 1905-1907 (1991).

91CPB2830 Y. Yokoyama, H. Suzuki, S. Matsumoto, Y. Sunaga, M. Tani and Y. Murakami, *Chem. Pharm. Bull.*, **39**, 2830-2836 (1991).

91H221 T. Kawasaki, A. Kodoma, T. Nishida, K. Shimizu and M. Somei, *Heterocycles*, **32**, 221-227 (1991).

91H1377 Y. Murai, G. Masuda, S. Inoue and K. Sato, *Heterocycles*, **32**, 1377-1386 (1991).

91H1199 M. Ohno, S. Shimizu and S. Eguchi, *Heterocycles*, **32**, 1199-1202 (1991)

91JA4230 D. L. Boger and M. Zhang, *J. Am. Chem. Soc.*, **113**, 4230-4234 (1991).

91JA4685 J. H. Tidwell, D. R. Senn and S. L. Buchwald, *J.Am.Chem.Soc.*, **113**, 4685-4686 (1991).

91JA6689 R. C. Larock and E. K. Yum, *J. Am. Chem. Soc.*, **113**, 6689-6690 (1991).

91JCS(P1)1721 G. B. Jones, C. J. Moody, A. Padwa and J. M. Kassir, *J. Chem. Soc. Perkin Trans. 1*, 1721-1727, (1991).

91JCS(P1)2445 T. Kawasaki, Y. Noraka, M. Kobayashi and M. Sakamoto, *J. Chem. Soc. Perkin Trans. 1*, 2445-2448 (1991).

91JCS(P1)2757 G. Bartoli, M. Bosco, R. Dalpozzo, G. Palmieri and E. Marcantoni, *J. Chem. Soc. Perkin Trans. 1*, 2757-2761 (1991).

91JCS(P1)3047 S. Horne, N. Taylor, S. Collins and R. Rodrigo, *J. Chem. Soc. Perkin Trans. 1*, 3047-3051 (1991).

91JCS(P1)3245 G. A. Hunter, H. McNab, L. C. Monahan and A. J. Blake, *J. Chem. Soc. Perkin Trans. 1*, 3245-3251 (1991).

91JFC61 J. Leroy, *J. Fluorine Chem.*, **53**, 61-70 (1991).

91JHC1293 J.-G. Rodriquez and A. San Andres, *J. Heterocycl. Chem.*, **28**, 1293-1299 (1991).

91JHC1569 J. E. Oliver, R. M. Waters, W. R. Lusby and J. L. Flippen-Anderson, *J. Heterocycl. Chem.*, **28**, 1569-1572 (1991).

91JHC1671 T. D. Lash and M. C. Hoehner, *J. Heterocycl. Chem.*, **28**, 1671-1676 (1991).

91JHC2053 N. Ono, E. Muratani and T. Ogawa, *J. Heterocycl. Chem.*, **28**, 2053-2055 (1991).

91JO1405	A. Gieseler, E. Steckhan, O. Wiest and F. Knoch, *J. Org. Chem.*, **56**, 1405-1411 (1991).
91JO1922	M. Iwasaki, Y. Kobayashi, J.-P. Li, H. Matsuzaka, Y. Ishii and M. Hidai, *J. Org. Chem.*, 56, 1922-1927 (1991).
91JO2166	N. Chatani and T. Hanafusa, *J. Org. Chem.*, **56**, 2166-2170 (1991).
91JO2256	Y. Naruse, Y. Ito and S. Inagaki, *J. Org. Chem.*, **56**, 2256-2258 (1991).
91JO3001	D. Zhao, D. L. Hughes, D. R. Bender, A. M. DeMarco and P. J. Reider, *J.Org.Chem.*, **56**, 3001-3006 (1991).
91JO3479	F. E. Ziegler and L. O. Jeroncic, *J. Org. Chem.*, **56**, 3479-3486 (1991).
91JO4706	R. R. Webb, II, M. C. Venuti and C. Eigenbrot, *J. Org. Chem.*, **56**, 4706-4713 (1991).
91JO6924	V. Amarnath, D. C. Anthony, K. Amarnath, W. M. Valentine, L. A. Wetterau and D. G. Graham, *J. Org. Chem.*, **56**, 6924-6931 (1991).
91S171	S.-S. Chou and T.-M. Yuan, *Synthesis*, 171-172 (1991).
91S613	S. Martina, V. Enkelmann, G. Wegner and A.-D. Schluter, *Synthesis*, 613 (1991).
91S701	T. Kawasaki, Y. Nonaka, M. Uemura and M. Sakamoto, *Synthesis*, 701-702 (1991).
91S863	A. R. Katritzky, G. J. Hitchings and X. Zhao, *Synthesis*, 863-867 (1991).
91S871	R. D. Clark, J. M. Muchowski, L. E. Fisher, L. A. Flippin, D. B. Repke and M. Souchet, *Synthesis*, 871-878 (1991).
91S1079	M. Gharpure, A. Stoller,F. Bellamy, G. Firnau and V. Snieckus, *Synthesis*, 1079-1082 (1991).
91SC611	D. Dobson, A. Todd and J. Gilmore, *Synth. Commun.*, 21, 611-617 (1991).
91SM403	S. Martina, V. Enkelmann, A.-D. Schluter and G. Wegner, *Synth. Meth.*, 41, 403-406 (1991).
91T1925	M. Haber and U. Pindur, *Tetrahedron*, **47**, 1925-1936 (1991).
91T4639	D. van Leusen, E. Flentge, and A. M. van Leusen, *Tetrahedron*, **47**, 4639-4644 (1991).
91T7911	J. Castells, Y. Troin, A. Diez, M. Rubiralta, D. S. Grierson and H.-P. Husson, *Tetrahedron*, **47**, 7911-7924 (1991)
91TL1093	E. M. Campi, W. R. Jackson and Y. Nilsson, *Tetraheron Lett.*, **32**, 1093-1094 (1991).
91TL1737	J. Inanaga, O. Ujikawa and M. Yamaguchi, *Tetrahedron Lett.*, **32**, 1737-1740 (1991).
91TL2671	P. S. Almeida, S. Prabhakar, A. M. Lobo and M. J. Marcelo-Curto, *Tetrahedron Lett.*, **32**, 2671-2674 (1991).
91TL2803	Y. Endo, S. Hizatate and K. Shudo, *Tetrahedron Lett.*, **32**, 2803-2806 (1991).
91TL3317	A. P. Kozikowski and D. Ma, *Tetrahedron Lett.*, **32**, 3317-3320 (1991).
91TL6695	A. Furstner, D. N. Jumbaum and H. Weidmann, *Tetrahedron Lett.*, **32**, 6695-6696 (1991).
91TL7195	J. E. Macor and J. M. Wehner, *Tetrahedron Lett.*, **32**, 7195-7198 (1991).
91TL7237	S. Blechert and T. Wirth, *Tetrahedron Lett.*, **32**, 7237-7240 (1991).

CHAPTER 5.3

Five-Membered Ring Systems: Furans and Benzo Derivatives

C. W. BIRD
King's College, University of London, UK

5.31 INTRODUCTION

The frenetic activity of the past few years, largely associated with the synthesis of avermectin, the milbemycins and assorted polether antibiotics, has abated and furan chemistry has apparently entered a relatively quiescent phase. The most obvious growth area is in the application of organometallic derivatives to elaboration of the furan ring. Emphasis has again been placed on presenting new developments rather than the noting of full accounts of work reviewed earlier, albeit in preliminary form.

5.32 REACTIONS

The major product of fluorination of furan is *syn*-2,5-difluoro-2,5-dihydrofuran <90G749>. Direct conversion of furan to the 2-bromo or 2,5-dibromo derivative can be effected satisfactorily by bromine in DMF <90SC3371>. The choice of solvent is also critical in the halodesilylation of 2-trimethylsilylfurans. The reagents of choice being sulfuryl chloride, bromine or iodine monochloride in acetonitrile <91JHC853; 91CE699>. Electrophilic substitution of dibenzofuran occurs selectively at the 2 and 3 positions, with the ratio of 2 to 3 substitution depending upon the electrophile <91JOC4671>. A novel electrophile is the tosylimine, $TsN{=}CHCO_2Me$, which reacts at room temperature with furan forming the amino acid derivative (1) <91SL29>. 2-Trimethylsilyloxyfuran reacts stereoselectively with *N*-acyl-*N*,*O*-acetals, $RO_2CNCHR^1OR^2$, at -78°C in the presence of boron trifluoride etherate, giving predominantly the *syn*-isomer (2) <91TL3795>.

Treatment of 3-silyloxycarbonylfurans with LDA provides 2-silyl-3-furoic acids <91SL33, which undergo regiospecific lithiation at C4 providing access to a variety of 4-substituted furoic acids <91JCS(P1)2600>.

A number of examples are available of the palladium catalysed coupling of 2- or 3-tributylstannylfurans and 2-tributylstannyl-4,5-dihydrofurans with aryl, vinyl or acyl halides which provide the corresponding aryl, vinyl or acylfurans <90BCJ2828; 91CJC1326; 91S242; 90TL6077>. Tetrakis(triphenylphosphine)palladium (0) mediates the reaction of aryl bromides with furan providing 2-arylfurans <90H(31)1951>. It has also been reported that the previously described arylation of 2,3-dihydrofuran at C4 witharyl triflates proceeds with excellent stereoselectivity when catalysed by palladous acetate and (R)-2,2'-bis(diphenylphosphino)-1,1'-binaphthyl <91JA1417>. The arylation of (3) with aryllead triacetate proceeds *via* (4), to give a convenient access to 2-arylbenzofuran-3-ones (5) <90TL6637>.

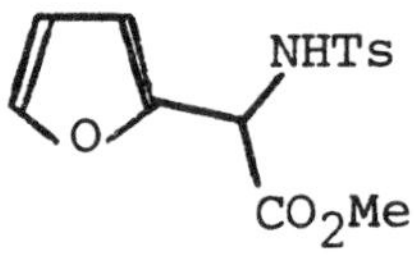

(1) 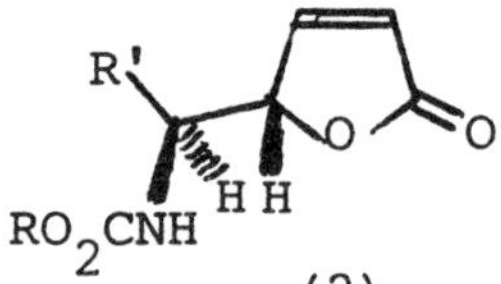 (2) 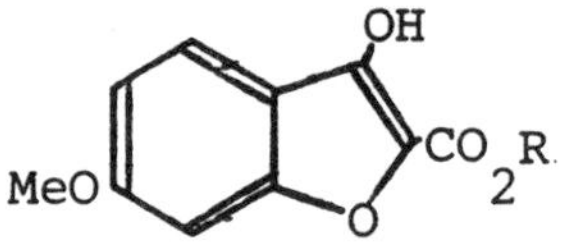 (3) R=Allyl

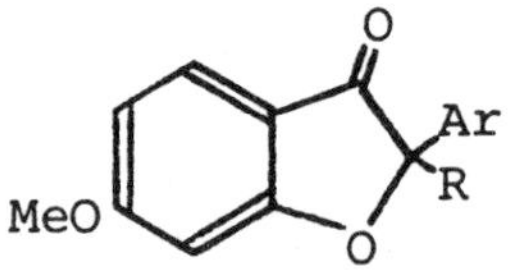

(4) R=CO_2Allyl (5) R=H (6) 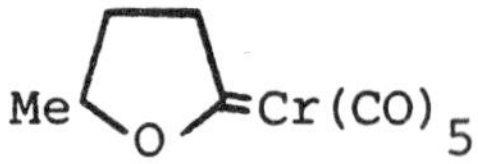 (7)

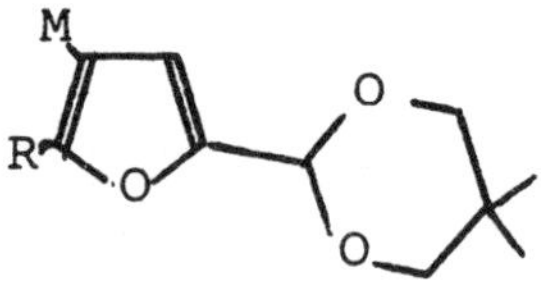

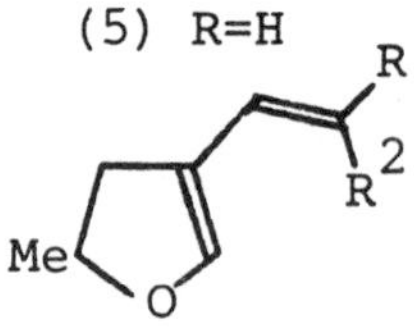

(8) 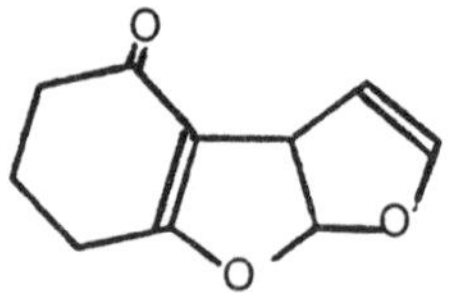 (9) 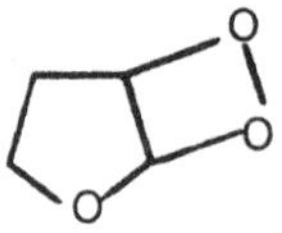 (10)

Good stereoselectivity is observed in the formation of furylcarbinols from either difurylzinc and aldehydes, or furfuraldehyde and dialkylzincs when the reactions are conducted in the presence of chiral β-aminoalcohols <90JCS(P1)3214; 91JCS(P1)25>. High chemoselectivity was obtained in the reaction of (6) with $MeO_2CC{\equiv}C(CH_2)_2CHO$ when M=$Ti(OPr^i)_3$, whereas with M=ZnCl, MgBr or Li the reagent proved unselective and/or too unreactive <91TL 903>. The base catalysed reaction of the carbene chromium complex (7) with aldehydes, R^1R^2CHCHO, provides 3-vinyl-4,5-dihydrofurans (8) <91CC437>.

Though acyclic diazoketones react with furans to give αβ,γδ-dienals or dienones, cyclic diazo-β-diketones react differently. Thus, 2-diazocyclohexane-1,3-dione and furan in the presence of rhodium (ii) diacetate generate the furanofuran derivative (9) <91JOC6269>.

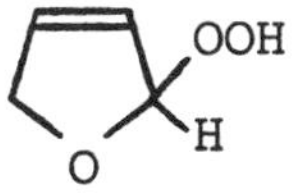

(11) 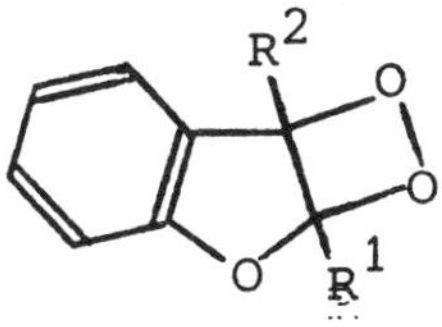 (12) 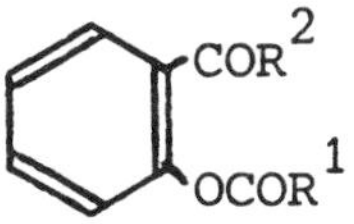 (13)

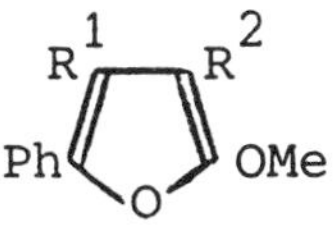

(14) 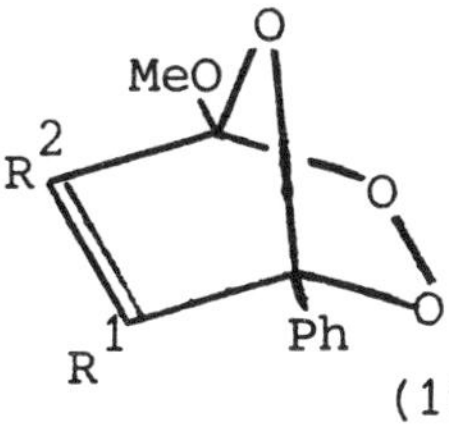 (15) 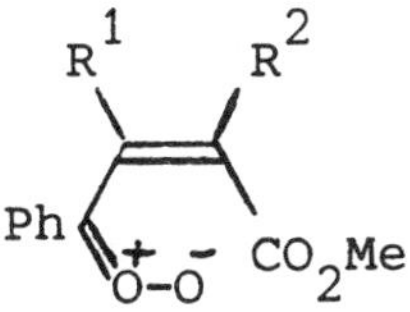 (16)

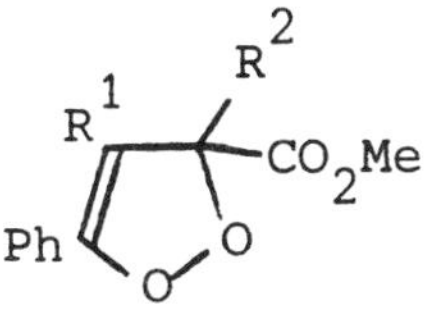

(17) 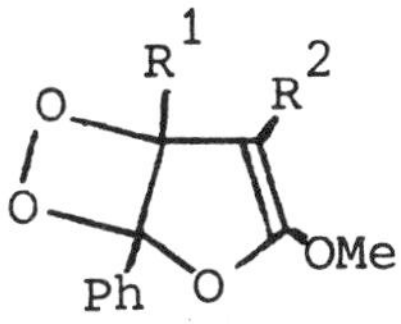 (18) 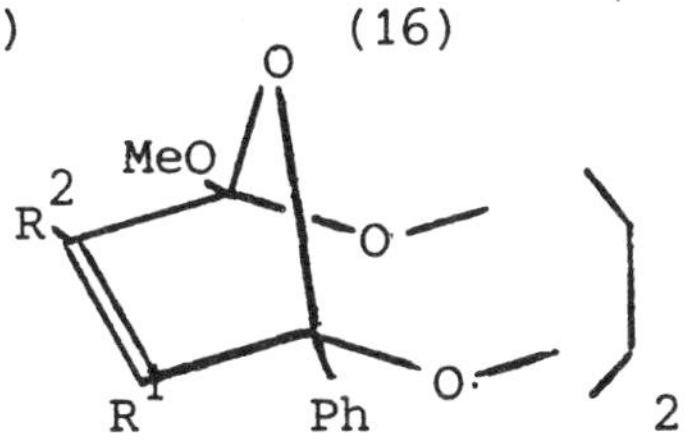 (19)

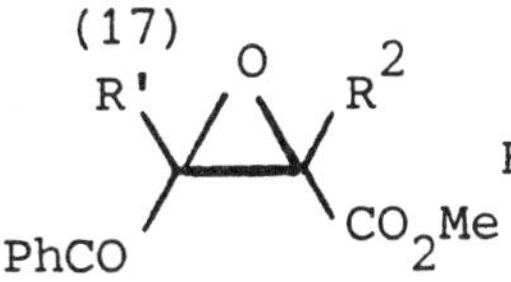

(20) 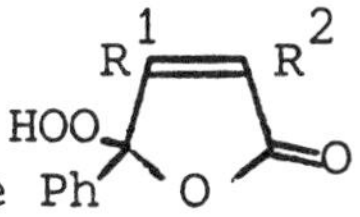 (21)

(a) R^1=H, R^2=Ph

(b) R^1=Me,Ph; R^2=CO_2Me

The reaction of furans with singlet oxygen continues to attract much attention. 2,3-Dihydrofuran forms a mixture of the dioxetane (10) and the hydroperoxide (11) which undergoes thermal decomposition to 3-formyloxypropanal and furan respectively. The proportions of (10) and (11) depend upon the solvent employed <91JOC4017>. Benzofurans give predominantly the dioxetanes (12) which decompose at 50°C to (13) <91LA33>. In contrast, at -78°, the furans (14) form the 2,5-adducts (15). Depending on the solvent polarity (15a) rearranges to the carbonyl oxide (16a) or the dioxole (17a) <91JCS(P1)1479>, while the adduct (15b) at -50° initially rearranges to the dioxetane (18b) and then more slowly to the dimer (19b). Warming this dimer at 70° generates the epoxide (20b). while mild acid hydrolysis provides the hydroperoxide

(21b) <91CC1061>. Optimum conditions have been establshed for the singlet oxygen mediated conversion of 4-substituted-2-trialkylsilylfurans to 4-substituted-5-hydroxyfuran-2(5H)-ones <91JOC7007>. The 2,5-addition of singlet oxygen to the furylcarbinol (22) followed by reduction with triphenylphosphine yields the dihydro pyrone (23) <90H(31)1941>. An imaginative application of such chemistry to the stereoselective synthesis of decalins has been described <90TL7201>. Thus reaction of the bisfuran (24) with singlet oxygen in methanol gave (25) which was reduced to the enedione (26) with triphenylphosphine. Subsequent zinc iodide catalysed intramolecular Diels-Alder addition provided (27). The

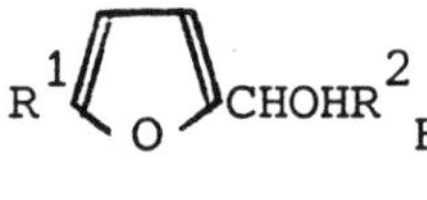

(22) 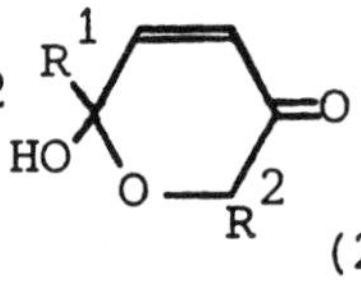 (23) 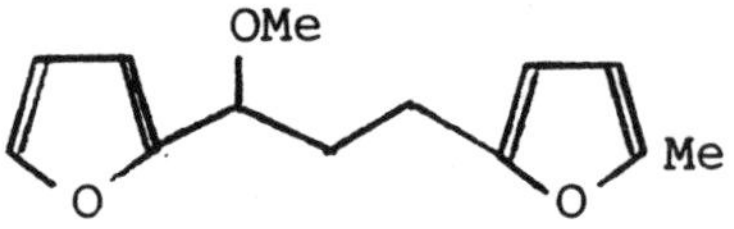 (24)

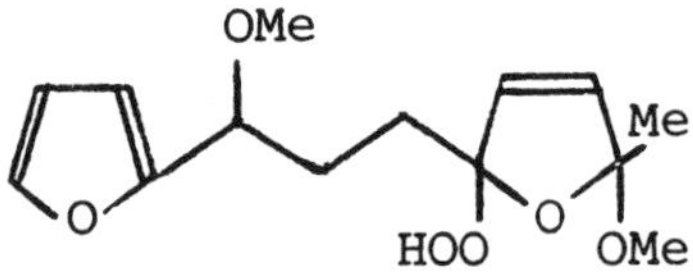

(25) 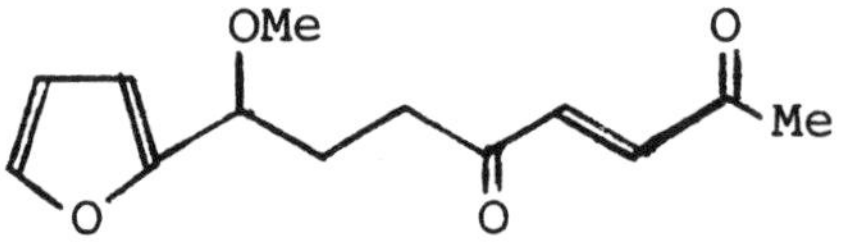 (26)

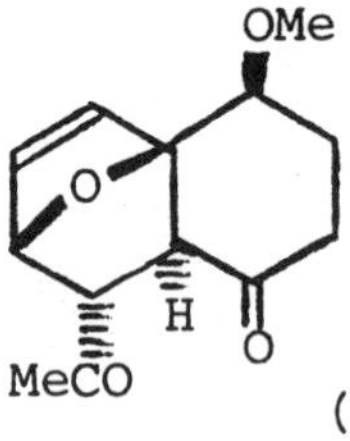

(27) 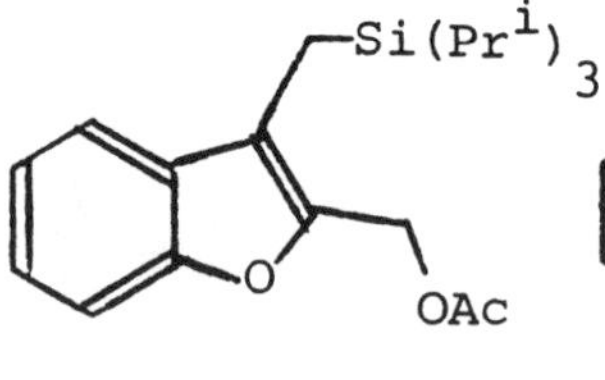 (28) 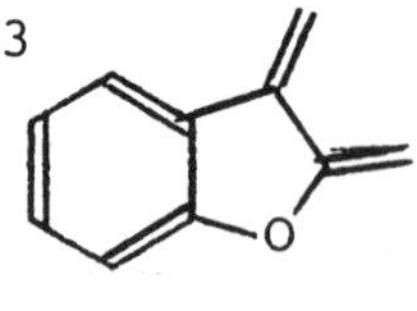 (29)

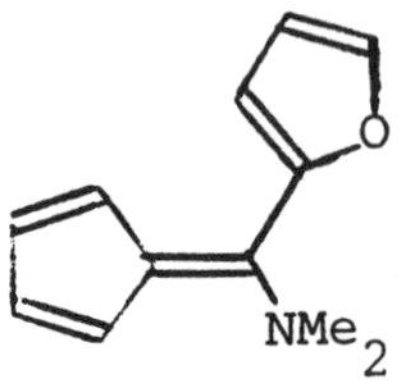

(30) 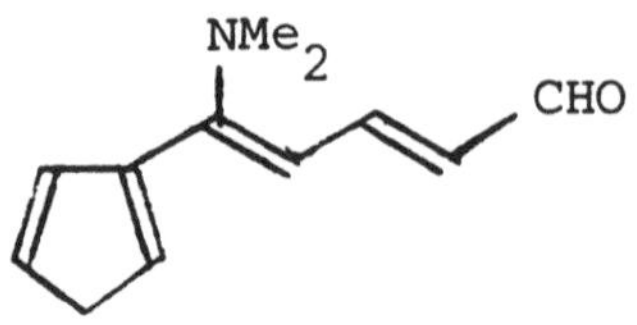 (31)

treatment of (28) with tetrabutylammonium fluoride generates 2,3-bismethylenebenzofuran (29), which has been intercepted by a variety of dienophiles yielding the corresponding tetrahydrodibenzofurans <91SL627>.

The most remarkable furan transformation reported this year is the conversion of the furylfulvene (30) by lithium naphthalenide to 4-dimethylaminoazulene. A possible intermediate is the aldehyde (31) which is a coproduct of the reaction <91TL3499>.

5.33 RING SYNTHESIS

5.331 Formation of the Carbon to Oxygen Bond

Triphenylphosphonium anhydride $[(Ph_3P^+)_2O.(TfO)_2^-]$, from triphenylphosphine oxide and triflic anhydride, effects the cyclisation of appropriate 1,4-diols to, *inter alia*, tetrahydrofuran, 2,3-dihydrobenzofuran and 1,3-dihydroisobenzofuran <90SL423>. The cyclisation of 2,2'-dihydroxybiphenyls to dibenzofurans is readily achieved by refluxing in *o*-xylene with Nafion-H resin

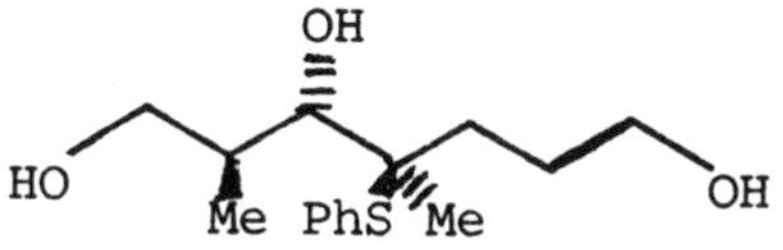

(32) 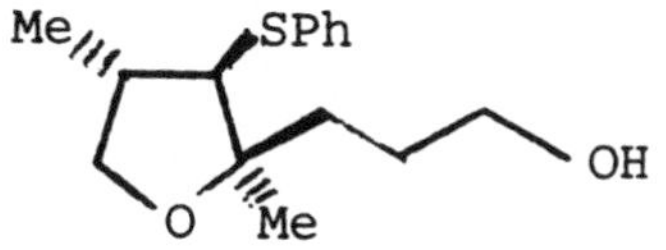 (33)

PhC≡C.CF_2CHOHR

(34) R=Ph,Et,vinyl

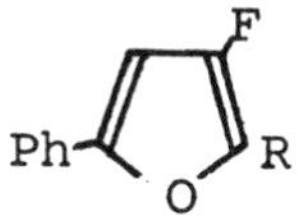

(35) 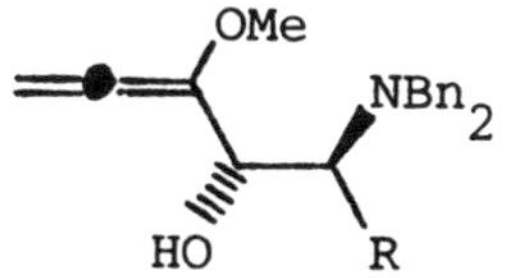 (36)

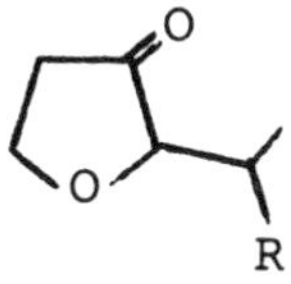

(37) 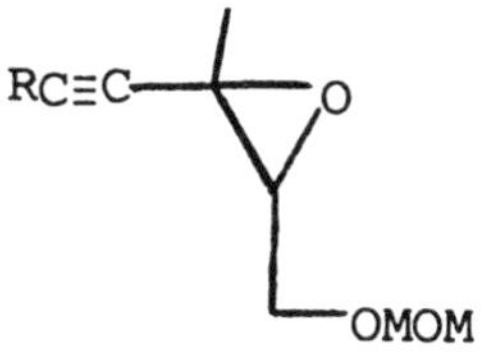 (38) 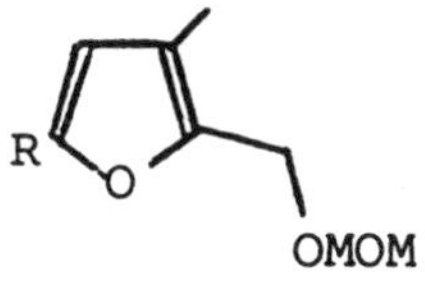 (39)

<91JOC3192>. The oxidative cyclisation of 2-hydroxystilbenes with dichlorodicyanobenzoquinone gives good yields of 2-arylbenzofurans <91CE323>. The factors controlling the stereospecific, acid catalysed cyclisation of compounds such as (32) to (33), which proceeds *via* an episulfonium ion, have been investigated <90TL3457>.

3-Fluorofurans (35) are formed by potassium *t*-butoxide mediated cyclisation of (34), which are obtained by a Reformatsky reaction of bromodifluoromethyl phenylacetylene with a range of aldehydes <91CC1134>. Analogous cyclisation of (36), synthesised from α-lithiomethoxyallene and α-aminoaldehydes, provides after acid hydrolysis the tetrahydrofuranones (37) <91SL179>. The rearrangement of (38) to (39) is effected by potassium hydride <91JOC1685>. The γδ-alkenylketones (40) are converted by iodine to the dihydrofurans (41) which undergo elimination of hydrogen iodide with DBU forming initially methylenefurans (42), which may be isomerised to the corresponding methylfurans <90H(31)1417>. Benzo[c]-furans (45) have been generated by base catalysed addition of methanol to (43), forming (44) and subsequent

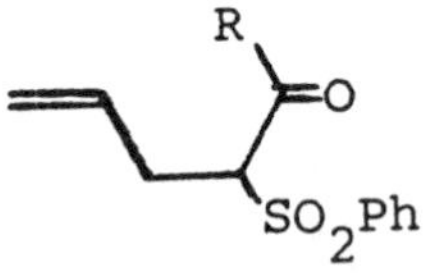

(40) R=alkyl,Ph 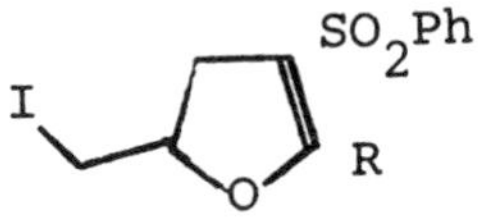 (41) 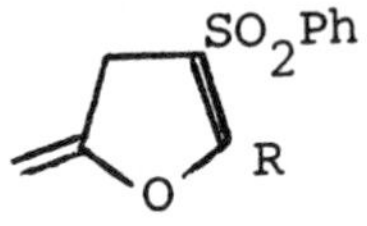 (42)

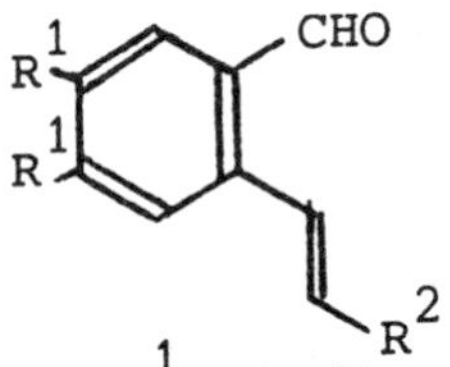

(43) R^1=OMe,H
R^2=CO_2Me,CN,NO_2

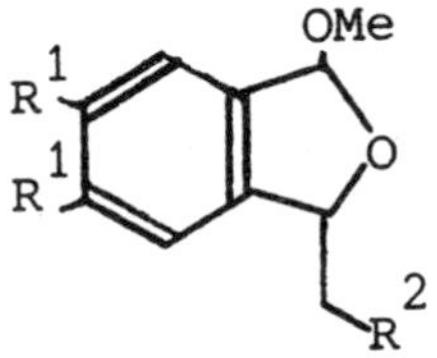

(44) 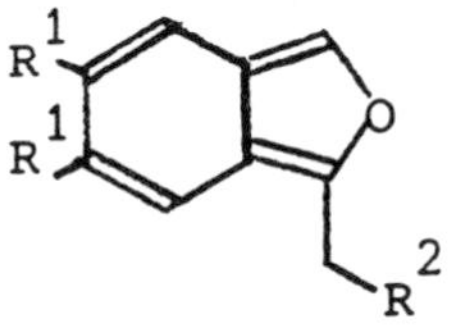 (45)

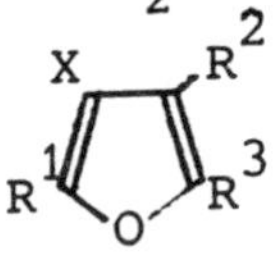

(46a) X=H (47) $R^1C{\equiv}C.CH_2CH(COR^2)_2$ (48)

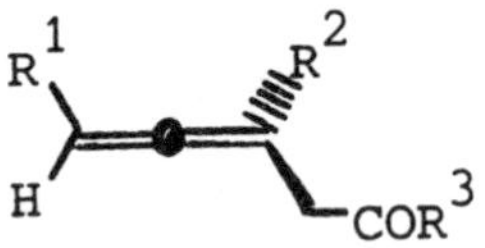

(46b) X=allyl

acid catalysed elimination of methanol <91JOC1882>.

β-Alkynylketones, $R^1C{\equiv}CCHR^2COR^3$, are cyclised to the furans (46a) by palladium (II) catalysts. If the reactionis carried out in the presence of allyl halides then the allylfuran (46b) results <91JOC5816>. Silver ion catalyses the cyclisation of allenylketones (47) to furans (46a) <91JOC960>. This chemistry has been applied to the synthesis of the 2,5-furanmacrocyclic sytems encountered in some macrocyclic diterpenes <91JOC6264>. Polyphosphoric acid converts the alkynes (48) into the furans (49), while the corresponding olefins give 4,5-dihydrofurans <90SL673>.

5.332 Formation of a Carbon to Carbon Bond

Intramolecular free-radical cyclisation reactions continue to receive exhaustive attention. The tributylstannane promoted cyclisation of (50) to (51) typefies the formation of a C1-C2 bond by this approach <91TL 1491; 91JOC5245>. The samarium (II) iodide mediated cyclisation of *o*-allyloxy or propargyloxyaryl iodides (bromides) initially yields an organosamarium species (eg. 52), which can be intercepted with a variety of electrophiles <90SL773; 91TL1737>. Sequential radical ring closure-radical ring opening is observed on treatment of (53) with tributylstannane and triethylboron which yields (54) <91JOC5285>. The photochemically initiated addition of diphenyl diselenide to allyl propargyl ether yields (55) <91JOC5721>.

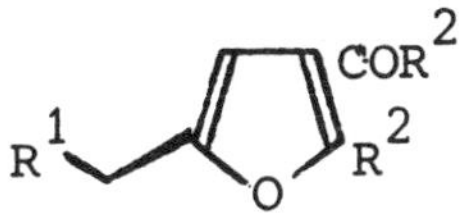

(49) 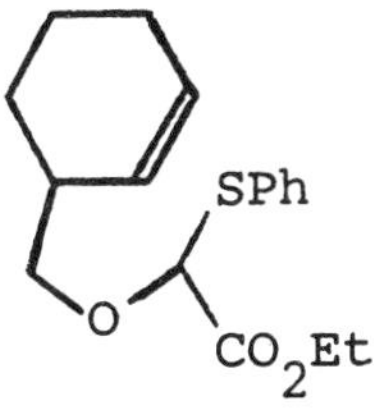 (50) 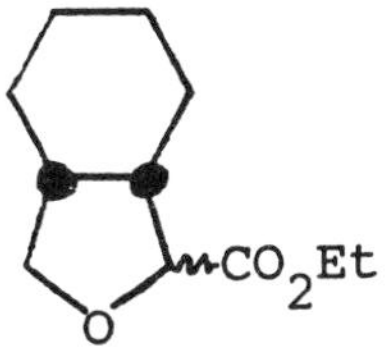 (51)

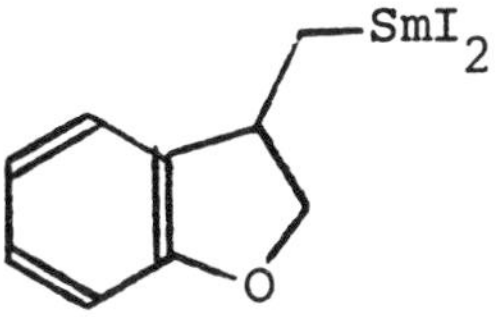

(52) 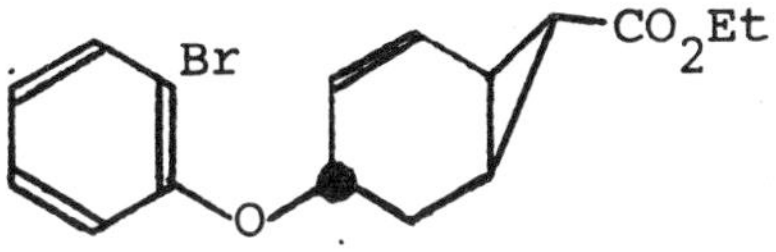 (53)

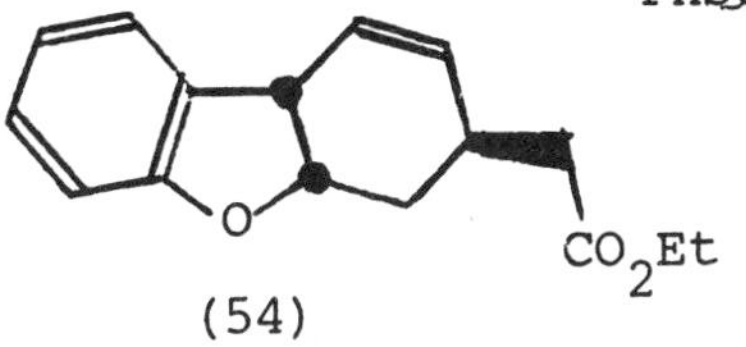

(54) 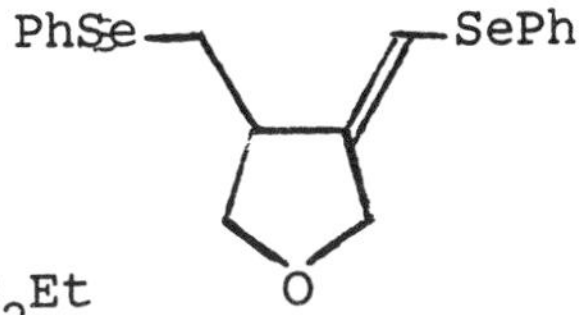 (55) 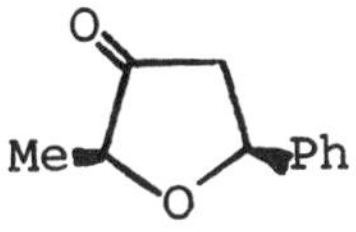 (56)

Copper acetylacetonate promoted cyclisation of diazo-3-benzyloxybutan-2-one gives predominantly the tetrahydrofuran-3-one (56) <90H(31)1745>.

The cyclisation of (57) to (58) by treatment with caesium fluoride employs a relatively novel method of carbanion generation <90JA7438>. The same workers also describe tetrakis(triphenylphosphine)palladium (0) catalysed cyclisations of, for example, (59) to (60). The conversion of (61) to the dihydrofuran (62) is initiated by vinylic deprotonation by LDA <91JOC3556>.

A Heck-type reaction of *o*-propargyloxyiodobenzene with arylzinc chlorides in the presence of a palladium (II) catalyst generates (63) <90H(31)2181>.

5.333 Formation of Two Bonds

2,3-Disubstituted furans may be obtained from Feist-Benary condensations using 1,2-dibromoethyl acetate as a coreactant <90SC1923>. Similarly substituted furans can be synthesised from an αβ-enal, $R^2CH=CHCHO$, and an aldehyde, R^1CHO, by a one pot procedure, which entails the intermediary generation of an arsonium ylide <91TL 2913>. Allenyldimethylsulfonium bromide, from dimethyl sulfide and propargyl bromide, has been used to annul-

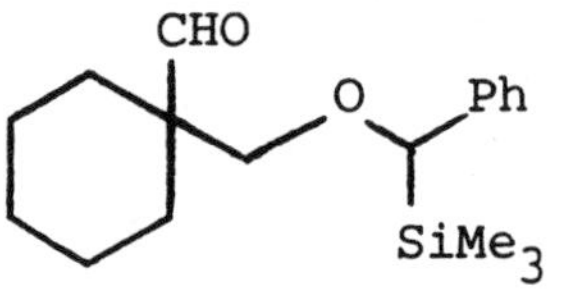

(57) 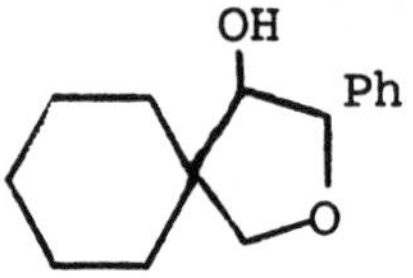 (58)

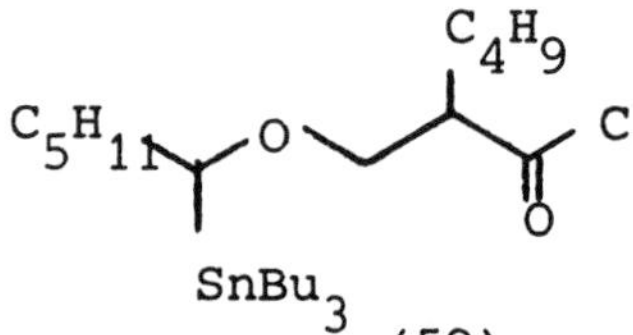

(59) 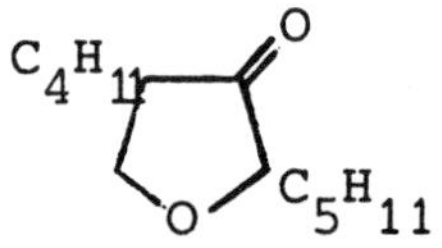 (60) 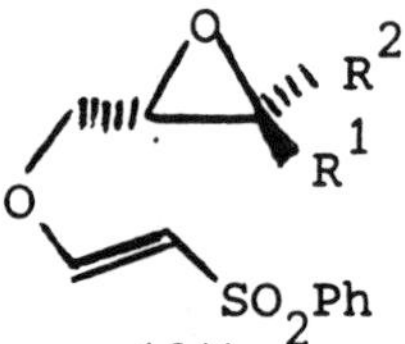 (61)

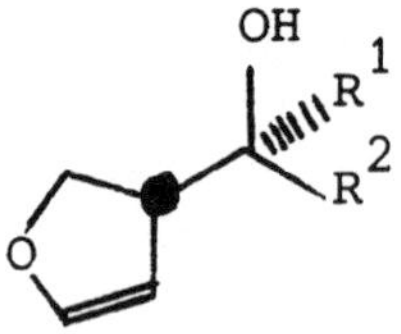

(62) 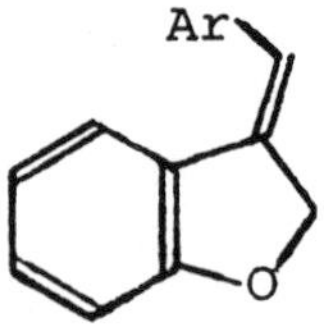 (63) 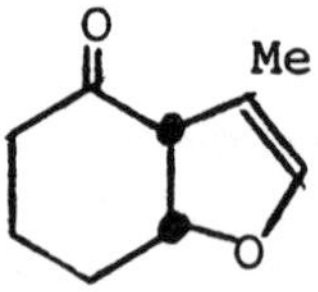 (64)

ate 1,3-cyclohexanediones to (eg.64) <90H(31)1003>.The application of the Vilsmeier reaction to arylpropane-1,2-diones yields the furans (65) <90JHC1843>. The reaction of dketene with styrenes, $PhCR'=CH_2$, promoted by $VO(OR)Cl_2$ provides the dihydrofurans (66) <91JOMC(407)C1>. Condensation of α-hydroxyketones, RCOCHOHR', with di-*tert*-butyl acetylenedicarboxylate has been used to generate the 4-hydroxydihydrofurans (67), which represent a structural sub-unit of the insect antifeedant azadirachtin <91TL4687>.

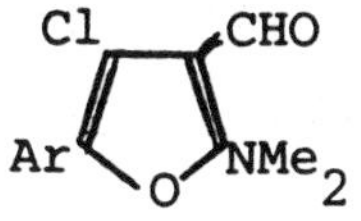

The stannic chloride catalysed rearrangement of allylic acetals such as (68) to the tetrahydrofurans (69) has been studied extensively <91JA5354,5365,5378>, The conversion of the vinyloxiranes (70) by trimethylsilyl iodide and hexamethyldisilazane into the dihydrofurans (71) clearly entails the intermediary cleavage of the oxirane carbon-carbon bond<91JOC4598>.

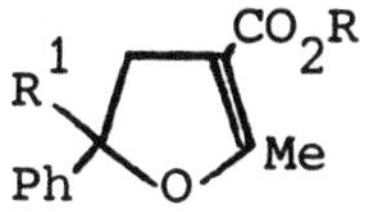

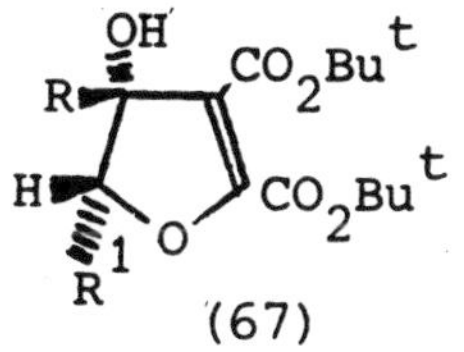

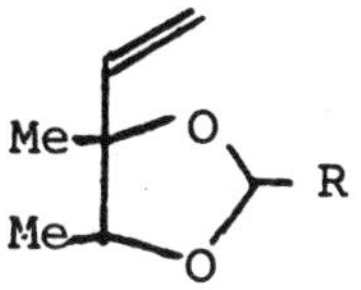

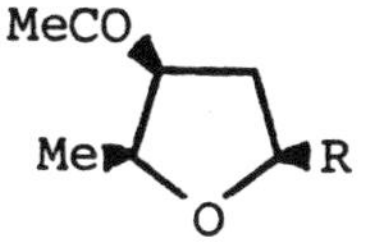

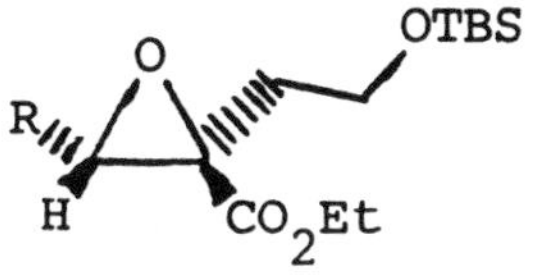

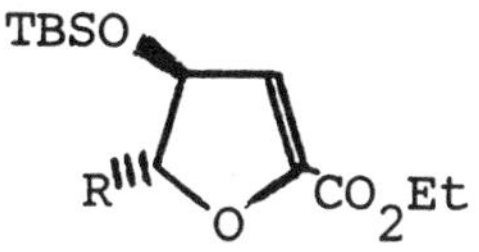

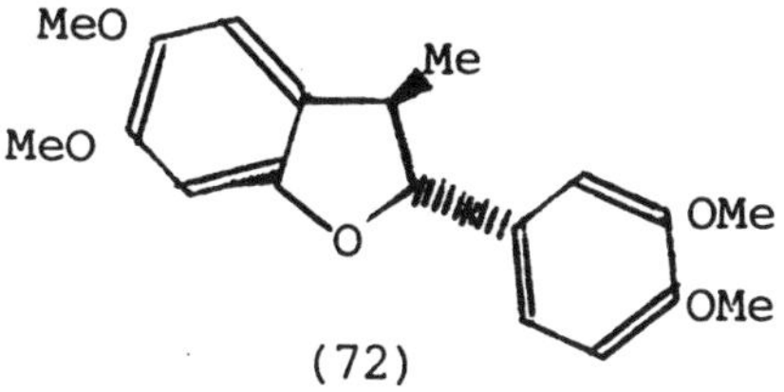

The oxidation of *p*-methoxysubstituted phenols with iodobenzene bi(trifluoroacetate) in the presence of electron-rich styrenes gives 2,3-dihydrobenzofurans. Thus, (72) is obtained from 3,4-dimethoxyphenol and 1,2-dimethoxy-4-propenylbenzene <91JOC1979>.

REFERENCES

90BCJ2828 H. Ishida, K. Yui, Y. Aso, T. Otsubo and F. Ogura, Bull, Chem. Soc. Japan, 1990, 63, 2828.

90G749 G. Cerichelli, M.E. Crestoni and S. Fornarini, Gazz. Chim. Ital., 1990, 120,749.

90H(31)1003 M. Aso, M. Sakamoto, N. Urakawa and K. Kanematsu, Heterocycles, 1990, 31, 1003.

90H(31)1417 J.W. Lee and D.Y. Oh, Heterocycles, 1990, 31, 1417.

90H(31)1745 Y.S. Hon, R.C. Chang and T.Y. Chau, Heterocycles, 1990, 31, 1745.

90H(31)1941 Y.H. Kuo and K.S. Shih, Heterocycles, 1990, 31, 1941.

90H(31)1951 A. Ohta, Y. Akita, T. Ohkuwa, M. Chiba, R. Fukunaga, A. Miyafuji, T. Nakata, N. Tani and Y. Aoyagi, Heterocycles, 1990, 31, 1951.

90H(31)2181 F.T. Luo and R.T. Wang, Heterocycles, 1990, 31, 2181.

90JA7438 R.J. Linderman, D.M. Graves, W.R. Kwochka, A.F. Ghannam and T.V. Anklekar, J. Amer. Chem. Soc., 1990, 112, 7438.

90JCS(P1)3214 K. Soai and Y. Kawase, J. Chem. Soc., Perkin 1, 1990, 3214.

90JHC1843 A.M. Jones, A.J. Simpson and S.P. Stanforth, J. Het. Chem., 1990, 27, 1843.

90SC1923 R.C. Cambie, S.C. Moratti, P.S. Rutledge and P.D. Woodgate, Synth. Commun., 1990, 20, 1923.

90SC3371 M.A. Keegstra, A.J.A. Klomp and L. Brandsma, Synth. Commun., 1990, 20, 3371.

90SL423 J.B. Hendrickson and M.S. Hussoin, Synlett,1990, 423.

90SL673 J. Barluenga, M. Thomas and A. Suarez-Sobrino, Synlett, 1990, 673.

90SL773 D.P. Curran, T.L. Fevig and M.J. Totleben, Synlett, 1990, 773.

90TL3457 S. McIntyre and S. Warren, Tetrahedron Lett., 1990, 31, 3457.

90TL6077 D. MacLeod, D. Moorcroft, P. Quayle, M.R.J. Dorrity, J.F. Malone and G.M. Davies, Tetrahedron Lett., 1990, 31, 6077.

90TL6637 D.H.R. Barton, D.M.X. Donnelly, J.P. Finet, P.J. Guiry and J.M. Kielty, Tetrahedron Lett., 1990, 31, 6637.

90TL7201 B.L. Feringa, O.J. Gelling and L. Meesters, Tetrahedron Lett., 1990, 31, 7201.

91CC437 L. Lattuada, E. Licandro, S. Maiorana and A. Papagni, J. Chem. Soc., Chem. Commun., 1991, 437.

91CC1061 M.R. Iesce, M.L. Graziano, F. Cermola and R. Scarpati,J. Chem. Soc., Chem. Commun., 1991, 1061.

91CC1134 H.L. Sham and D.A. Betebenner, J. Chem. Soc., Chem. Commun., 1991, 1134.

91CE323 K. Imafuku and R. Fujita, Chem. Express, 1991, 6, 323.

91CE699 K. Nakayama and A. Tanaka, Chem. Express, 1991, 6, 699.

91CJC1326 B.A. Keay and J.L.J. Bontront, Can. J. Chem., 1991, 69, 1326.

91JA1417 F. Ozawa, A. Kubo and T. Hayashi, J. Amer. Chem. Soc., 1991, 113, 1417.

91JA5354 M.H. Hopkins, L.E. Overman and G.M. Rishton, J. Amer. Chem. Soc., 1991, 113, 5354.

91JA5365 M.J. Brown, T. Harrison, P.M. Herrinton, M.H. Hopkins, K.D. Hutchinson, P. Mishra and L.E. Overman, J. Amer. Chem. Soc., 1991, 113, 5365.

91JA5378 M.J. Brown, T. Harrison and L.E. Overman, J. Amer. Chem. Soc., 1991, 113, 5378.

91JCS(P1)25 M. Hayashi, T. Kaneko and N. Oguni, J. Chem. Soc., Perkin 1, 1991, 25.

91JCS(P1)1479 M.L. Graziano, M.R. Iesce, F. Cermola, G. Cimminiello and R. Scarpati, J. Chem. Soc., Perkin 1, 1991, 1479.

91JCS(P1)2600 S. Yu and B.A. Keay, J. Chem. Soc., Perkin 1, 1991, 2600.

91JHC853 K. Nakayama, Y. Harigaya, H. Okamoto and A. Tanaka, J. Het. Chem., 1991, 28, 853.

91JOC960 J.A. Marshall and X.J. Wang, J. Org. Chem., 1991, 56, 960.

91JOC1685 J.A. Marshall and W.J. DuBay, J. Org. Chem., 1991, 56, 1685.

91JOC1882 S.K. Meegalla and R. Rodrigo, J. Org. Chem., 1991, 56, 1882.

91JOC1979 S. Wang, B.D. Gates and J.S. Swenton, J. Org. Chem., 1991, 56, 1979.

91JOC3192 T. Yamato, C. Hideshima, G.K. Surya Prakash and G.A. Olah, J. Org. Chem., 1991, 56, 3192.

91JOC3556 A. Padwa, W.H. Bullock, A.D. Dyszlewski, S.W. McCombie, B.B. Shankar and A.K. Ganguly, J. Org. Chem., 1991, 56, 3556.

91JOC4017 K. Gollnick and K. Knutzen-Mies, J. Org. Chem., 1991, 56, 4017.

91JOC4598 T. Hudlicky and G. Barbieri, J. Org. Chem., 1991, 56, 4598.

91JOC4671 T. Keumi, N. Tomioka, K. Hamanaka, H. Kakihara, M. Fukushima, T. Morita and H. Kitajima, J. Org. Chem., 1991, 56, 4671.

91JOC5245 V.H. Rawal, S.P. Singh, C. Dufour and C. Michoud, J. Org. Chem., 1991, 56, 5245.

91JOC5285 D.L.J. Clive and S. Daigneault, J. Org. Chem., 1991, 56, 5285.

91JOC5721 A. Ogawa, H. Yokoyama, K. Yokoyama, T. Masawaki, N. Kambe and N. Sonoda, J. Org. Chem., 1991, 56, 5721.

91JOC5816 Y. Fukuda, H. Shiragami, K. Utimoto and H. Nozaki, J. Org. Chem., 1991, 56, 5816.

91JOC6264 J.A. Marshall and X. Wang, J. Org. Chem., 1991, 56, 6264.

91JOC6269 M.C. Pirrung, J. Zhang and A.T. McPhail, J. Org. Chem., 1991, 56, 6269.

91JOC7007 G.C.M. Lee, E.T. Syage, D.A. Harcourt, J.M. Holmes and M.E. Garst, J. Org. Chem., 1991, 56, 7007.

91JOMC(407)C1 T. Hirao, T. Fujii and Y. Ohshiro, J, Organomet. Chem., 1991, 407, C1.

91LA33 W. Adam, O. Albrecht, E. Feineis, I. Reuther, C.R. Saha-Möller, P. Seufert-Baumbach and D. Wild, Liebigs Ann. Chem., 1991, 33.

91S242 T.R. Bailey, Synthesis, 1991, 242.

91SL29 P. Hamley, A.B. Holmes, A. Kee, T. Ladd-uwahetty and D.F. Smith, Synlett, 1991,29.

91SL33 G. Beese and B.A. Keay, Synlett, 1991, 33.

91SL179 S. Hormuth and H.U. Reissig, Synlett, 1991, 179.

91SL627 S.B. Bedford, M.J. Begley, P. Cornwall and D.W. Knight, Synlett, 1991, 627.

91TL903 H. Haarmann and W. Eberbach, Tetrahedron Lett., 1991, 32, 903.

91TL1491 L.D.M. Lolkema, H. Hiemstra, A.A.A. Ghouch and W.N. Speckamp, Tetrahedron Lett., 1991, 32, 1491.

91TL1737 J. Inanaga, O. Ujikawa and M. Yamaguchi, Tetrahedron Lett., 1991, 32, 1737.

91TL2913 S. Kim and Y.G. Kim, Tetrahedron Lett., 1991, 32, 2913.

91TL3499 T. Kawase, S. Fujino and M. Oda, Tetrahedron Lett., 1991, 32, 3499.

91TL3795 K.E. Harding, M.T. Coleman and L.T. Liu, Tetrahedron Lett., 1991, 32, 3795.

91TL4687 J. Jauch and V. Schurig, Tetrahedron Lett., 1991, 32, 4687.

CHAPTER 5.4
Five-Membered Ring Systems: With More Than One N Atom

S. A. LANG, Jr. and V. J. LEE
American Cyanamid Company, Pearl River, NY, USA

5.4.1 INTRODUCTION

These heterocycles continue to be the focal point of many of the reports in the literature and they play an important part in many new metal complexes and as the structural basis for a variety of novel heterocycles. The biological significance of the polyaza heterocycles is well documented in the agricultural and medicinal chemistry fields. The diversity of biological activity imparted by these heterocycles is shown in recent publications on a) conformationally constrainted renin inhibitory peptides [91JMC(34)1276], b) HIV protease inhibitors [91JMC(34)2852], c) angiotensin II analogues [91JMC(34)2402] and AT_2 subtype nonpeptidic angiotensin II inhibitors [91JMC(34)3248]. While this is not a comprehensive review, the examples cited are representative of the tremendous activity in this area.

5.4.2 PYRAZOLES AND FUSED-RING PYRAZOLES

New methodology or variations of existing methodology continue to be reported. The direct functionalization or modification of substituents on pyrazoles continues to be a synthetic challenge. The *cine*-substitution of 1,4-dinitropyrazoles (**1**) provides an efficient entry to highly substituted pyrazoles [91JCS(P1)1077] and is the cornerstone for an efficient synthesis of formycin (**2**) and pyrazofurin (**3**). Desilylation of 1-(dimethylsulfamoyl)-5-trimethylsilyl-pyrazole (**4**), in the presence of electrophiles, affords 5-substituted pyrazoles (**5**) [91CB(124)1639]. Select pyrazole-4-carboxylic acids (**7**) are obtained by a cobalt-manganese bromide mediated oxidation of 4-methylpyrazoles (**6**) [91CL585]. 5-Hydroxy-1H-pyrazole-4-carbodithioates (**9**) are readily obtained from 5-pyrazolones (**8**) and carbon disulfide [91S481]. 5-Amino-1-substituted-1H-pyrazole-4-carbothialdehyde (**11**) are obtained from 5-azido-1-substituted-1H-pyrazole-4-carboxyaldehyde (**10**) and hydrogen sulfide [91S609].

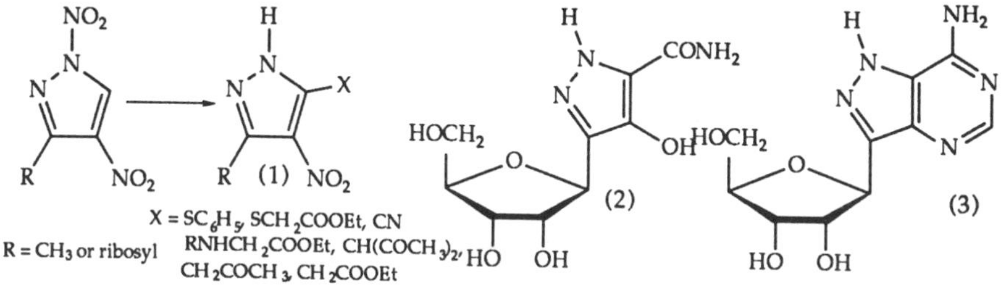

(1) (2) (3)

$X = SC_6H_5, SCH_2COOEt, CN, RNHCH_2COOEt, CH(COCH_3)_2, CH_2COCH_3, CH_2COOEt$

$R = CH_3$ or ribosyl

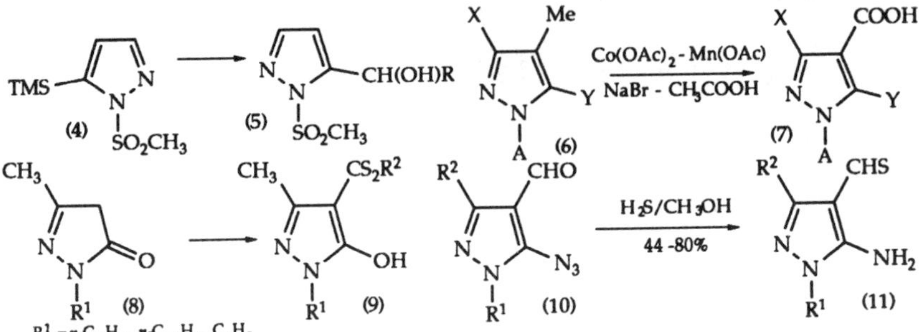

$R^1 = n\text{-}C_8H_{17}, n\text{-}C_{12}H_{25}, C_6H_5$
$R^2 = n\text{-}C_4H_9, C_6H_5CH_2, n\text{-}C_8H_{17}, n\text{-}C_{10}H_{21}, n\text{-}C_{12}H_{25}$

$R^1 = CH_3, C_6H_5$ $R^2 = CH_3, C_6H_5, t\text{-Bu}$

New hydrazine-based annulations with heterocyclic or non-heterocyclic 1,3-dicarbonyl equivalents continued to be reported. For example, the α-(heterocycloalkylidene)malononitriles **(12)** afford various δ-(functionalized alkyl)pyrazoles **(13)** [91JHC(28)1257]. Similarly, chromones **(14)**, 3-acyl-4-hydroxy-2-pyrones **(16)** and 6-aryl-4-(methylthio)-2-oxo-2H-pyran-3-carbonitriles **(18)** afford 2-(pyrazolyl)phenols **(15)**[91JHC(28)667], pyrano[2,3-c]pyrazol-4(1H)-ones **(17)** [91SC(21)1189] and 5(3)-aryl-3(5)-(cyanomethyl)pyrazoles **(19)** [91LA1229], respectively. The chiral heterocyclic glutamate agonist probe **(20)** was efficiently obtained from benzyl pyroglutamate [91JCS(CC)314]. Anomeric mixtures of the pyrazolyl nucleosides **(23,24)** were obtained from the glycosidic hydrazines and α-oxo-ketene dithioacetals **(21)** [91BCJ(64)2306] or β-ethoxy-β-(ethoxycarbonyl)-acrylonitrile **(22)** [91NN(10)625].

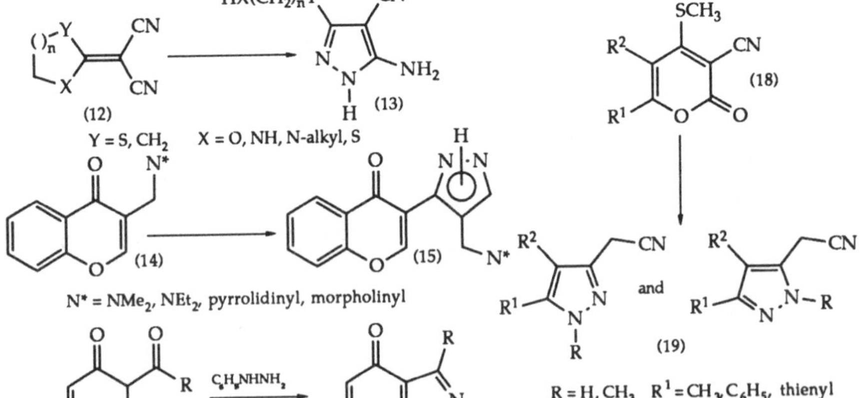

$Y = S, CH_2$ $X = O, NH, N\text{-alkyl}, S$

$N^* = NMe_2, NEt_2$, pyrrolidinyl, morpholinyl

$R = H, CH_3$ $R^1 = CH_3, C_6H_5$, thienyl
$R^2 = H, 3,4\text{-}(CH_3O)_2C_6H_3\text{-}$

Other complex 1,3-dicarbonyl equivalents have also found use in these annulations; for example, 2-(dimethylaminomethylene)-1,3-bis(dimethylimmo-nio)propane bisperchlorate **(25)** affords 1-arylpyrazole-4-carboxaldehyde hydra-zones **(26)** [91JHC(28)1281]. In contrast, the reaction of the bifunctional α-oxoketene dithioacetals **(27)** with hydrazines is pH sensitive, with acidic conditions affording 3(or 5)-styryl-5(or 3)-(methylthio)pyrazoles **(28)**. Under non-acidic conditions, formation of the 3-[β,β-di(methylthio)vinyl]-pyrazolines **(29)** occurs predominantly [91JHC(28)1341]. A one-pot synthesis of withasomnine is reported [91SC(21)1667].

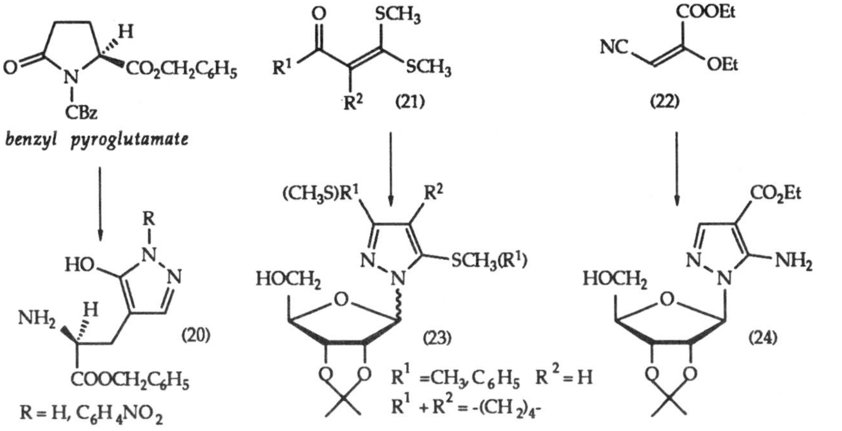

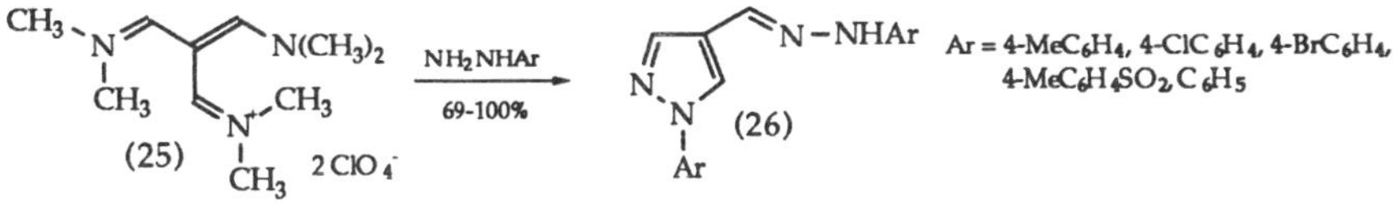

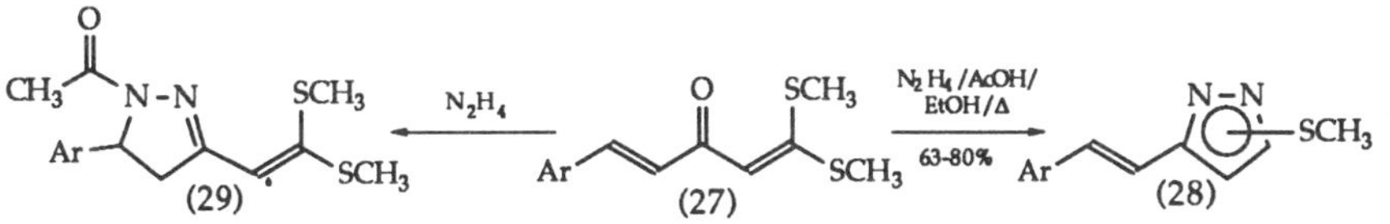

The conversion of α-halo(amino or alkylthio)-β-cyano-azaheterocycles to annelated (2-aminopyrazolo)heterocycles continues to be a cornerstone. For example, aminocyanoquinoline (**30**) and 6-(methylthio)-5-cyanouracil (**32**) afford 3-amino-1H-pyrazolo[3,4-b]quinoline (**31**)[91SC(21)1611] and 3-amino-2H-pyrazolo-[3,4-d]pyrimidine-4,6(5H,7H)-dione (**33**) [91JHC(28)1039]. However, regioselectivity may be controlled by the electronic nature of the hydrazine substituents; for example, either 3-amino-1-aryl-6-dimethylamino-4-oxo-1,4-dihydropyrazolo[3,4-d]- [1,3]oxazines (**35**) or 3-amino-2-aryl-6-dimethylamino-4-oxo-2,4-dihydropyrazolo- [3,4-d][1,3]oxazines (**36**) maybe obtained from (**34**). When R is electron withdrawing, isomer (**36**) is preferred [91S658].

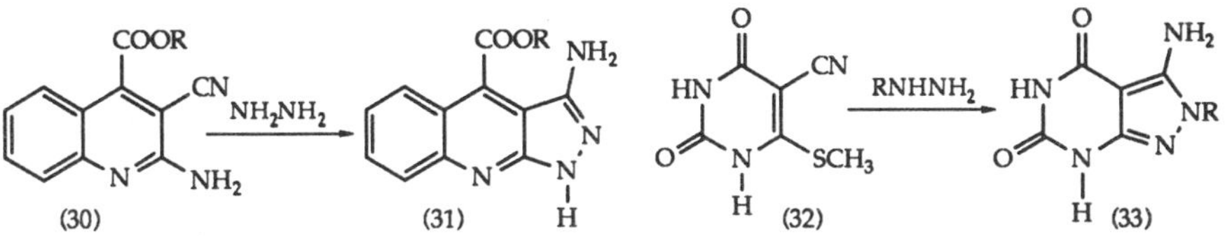

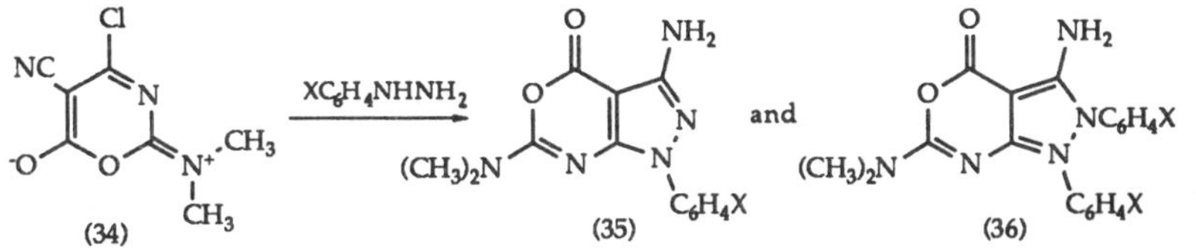

Two nonconventional intramolecular cyclization approaches to pyrazoles have been disclosed. α-Keto(arylhydrazones), on treatment with α-(alkylthio)phosphonates, afford 1-aryl-3,4-disubstituted pyrazoles (**37**) in excellent yields [91SL483]. An alternative oxidative cyclization of α-ketohydrazones with phenyl iodo(tosylate)hydroxide affords 1-aryl-3-alkoxycarbonyl-4-hydroxypyrazoles (**38**) via an intermediate tosyloxyhydrazone [91SC(21)1583].

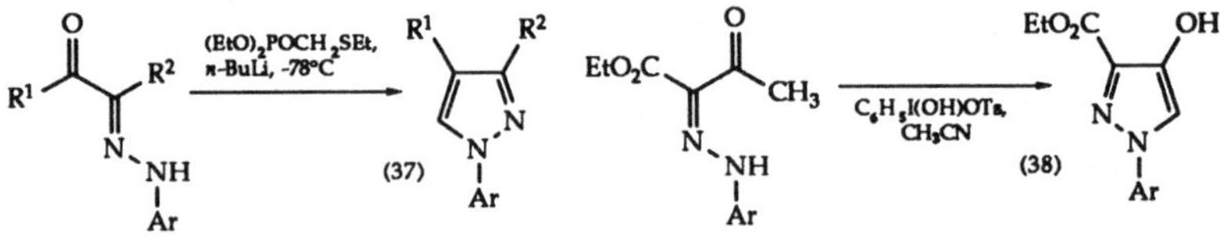

[3+2] Cycloadditions continue to be the synthetic approach of choice for pyrazolyl C-nucleosides. 4-Deoxypyrazofurin (**40**) was obtained by the cycloaddition of diazoalkane (**39**) to a propiolic ester [91S747]. Alternatively, the cycloaddition of hydrazones to glycosyl nitroalkenes (**41**) is operationally simpler and provides more latitude for substituents [91CR(210)175; 91CR(211)287]. The "normal" regiochemistry (**42**) is reversed (*cf.* **43**) when an aryl hydrazone of formaldehyde is employed.

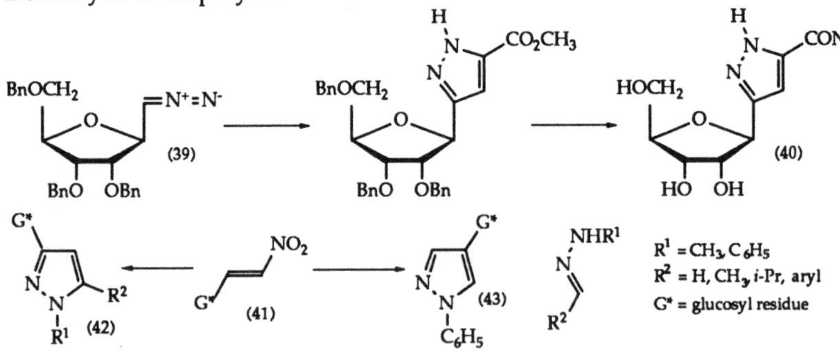

In contrast to the above methods, few syntheses of pyrazoles via intramolecular rearrangements are reported. The photochemical or thermal rearrangement of 2-alkenyl-5-substituted tetrazoles to 3-substituted pyrazoles is reported to occur via an intermediate 4H-pyrazole and is sensitive to electronic effects[91JCS(P1)329]. This promising rearrangement is exemplified in the conversion of the tetrazole C-glycoside (**44**) to a pyrazole C-nucleoside (**45**)[91NN(10)729]. The rearrangement of 6-diazidomethyl-1,2,3,4-tetrahydro-2-oxopyrimidine-5-carboxylates (**46**) to pyrazolopyrimidines (**48**) occurs via 1,5-electrocyclization (vinyl-diazomethane-->3H-pyrazole interconversion) of an intermediate diazomethylpyrimidine (**47**). When R = H (**46b**), rearrangement in the presence of oxygen affords the pyrimidinones (**49**) [91JCS(P1)1342]. The thermal rearrangement of 5-azido-4-iminomethylpyrazoles (**50**) via a nitrene intermediate to the corresponding 2-amino (or hydrazino)-3-cyanopyrazoles (**51**) is temperature and solvent sensitive. Further studies of a ring contraction of 1,2,4-triazepino[2,3-a]benzimidazol-4-ones (**52**) to pyrazolobenz-imidazoles (**53**), first reported by Claramunt [74JHC(11)751], is found to occur via a tetracyclic β-lactam intermediate [91JOC(56)74].

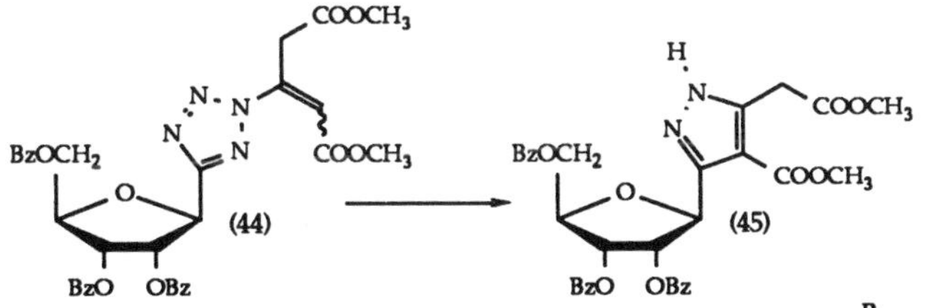

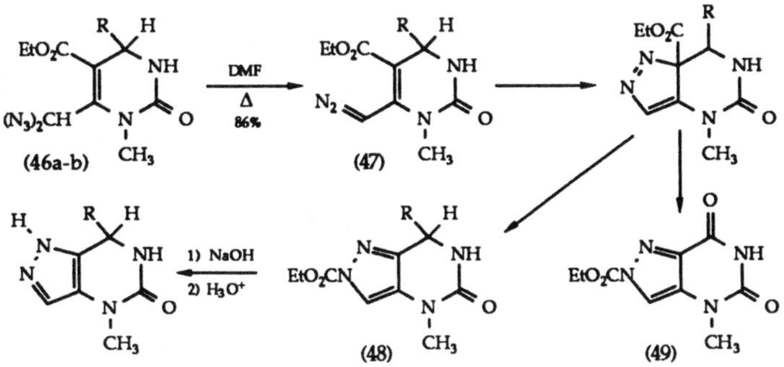

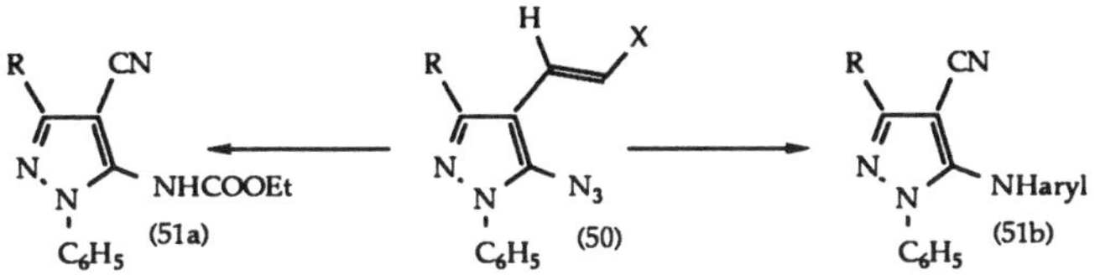

R = CH_3, C_6H_5 X = NHCOOEt, 4-$CH_3C_6H_4$-, 4-(CH_3O)C_6H_4-

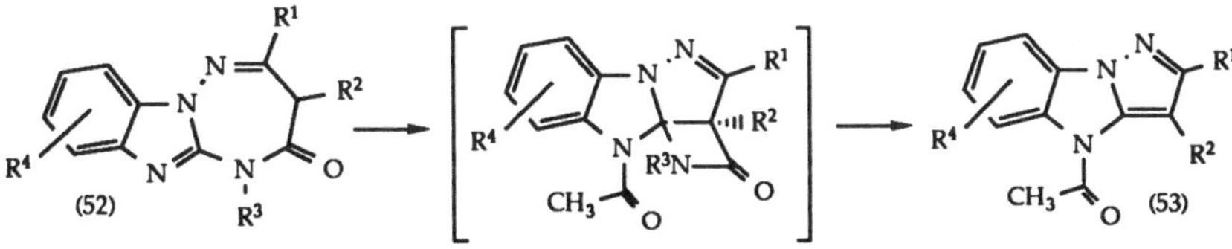

R^1 = CF_3, alkyl R^2 = H, CH_3 R^3 = H, CH_3 R^4 = H, CH_3

5.4.3 IMIDAZOLES AND FUSED-RING IMIDAZOLES

The multifaceted biological properties of imidazoles continue to provide ample opportunities for mechanistic and synthetic studies. Research on nuclear modifications by a) N-alkylation or glycosidation, b) C-2 arylation or glycosidation, c) regioselective metallation and d) intramolecular rearrangements have been reported during this period. The direct functionalization of imidazolide anions continues to be a favorite approach. Reaction conditions have been described for the direct formation of the thermodynamically less stable C-4(5) imidazolide anion vs. the thermodynamically stable C-2 anion. For example, 1-substituted-4,5-dicyano-imidazoles (**54**), on lithiation and oxidative dimerization, afford the biimidazoles (**55**) [91JACS(113)6178]. In contrast, the 1-substituted-4-iodoimidazoles (**56**), on metallation with ethyl magnesium bromide in methylene chloride afford C-4 adducts (**57**). The proper choice of solvent suppresses the C-2 reactivity [91JOC(56)5739]. The use of crown ethers (*e.g.* 15-crown-5 or 18-crown-6) and acetonitrile with sodium (or potassium) imidazolides is highly effective for N-alkylation, as exemplified for the synthesis of the N-(imidazolylalkyl)purine (**58**) [91S709] and the acyclic nucleoside (**59**) [91NN(10)1285]. Similarly, selective additions to epoxides have also been realized [91TL(32)611; 91S769; 91JCS(P1)2479].

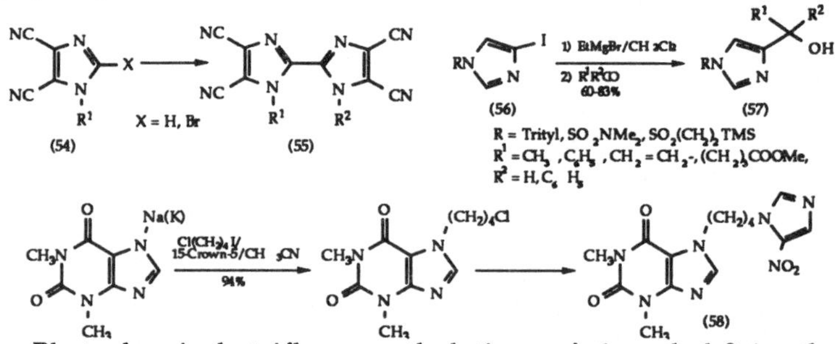

Photochemical trifluoromethylation of 1-methyl-2-(methyl-thio) imidazole (**60**), while a highly specific method, provides an entry into 4-(trifluoromethyl)-1-methylimidazole (**61**) [91BCJ(64)2255].

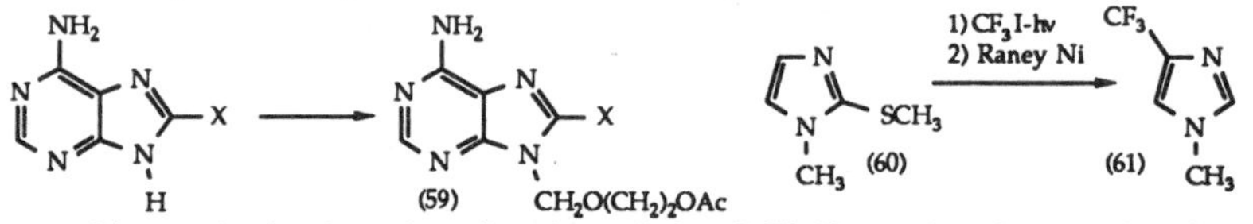

Alternatively, functionalization of the C-TMS imidazoles, under fluoride desilyation or electrophilic reaction conditions, affords another route to nuclear modifications. Based on studies with 1-methyl-2-(trimethylsilyl)-imidazole (**62a**), 1-methyl-2,5-bis(trimethylsilyl)imidazole (**62b**) and 1-methyl-4,5-bis-(trimethylsilyl)-imidazole (**62c**), the order of site reactivity is C-2> C-5> C-4 [91CB(124)1639]. The selective N-(or C) glycosidation and C-ethynylation of imidazoles continues to provide novel chemotherapeutic agents. The palladium catalyzed addition of allyl acetates and vinyl epoxides to imidazoles bases provides entry to 2',3'-dehydro-2',3'-dideoxynucleoside analogues (**63-64**) [91JCS(P1)2603; 91JOC(56)4990]. A modified Stille coupling [Pd(0)] of the iodouracil (**65**) with a C-2 zinc imidazolide affords the 5-(imidazolyl)uracil (**66**) [91JMC(34)2383]. Similarly, free-radical based C-glycosidation of caffeine affords the C-glycoside (**67**) [91TL(32)3377]. In a different tact, ethynylation of the bromoguanine (**68**) affords an efficient entry to the DDATHF analogue (**69**) [91JOC(56)6937].

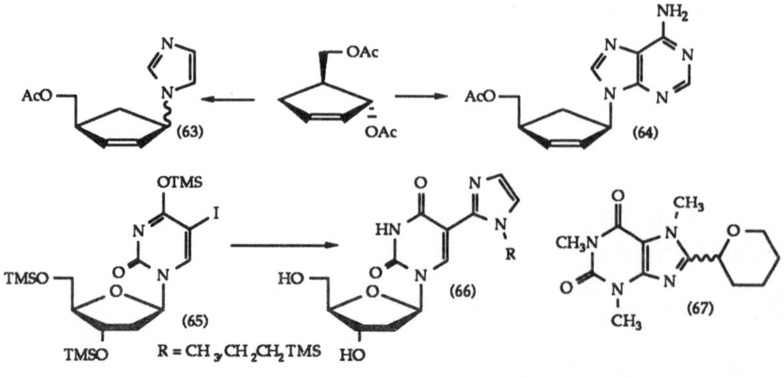

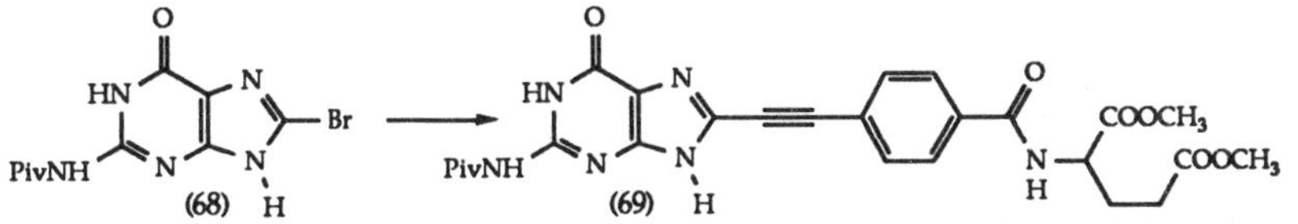

New synthesis pathways for imidazoles and bicyclic imidazoles continue to be described. The cycloaddition of inner ionic (**70**) with N-sulfonylimines provides a versatile synthesis of 1,4,5-trisubstituted and 1,2,4,5-tetrasubstituted imidazoles [91JCR(S)188]. The intramolecular cyclization of N-cyanolactam 2-imines (**71**) affords 2-aminoazacycloalkyl[1,2-a]imidazoles (**72**) [91LA975]. Similarly, pyridinium methides (**73**) react with perfluoroalkylnitriles to afford 2-(perfluoroalkyl)imidazo[1,2-a]pyridines (**74**) [91JFC(51)407]. Imidazolo-pyridines (**75**) and imidazo[5,1-a]isoquinolines (**76**) are available in a "one-pot" synthesis (phase-transfer catalysis, $CHCl_3$, KOH) from α-(aminomethyl) pyridines or α-(aminomethyl)isoquinolines [91JOC(56)2400]. An alternative annulation procedure is illustrated for the synthesis of 3-cyano-8-fluoro-6-trifluoromethylimidazo[1,2-a]pyridine (**77**) from 2,3-difluoro-5-trifluoromethylpyridine [91JHC(28)971]. A novel synthesis of imidazoles from isoxazol-5(4H)-ones via 4-aminoisoxazol-5(4H)-ones (**78**), a 1,2-diaminoethylene synthon, has recently been reported [91S127].

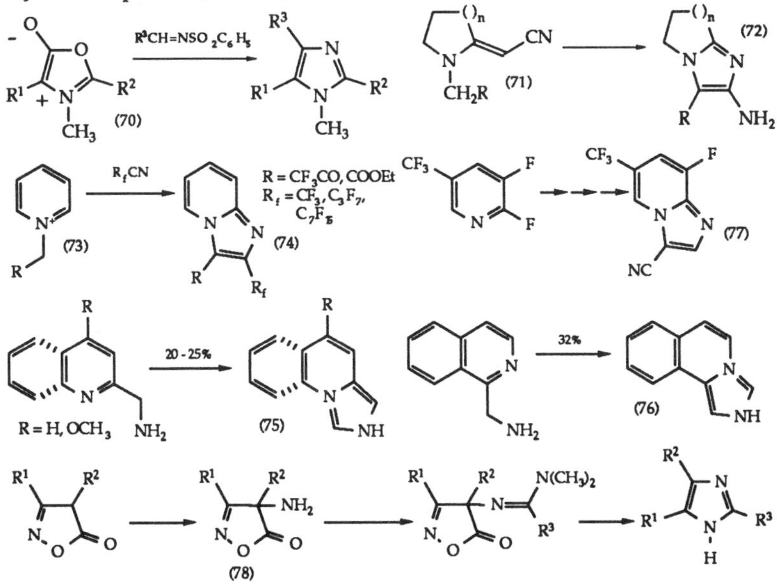

Photochemical rearrangement of 2-alkenyl-5-phenyltetrazoles is reported to occur via an intermediate 4H-imidazole (**79**) which thermally rearranges to the 1H-imidazoles [91JCS(P1)335] (*cf.* **45**). The thermal rearrangement of bicyclic N-cyano-N'-vinylhydrazines (**81**) and bicyclic N-cyano-N'-aryhydrazines (**82**) provides an interesting entry to imidazolodiazepines (**83-84**) [91TL(32)907].

(79a) (80a)

$R^* = 4\text{-}CH_3C_6H_4$

(79b) (80b)

(81) (83) (82) (84) X = H, Cl

5.4.4 1,2,3-TRIAZOLES AND DERIVATIVES

Triazoles, 1,2,3-benzotriazoles in particular, are important in catalytic roles in the synthesis of a variety of aliphatic compounds. A general method for the transformation of ketones into their enol ethers in good yield employes 1,2,3-benzotriazole adducts as an intermediate compound [91S279]. Benzotriazole was reacted with ketals (acetals) in the presence of H_2SO_4 to form a hemiaminal **(85)** which was decomposed at low pressure in the presence of NaH to generate isomeric enol ethers in 60-95% yields.

benzotriazole, H_2SO_4(cat)/CH_2Cl_2, 25°C, 5 hr, 65-88% → (85); NaH, 1-5 Torr, 140-160°C, 72-92% → (86a) + (86b)

The adduct **(85)** at elevated temperatures can reversibly exist in an anionic/catonic form which on recombination and eventual splitting generates the isomeric olefins **(86a)** and **(86b)**. Benzotriazolo-1-ylalkylamides **(87)** react with CH_2 acids to produce α-amidoalkylation products **(88)** in 55-70% yields. Alkylation of **(89)** with methyl iodide generated benzotriazoloalkanes **(90)** in good yields [91JOC(56)4439].

(87) $O_2NCH_2CO_2Et$, $AlCl_3/CH_2Cl_2$ → (88)

BT = benzotriazole

(89) + $CH_2R^2R^3$ $\xrightarrow[N_2,\ rt]{CH_3I}$ BTCHCHR^3 (90) + morpholinium iodide

Substituted diarylmethanes and their heterocyclic analogues with electron rich donating groups can, for example, be used to prepare pyrrole analogue (**91**) in moderate yields by lithiation and reaction with an appropriate electrophile [91JOC(56)4397].

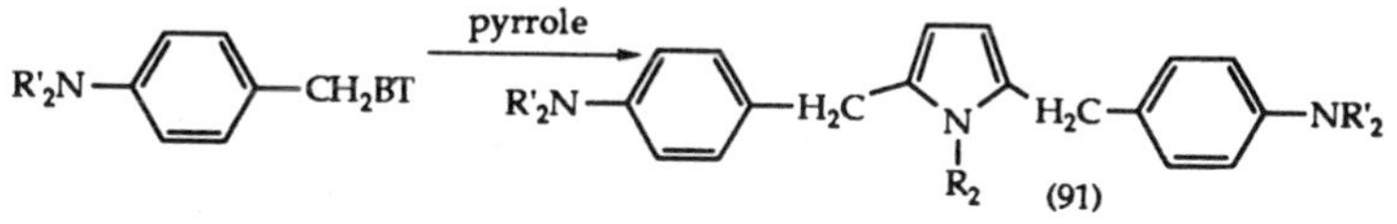

Methylnaphthylamines, which are useful intermediates to prepare additional derivatives via lithiation and subsequent treatment, can be easily synthesized from the intermediate Mannich base prepared from 1,2,3-benzotriazole and naphthylamines [91JOC(56)5045]. Esters can be prepared by substitution of a benzotriazole group (**92**) and subsequent displacement with organozinc reagents as shown [91S69].

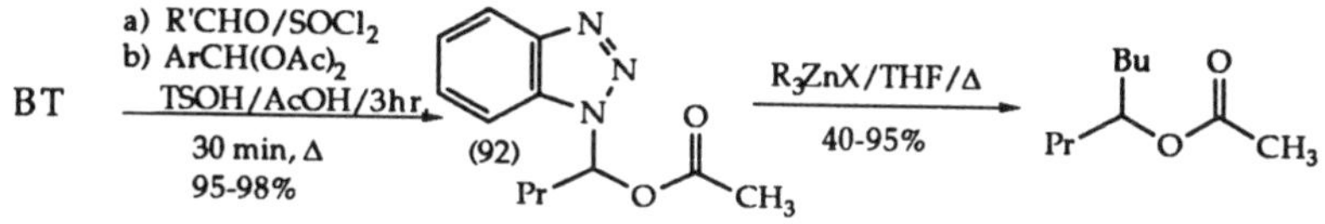

Benzotriazole-1-methylaryl derivatives (**93**) form anions with n-BuLi and these react with electrophiles which are cleaved with organometallics to displace the BT in 30-95% yields [91CB(124)1819]. Benzotriazole imines can also be utilized to prepare a variety of secondary amine possessing primary secondary alkyl groups [91S703].

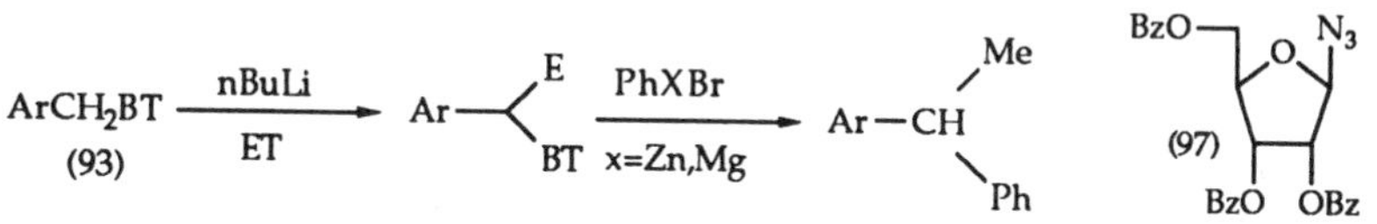

Benzotriazolo esters have been employed as N-protecting amino acids in the synthesis of dipeptides [91S689]. 1H-Benzotriazoles react with aldehydes or thioureas to provide intermediate Mannich-type adducts (**94**) which on reaction with $NaBH_4$ or benzyl magnesium chloride generate thioureas or carbodiimides [91JCS(PI)2199]. Benzotriazolo methyl anion equivalents can also be used to prepare ketones [91JOC(56)6917].

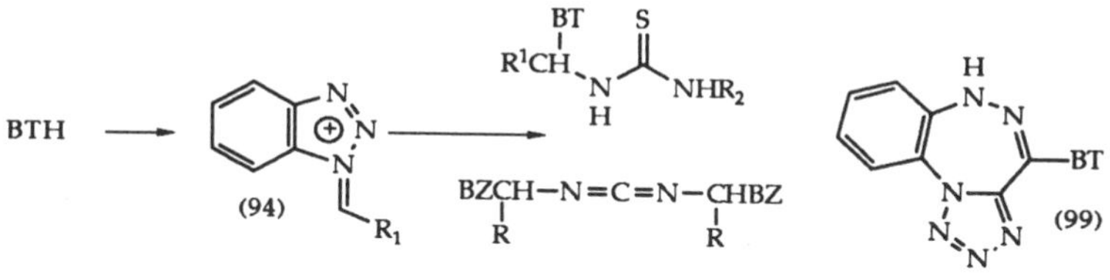

1H-1,2,3-Triazoles can easily by prepared from benzyl azides under Dimroth conditions while acetonitrile or diethylmalonate derivatives generate 5-amino or 5-hydroxy analogs respectively [91JHC(28)301]. Other active methylene compounds such as ethyl cyanoacetate, fluoroacetate or nitroacetate failed to react. Oxidative cyclization with $Cu(OAc)_2$ on **(95)** produced 5-dialkylamino-2-aryl-1,2,3-triazole-4-carbonitriles in moderate yields [91GEP289044].

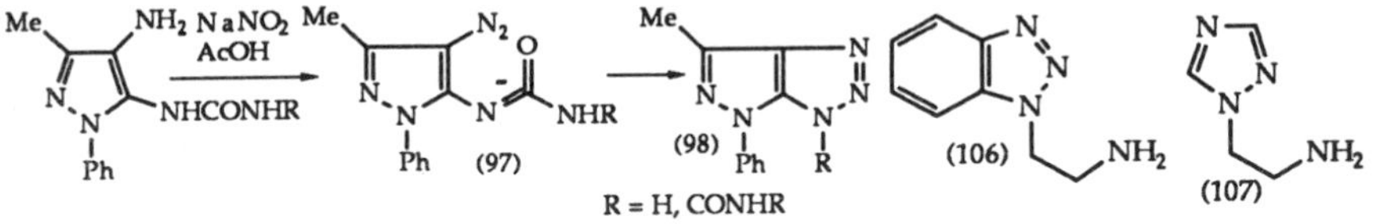

Azide **(97)** was cyclized to produce nucleoside analogs [91MI(10)727]. Intermediate diazo pyrazole **(97)** produced with hydroxide or triethylamine fused triazolopyrazoles **(98)** [91H(32)727]. Other condensed heterocycles containing the 1,2,3-triazole unit synthesized include triazolo[4,5-d][1,2]diazocines [91CPB(39)1713], benzotriazolo-6H-benzo[c]tetrazolo[1,5-e]triazepine **(99)** the latter produced by series of novel rearrangements [91JOC(56)1299], and fused triazolo polycyclic aromatic hydrocarbons [91JOC(56)3219].

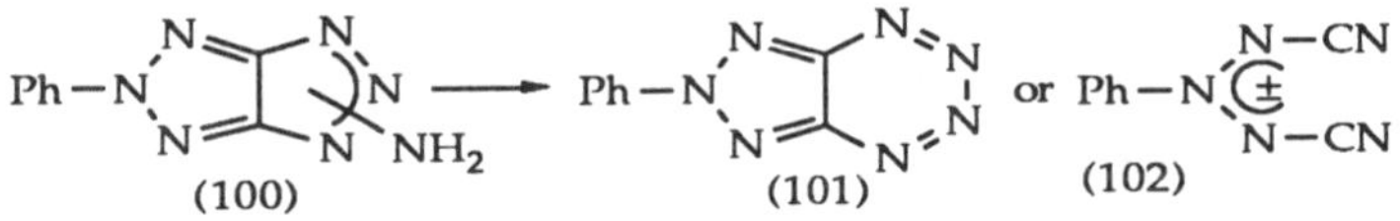

Lead tetracetate oxidation of 1- and 2-amino-5-phenyl[1,2,3]triazolo[4,5-d][1,2,3]triazoles **(100)** produced the novel fused tetrazine **(101)** or azide **(102)**. X-ray structure determination of the products confirmed the assignments. Degradation products of the tetrazine were also reported [91JCS(PI)2045].

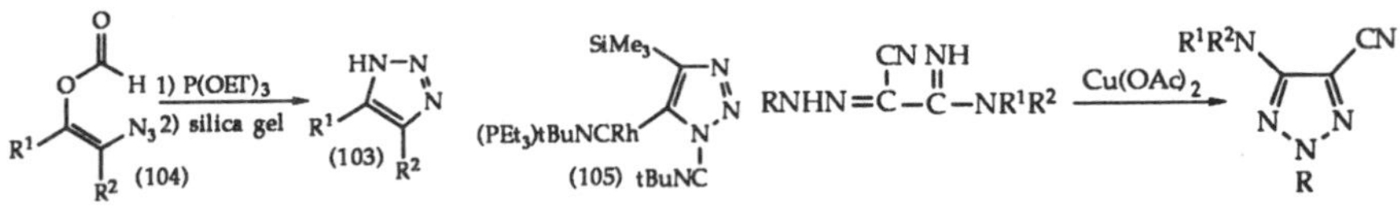

1,2,3-Triazoles **(103)** are also conveniently synthesized from formyloxyvinyl azides **(104)** and triethyl phosphite [91JOC(56)2680]. 1H-1,2,3-triazole was prepared in 92% by autoclaving the 5-thiol sodium salt with RaNi [91JAP03047173]. The rhodium complex of tris(triethylphosphine) **(105)** was determined by X-ray [91JCS(CC)809]. Triazolo and benzotriazolo ethylamines **(106)** and **(107)** are potentially useful synthetic intermediates [91MI(21)535].

5.4.5 1,2,4-TRIAZOLES AND DERIVATIVES

Proton transfer chemistry on 1,2,4-triazolidine-3,5-dione **(108)** generated pK_a constants in a variety of solvents. The deprotonation is consistant with hydrazine proton removal and suggest that the hydrogen in this system is remarkably acidic and its acidity is due to the cyclic diacyl hydrazide structure [91JOC(56)5643]. The thiadiazaphosphol-2-ones **(109)**, prepared by the condensation of amino-5-mercapto-1,2,4-triazole **(110)** with alkyl dichlorophosphates or dichlorothiophosphates represents a heterocycle with heteroatoms at 6 of the 8 ring positions [91H(32)269].

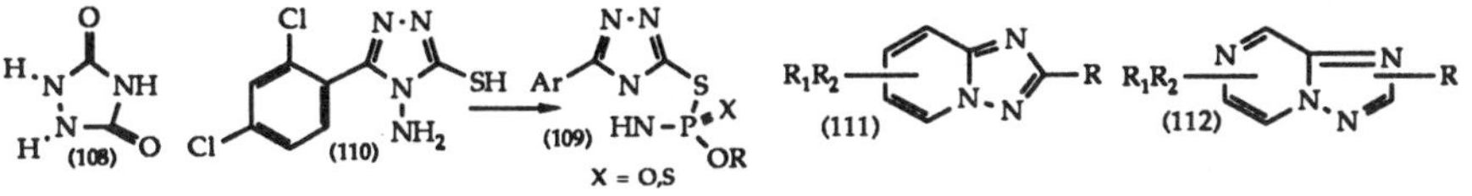

Cesium fluoroxysulfate was demonstrated to be a useful N-fluorinating agent for a number of heterocycles. This reagent was less useful than elemental fluorine and fluoromethyl hypofluoride but superior to perchloryl fluoride. 1-Fluorobenzo-1,2,3-triazole, stable at room temperature, was prepared in 25% yield [91T(47)7447]. Fused derivatives comprise the bulk of literature references with 1,2,4-triazolo[1,5-a]pyridines (**111**) and derivatives [91JHC(28)797] being the most reported [91GER3926770] followed by triazolopyrimidines (**112**) in 70-90% yields [91KGS105], [91KGS700].

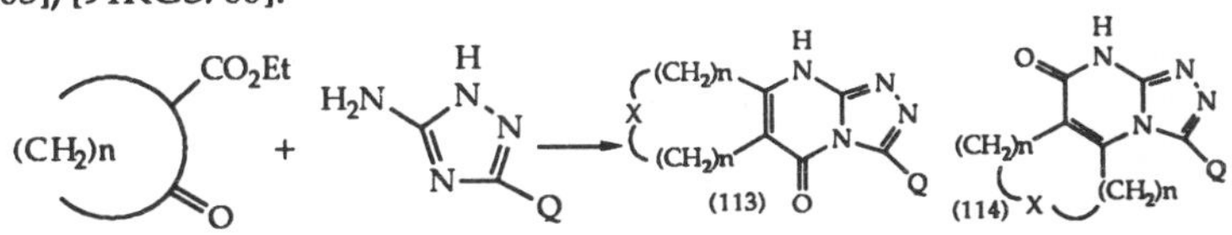

For example, triazolopyrimidinones (**113,114**) were prepared by the condensation of cyclic keto esters and substituted amino-1,2,4-triazoles [91JHC(28)721]. The synthesis was also extended to additional heterocyclic fused rings [91JHC(28)561] or aryl fused analogs [91JMC(34)281]. Dehydrotriazolopyrimidines (**115**) were prepared by the reaction of aminotriazole with (**116**) in the presence of $ZnCl_2$ [91S189]. Pyridazine analogs (**117**) were prepared from pyridazone (**118**) as shown [91H(32)449].

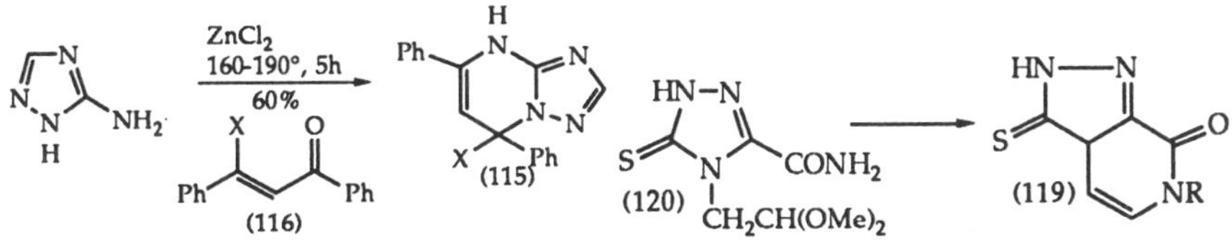

Fused triazolopyrazinones-thiones (**119**) were prepared in 79% by first the reaction of thiosemicarbazides with NH_4OH to generate thiones (**120**) with subsequent cyclization in refluxing aq. HCl [91JAP0314828].

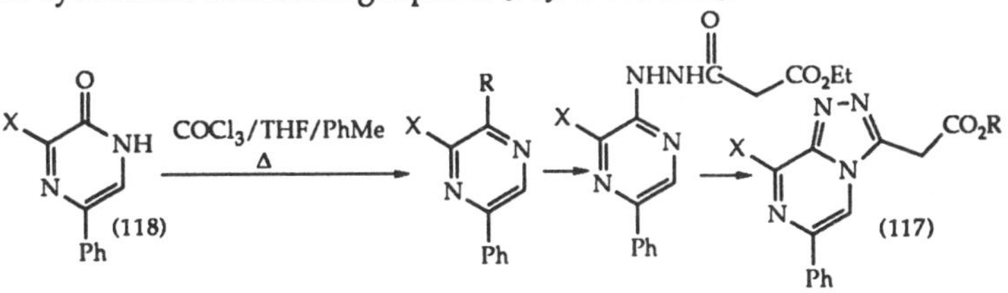

Triazolo-thiadiazines (**121**) were prepared in 2 steps in 60-80% yields by condensation of aminothiazolo-3-thione with carbonothionic dihydrazine and HOAc and subsequent cyclization [91CI(9)319].

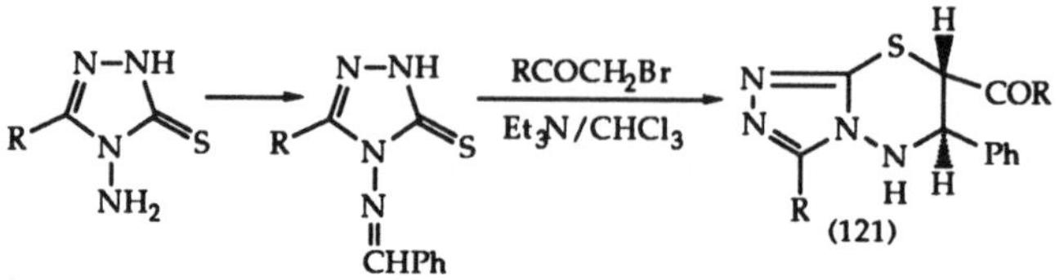

Mercaptotriazoles (**122**) were useful in generation of a series of unique triazolo fused ring systems (**123,124**) under a variety of conditions and in >50% yields [91H(32)231]. 3-Methyl-1,2,4-triazole-5-thione reacts with dimethyl acetylenedicarboxylate to generate, based on spectral analysis, fused derivative (**125**) while reaction with diethyl diazocarboxylate generated the disulfide (**126**) [91JHC(28)325].

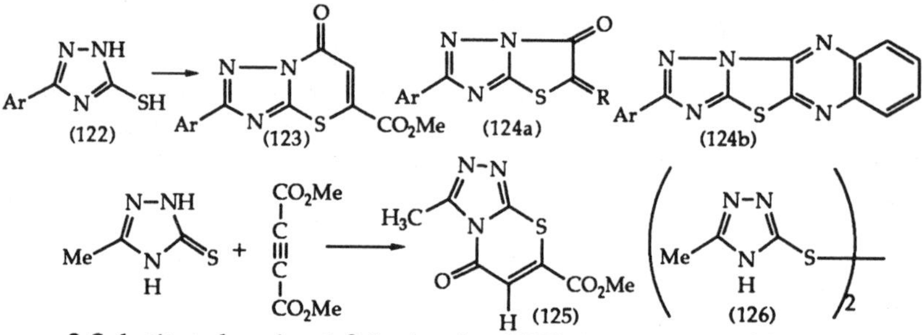

2-Substituted-amino-1,2,4-triazoles (**127**) were prepared by the reaction of 2-imino-1,3-thiazetidines with hydrazine in 25-90% yields. This particular reagent was also highly versatile in the production of a number of other heterocyclic systems [91JHC(28)177]. Additional fused 1,2,4-triazole analogs include [1,2,4]triazolo[3',2':2.3]triazolo[5,4-b]pyridines (**128**) prepared by the condensation of aminothiazolopyridines (**128**) with arylnitriles and $AlCl_3$ and subsequent dehydrogenation-cyclization with lead tetracetate [91IJC(30B)517].

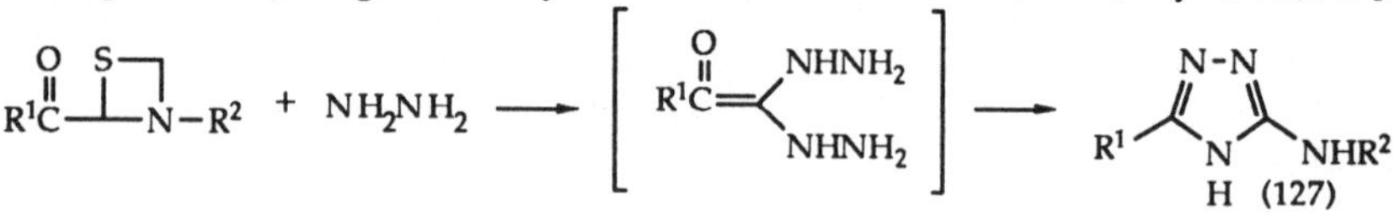

R^1 = aryl, CF_3, CH_3
R^2 = substituted phenyl

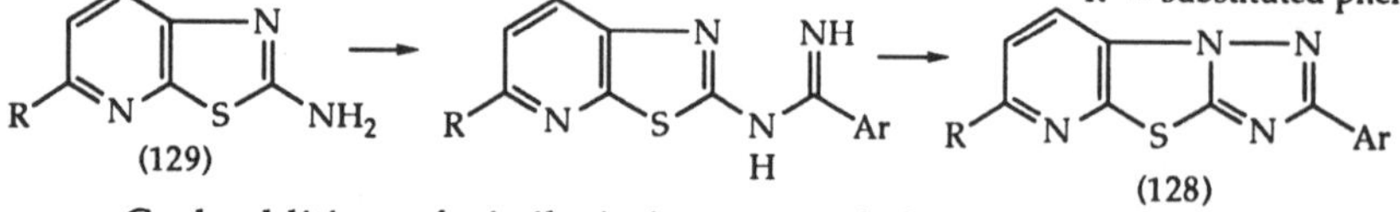

Cycloaddition of nitrile imines to aryl dicyanoacetates at 100-110°C generated dimers (**130**) while at slightly higher temperatures (120°) the diaryl aryl derivatives (**131**) were formed. At temperatures >130°, the imine dimerized to form the 1,2,3-triazole (**132**) [91HCA(74)501]. Fused 4H-[1,2,4]triazolo[4,3-a[1,5]benzo-diazepin-5-amines were prepared and evaluated for analgesic and/or antiinflammatory activity [91MI(26)489]. Related thienofused analogs were also prepared and possessed PAF inhibiting properties [91JMC(34)1440]. Additional extended pyrimido[2,1-a]phthalazines fused with a 1,2,4-triazolo unit to form new novel polycyclic heterocycles appeared [91JHC(28)545]. Pentacyclic triazolodiazepines were prepared as PAF antagonists [91JHC(28)1121]. Triazolo-

[3,4-c]-as-triazino[5,6-b]indoles represent a unique ring system in the 1991 literature [91H(32)1081]. 1,2,4-Triazoles form the basis of proton ionizable crown compounds as represented by structure **(133)** [91JHC(28)773] or are present as a part of an interesting inorganic iridium complex **(134)** [91ICA(181)245].

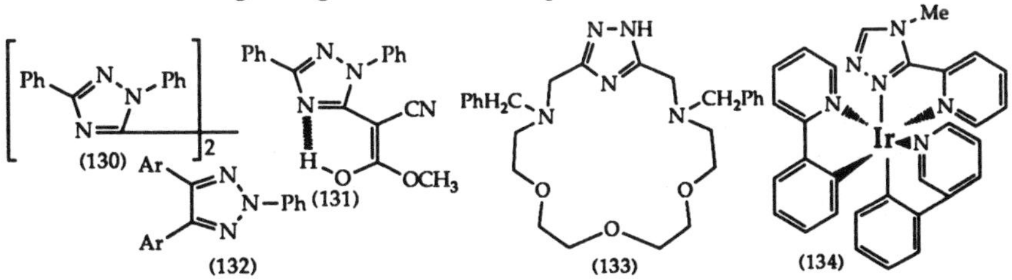

(130) (131) (132) (133) (134)

5.4.6 TETRAZOLES

The cyanotetrazole **(135)** was generated by degradation of azo compound **(136)**. The distribution of **(135)** to other products was less due to first-order isotope effects present in the hydrogen-abstraction step in the diazo radical [91JCS(PI)2045].

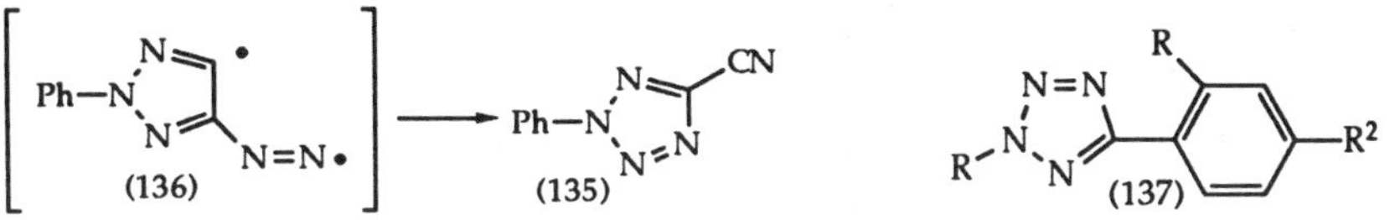

(136) (135) (137)

Ortho lithiation of phenyltetrazoles **(137)** generated o-substituted analogs containing ketone adducts [91USP183324]. The tetrazole unit continues to an important role in medicinal chemistry, often as an isosteric replacement for carboxylic acid functionalities [91JMC(34)1125] as in cholecystokinin conger synthesis. Studies on synthetic methodologies to sterically hindered tetrazoles has yielded a new one-step mild conversion of amides into tetrazoles **(138)** [91JOC(56)2395] which are useful as nonpeptide angiotension II receptor antagonists.

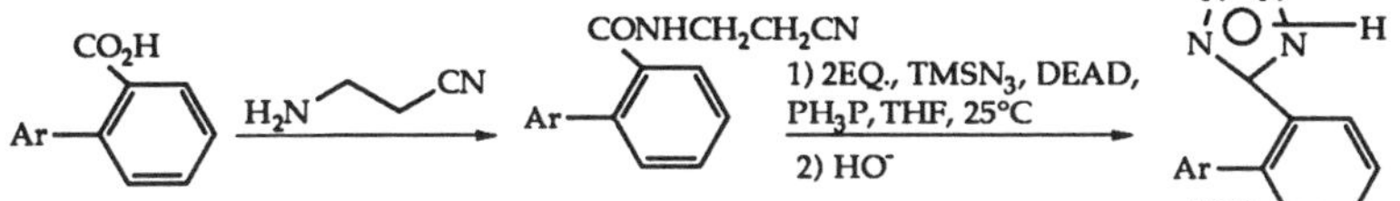

1) 2EQ., $TMSN_3$, DEAD, PH_3P, THF, 25°C
2) HO^-

(138)

6-Azidotetrazolo[1,5-b]pyridazine **(139)** reacts with phosphanes or phosphites at ambient temperature to produce phosphazene **(140)** while at higher temperatures 3,6-bis(phosphoranylideneamino)pyridizanes were produced. The use of phosphites at reflux produced **(141)** by a "Michaelis-Arbuzov" type rearrangement. Heating of (**140** R=CH_3O) in refluxing 1,2-dichlorobenzene generated methylamino analog (**142**) [91LA1225]. Bis(1-methyl-1H-tetrazole-5-yl)disulfide has been employed in the synthesis of unsymmetrical disulfides which are useful for a mild prep of other disulfides [91 JOC(56)5475]. A series of 3-(1H-tetrazol-5-yl)-4(3H)-quinazolinone derivatives **(143)** were prepared in 35% yields by refluxing with 5-aminotetrazole and triethyl or thioformate in DMF for 1 hr. followed by treatment with anthranylamides at 80° [91JAP03145489].

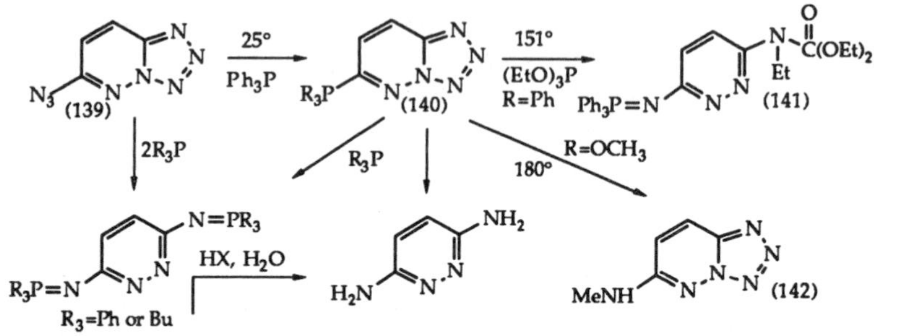

Sodium azide cyclocondensed with PhCCl=NPh-X in 90% yields to form 1,5-diaryltetrazoles [91ZOK(61)1483]. Aminosubstituted tetrazoles react with ethyl (arylazo) benzoylacetates in refluxing ethanol to generate tetrazolo-pyrimidines (**144**) in 45-65% yields [91CCC(56)1560]. N_1-Cyanodiazene-N-oxides and azide produces tetrazole azine oxides (**145**) [91MI1647]. N-(α-Lithioalkyl) tetrazoles (**146**) react with a variety of electrophiles to produce products in 30-90% yields [91JCS(PI)323].

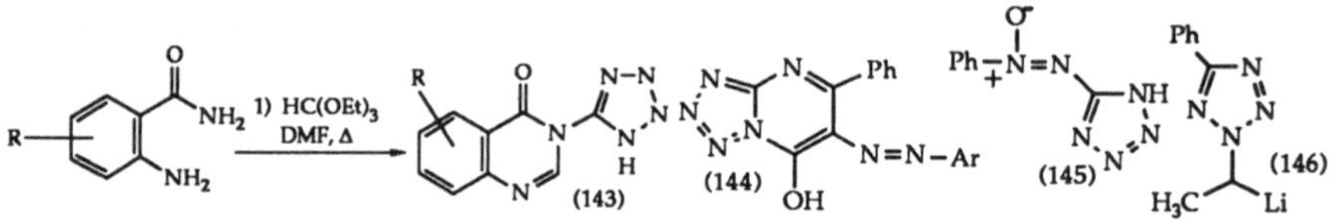

Acknowledgements: We wish to thank **Ms. Elizabeth Fromwiller** and **Ms. Lori Connolly** for their preparation and finalization of this manuscript.

5.4.7 REFERENCES

74JHC(11)751 R. M. Claramunt, J. M. Fabregá and J. Elguero, J. Heterocycl. Chem., 1974, 11, 751.

90USP50398814 R. F. Shuman and A. O. King, U.S. Patent 1990, 5039814; Chem. Abstr., 1991, 115, 183324u.

91BCSJ(64)2255 M. Nishida, H. Kimoto, S. Fujii, Y. Hayakawa and L. A. Cohen, Bull. Chem. Soc., Japan, 1991, 64, 2255.

91BCSJ(64)2306 M. Yokoyama, T. Ikuma, S. Sugasawa and H. Togo, Bull. Chem. Soc. Japan, 1991, 64, 2306.

91CB(124)1639 F. Effenberger, M. Roos, R. Ahmad and A. Krebs, Chem. Ber., 1991, 124, 1639.

91CB(124)1819 A. R. Katritzky, X. Lan and J. N. Lam, Chem. Ber., 1991, 124, 1819.

91CCC(56)1560 A. Deeb, Collect. Czech. Chem. Comm., 1991, 56, 1560.

91CI(9)319 S. G. Manjunatha, Chem. Ind. (London), 1991, 9, 319.

91CL585 N. Tanaka, S. Shinke and S. Takigawa, Chem. Letters, 1991, 585.

91CPB(39)1713 E. Oishi, N. Taido, A. Miyashita, S. Sato, S. Ohta, H. Ishida, K-i. Tanji and T. Higashino, Chem. Pharm. Bull., 1991, 39, 1713.

91CR(210)175 M. Gómez-Guillén and J. M. Lassaletta Simon, Carbohydrates Res., 1991, 210, 175.

91CR(211)287 M. Gómez-Guillén and J. M. Lassaletta Simon, Carbohydrates Res., 1991, 211, 287.

91GEP289044 H. Schaefer, K. Gewarld and P. Bellmann, German Patent (East) 18 April 1991, 289044, Chem. Abstr., 1991, 115, 159149v.

91GEP3926770 E. Schetezik and H. Reichelt, German Patent Application, 14 Feb. 1991, 3926770; Chem. Abstr., 1991, 114, 164253h.

91H(32)231 P. G. Gogoi and J. C. S. Kataky, Heterocycles, 1991, 32, 231.

91H(32)269 Gogoi & Kat, Heterocycles, 1991, 32, 269.

91H(32)449 R. H. Bradbury, Heterocycles, 1991, 32, 449.

91H(32)727 C. B. Vicentini, A. C. Veronese, T. Poli, M. Guarneri and P. Gioro, Heterocycles, 1991, 32, 727.

91H(32)1081 B. S. Holla and K. V. Udupa, Heterocycles, 1991, 32, 1081.

91HCA(74)501 R. Neidlein and Z. Sui, Helv. Chim. Acta, 1991, 74, 501.

91ICA(181)245 J. H. VanDiemen, J. G. Haasnoot, R. Hage, E. Muller and J. Reedijk, Inorg. Chim. Acta., 1991, 181, 245.

91IJC(30B)517 V. S. R. Prassad, T. Sambaiah and R. K. Kondal, Indian J. Chem., Section B, 1991, 30B, 517.
91JACS(113)6178 P. G. Apen and P. G. Rasmussen, J. Amer. Chem. Soc., 1991, 113, 6178.
91JAP03047173 T. Ara, K. Kurata, H. Fujisawa and O. Kunihide, Japanese Patent Application, 1991, 03047173; Chem. Abstr., 1991, 115, 8815.
91JAP03145489 C. Jiiiwan, R. Kannshen and Chiaaien, Japanese Patent, 03145489, 1991; Chem. Abstr., 1991, 115, 183350z.
91JAP03148281 Japanese Patent, 1991, 03148281; Chem. Abstr., 1991, 115, 159180y.
91JCR(S)188 R. Consonni, P. D. Croce, R. Ferraccioli and C. LaRosa, J. Chem. Res. (S), 1991, 188.
91JCS(CC)314 A. N. Bowler, P. M. Doyle and D. W. Young, J. Chem. Soc. Chem. Comm., 1991, 314.
91JCS(CC)809 E. Deydeer, M-J. Menu, M. Dortiguenave and Y. Dortiguenave, J. Chem. Soc. Chem. Comm., 1991, 809.
91JCS(P1)323 C. J. Moody, C. W. Rees and R. G. Young, J. Chem. Soc., Perkin Trans. 1, 1991, 323.
91JCS(P1)329 C. J. Moody, C. W. Rees and R. G. Young, J. Chem. Soc., Perkin Trans. 1, 1991, 329.
91JCS(P1)335 C. J. Moody, C. W. Rees and R. G. Young, J. Chem. Soc., Perkin Trans. 1, 1991, 335.
91JCS(P1)1077 J. G. Buchanan, A. O. Jumaah, G. Kerr, R. R. Talekar and R. H. Wightman, J. Chem. Soc, Perkin Trans. 1, 1991, 1077.
91JCS(P1)1342 C. O. Kappe and G. Färber, J. Chem. Soc., Perkin Trans. 1, 1991, 1342.
91JCS(P1)2045 T. Kaihoh, T. Itoh, K. Yamaguchi and A. Ohsawa, J. Chem. Soc., Perkin Trans. 1, 1991, 2045.
91JCS(P1)2199 A. R. Katritzky and M. F. Gordeev, J. Chem. Soc., Perkin Trans. 1, 1991, 2199.
91JCS(P1)2479 M. F. Jones, P. L. Myers, C. A. Robertson, R. Storer and C. Williamson, J. Chem. Soc., Perkin Trans. 1, 1991, 2479.
91JCS(P1)2603 E. A. Saville-Stones, S. D. Lindell, N. S. Jemmings, J. C. Head and M. J. Ford, J. Chem. Soc., Perkin Trans. 1, 1991, 2603.
91JFC(51)407 R. E. Banks and S. N. Khaffaff, J. Fluorine Chem., 1991, 51, 407.
91JHC(28)177 N. Okajima and Y. Okada, J. Heterocyclic Chem., 1991, 28, 177.
91JHC(28)301 I. F. Cottrell, D. Hands, P. G. Houghton, G. R. Humphrey and H. B. Stanley, J. Heterocycl. Chem., 1991, 28, 301.
91JHC(28)325 G. Giammona, M. Neri, B. Carlisi, A. Palazzo and C. LaRosa, J. Heterocycl. Chem., 1991, 28, 325.
91JHC(28)545 A. Santagati, M. Santagati and F. Russo, J. Heterocyclic Chem., 1991, 28, 545.
91JHC(28)561 J. Reiter and K. E-Reiter, J. Heterocycl. Chem., 1991, 28, 561.
91JHC(28)667 M.-C. Sacquet, M.-C. Fargeau-Bellassoued and B. Graffe, J. Heterocycl. Chem., 1991, 28, 667.
91JHC(28)721 J. Reiter, G. Berecy and I. Pallagi, J. Heterocycl. Chem., 1991, 28, 721.
91JHC(28)773 J. S. Bradshaw, K. E. Krakowick, P. Huszthy and R. I. Izatt, J. Heterocycl. Chem., 1991, 28, 773.
91JHC(28)797 M. T. Cocco, C. Congiu, V. Onnis and A. Maccioni, J. Heterocycl. Chem., 1991, 28, 797.
91JHC(28)971 D. E. Podhorez, J. Heterocycl. Chem., 1991, 28, 971.
91JHC(28)1039 Y. Tominaga, S. Ohno, S. Kohra, H. Fujito and H. Mazume, J. Het. Chem., 1991, 28, 1039.
91JHC(28)1121 A. Walser, F. Flynn and C. Mason, J. Heterocycl. Chem., 1991, 28, 1121.
91JHC(28)1257 M. Pätzel and J. Liebscher, J. Heterocycl. Chem., 1991, 28, 1257.
91JHC(28)1281 J. T. Gupton, J. E. Gall, S. W. Riesinger, S. Q. Smith, K. M. Bevirt, J. A. Sikorski, M. L. Dahl and Z. Arnold, J. Heterocycl. Chem., 1991, 28, 1281.
91JHC(28)1341 M. L. Purkayastha, B. Patro, H. Illa and H. Junijappa, J. Heterocycl. Chem., 1991, 28, 1341.
91JMC(34)281 J. E. Frances, W. D. Cash, B. Barbay, P. S. Bernard, R. A. Lovell, G. C. Mazzenga, R. C. Friedmann, J. L. Hyun, A. F. Braunwald, P. S. Loo and D. A. Bennett, J. Med. Chem., 1991, 34, 281.
91JMC(34)1125 J. W. Tilly, W. Danho, K. Lovey, R. Wagner, J. Swistock, R. Makofski, J. Michalewsky, J. Triscau, D. Nelson and S. Weatherford, J. Med. Chem., 1991, 34, 1125.
91JMC(34)1276 S. Thaisrivongs, J. R. Blinn, D. T. Pals and S. R. Turner, J. Med. Chem., 1991, 34, 1276.
91JMC(34)1440 A. Walser, T. Flynn, C. Mason, H. Crowley, C. Maresca and M. O'Donnell, J. Med. Chem., 1991, 34, 1440.
91JMC(34)2383 P. Wigerinck, C. Pannecouque, R. Snoeck, P. Claes, E. DeClerq and P. Herdewijn, J. Med. Chem., 1991, 34, 2383.
91JMC(34)2402 R. Mohan, Y.-L. Chou, R. Bihovsky, W. C. Lumma, Jr., P. W. Erhardt and K. J. Shaw, J. Med. Chem., 1991, 34, 2402.
91JMC(34)2852 S. J. deSolms, E. A. Giuliani, J. P. Guare, J. P. Vacca, W. M. Sanders, S. L. Graham, J. M. Wiggins, P. L. Darke, I. S. Sigal, J. A. Zugay, E. A. Emini, W. A. Schleif, J. C. Quintero, P. S. Anderson and J. R. Huff, J. Med. Chem., 1991, 34, 2852.
91JMC(34)3248 C. J. Blankley, J. C. Hodges, S. R. Klutchko, R. J. Himmelsbach, A. Chucholowski, C. J. Connolly, S. J. Neergaard, M. S. Van Nieuwenhze, A. Sebastian, J. Quin III, A. D. Essenberg and D. M. Cohen, J. Med. Chem., 1991, 34, 3248.
91JOC(56)74 C. Romano, E. de la Cuesta and C. Avendano, J. Org. Chem., 1991, 56, 74.
91JOC(56)1299 A. R. Katritzsky, W-Q. Fan and J. V. Greenhill, J. Org. Chem., 1991, 56, 1299.

91JOC(56)2395	J. V. Duncia, M. E. Pierce and J. B. Santella, J. Org. Chem., 1991, 56, 2395.
91JOC(56)2400	K. C. Langry, J. Org. Chem., 1991, 56, 2400.
91JOC(56)2680	S. V. D'Andrea, A. Ghosh, W. Wang, J. P. Freeman and J. Szmusykovicz, J. Org. Chem., 1991, 56, 2680.
91JOC(56)3219	B. F. Plummer, S. J. Russell and W. G. Reese, J. Org. Chem., 1991, 56, 3219.
91JOC(56)4397	A. R. Katritzsky, X. Lan and J. N. Lam, J. Org. Chem., 1991, 56, 4397.
91JOC(56)4439	A. R. Katritzsky, J. Pernak, W-Q Fan and F. Sacyewski, J. Org. Chem., 1991, 56, 4439.
91JOC(56)4990	M. R. Peel, D. D. Sternbach and M. R. Johnson, J. Org. Chem., 1991, 56, 4990.
91JOC(56)5045	A. R. Katritzsky, M. Black and W-Q. Fan, J. Org. Chem., 1991, 56, 5045.
91JOC(56)5475	M. Ohtani and M. Narisada, J. Org. Chem., 1991, 56, 5475.
91JOC(56)5643	M. J. Bausch, B. David, P. Dobrowolski, C. G-Fasano, R. Gostowski, D. Selmartin, V. Prasad, A. Vaughn and L-H. Wang, J. Org. Chem., 1991, 56, 5643.
91JOC(56)5739	R. M. Turner, S. D. Lindell and S. V. Ley, J. Org. Chem., 1991, 56, 5739.
91JOC(56)6917	A. R. Katritzsky, Z. Yang and J. N. Lam, J. Org. Chem., 1991, 56, 6917.
91JOC(56)6937	E. C. Taylor, D. Kuhnt and Z.-y. Chang, J. Org. Chem., 1991, 56, 6937.
91KGS105	S. M. Desenko, H. Estrada, V. D. Orlov and O. A. Ponomarev, Khim. Getero. Soedin., 1991, 105.
91KGS700	V. L. Rusinov, E. N. Vlomskii, G. G. Aleksandrov, V. E. Parshin and O. N. Chupakhin, Khim. Getero. Soedin., 1991, 700.
91LA921	P. Gmeiner and J. Sommer, Liebigs Ann. Chem., 1991, 921.
91LA975	M. Patzel, J. Bohrisch and J. Liebscher, Liebigs Ann. Chem., 1991, 975.
91LA1225	A. Deeb, H. Sterk and T. Kappe, Liebigs Ann. Chem., 1991, 1225.
91LA1229	V. J. Ram, M. Verma, F. A. Hussaini and A. Shoeb, Liebigs Ann. Chem., 1991, 21, 1229.
91MI(21)535	A. M. Cuardro, M. P. Matua, J. L. Garcia, J. J. Vaquero and J. Alvarez-Builla, Synthetic Comm., 1991, 21, 535.
91MI(26)489	G. Roma, G. C. Gross, M. DiBraccio, M. Ghia and F. Mattioli, Eur. J. Med. Chem., 1991, 26, 489.
91MI1647	S. G. Zoltin, A. I. Podgurskii, N. V. Zelinskogo and O. A. Luk'yanov, Izv. Akad, Nauh SSSR, Ser. Khim, 1991, 1647.
91NN(10)625	Y. S. Sanghvi, R. K. Robins and G. R. Revenkar, Nucleosides Nucleotides, 1991, 10, 625.
91NN(10)727	A. Stimac, L. B. Townsend and J. Kobe, Nucleosides Nucleotides, 1991, 10, 727.
91NN(10)729	M. Prhavc, L. B. Townsend and J. Kobe, Nucleosides Nucleotides, 1991, 10, 729.
91NN(10)1285	H. B. Lazrek, M. Taouritte, J.-L. Barascut and J.-L. Imbach, Nucleosides Nucleotides, 1991, 10, 1285.
91S69	A. R. Katritzky, S. Rachwal and B. Rachwal, Synthesis, 1991, 69.
91S127	E. M. Beccalli, A. Marchesini and T. Pilati, Synthesis, 1991, 127.
91S189	K. Wermann and M. Hartmann, Synthesis, 1991, 189.
91S279	A. R. Katritzky, S. I. Bazzuk and S. Rachwal, Synthesis, 1991, 279.
91S481	A. Oliva, I. Castro, C. Castillo and G. León, Synthesis, 1991, 481.
91S609	J. Becher, P. Lange Jorgensen, H. Frydendahl and B. Fält-Hansen, Synthesis, 1991, 609.
91S658	R. Neidlein and Z. Sui, Synthesis, 1991, 658.
91S689	K. Takeda, A. Ayabe, M. Suzuki, Y. Konda and Y. Harigaya, Synthesis, 1991, 689.
91S703	A. R. Katritzky, X. Zhao and G. J. Hitchings, Synthesis, 1991, 703.
91S709	A. Long, J. Parrick and R. J. Hodgkiss, Synthesis, 1991, 709.
91S747	D. R. Sauer and S. W. Schneller, Synthesis, 1991, 747.
91S769	V. Jäger, W. Hümmer, U. Stahl and T. Gracza, Synthesis, 1991, 769.
91SC(21)1189	Y. Rachedi, M. Hamdi, R. Sakellariou and V. Speziale, Syn. Comm., 1991, 21, 1189.
91SC(21)1583	H. V. Patel, K. A. Vyas, S. P. Pandey, F. Tavares and P. S. Fernandes, Syn. Comm., 1991, 21, 1583.
91SC(21)1611	B. Narsaiah, A. Sivaprasad and R. V. Venkataratnam, Syn. Comm., 1991, 21, 1611.
91SC(21)1667	A. Guzmán-Pérez and L. A. Maldonado, Syn. Comm., 1991, 21, 1667.
91SL483	H. V. Patel, K. A. Vyas, S. P. Pandey, F. Tavares and P. S. Fernandes, Syn. Lett., 1991, 483.
91T(47)7447	A. Gakh, S. V. Romaniko, B. I. Urgak and A. A. Fainzilberg, Tetrahedron, 1991, 47, 7447.
91TL(32)611	H. Baumgartner, C. Marschner, R. Pucher and H. Griengl, Tetrahedron Lett., 1991, 32, 611.
91TL(32)907	D. Rousselle, C. Musick, H. G. Viehe, B. Tinant and J.-P. Declercq, Tetrahedron Lett., 1991, 32, 907.
91TL(32)3377	H. Togo, M. Fujii, T. Ikuma and M. Yokohama, Tetrahedron Lett., 1991, 32, 3377.
91TL(32)3565	W. Dehaen and J. Becher, Tetrahedron Lett., 1991, 32, 3565.
91ZOK(61)1483	I. V. Nikonova, G. I. Koldobsii, A. B. Zhivich and V. A. Ostrovski, Zh. Obshch, Khim., 1991, 61, 1483; Chem. Abstr., 115, 279927u (1991).

CHAPTER 5.5

Five-Membered Ring Systems: With N & S (Se) Atoms

RIE TANAKA
Suntory Institute for Biomedical Research, Osaka, Japan

and

ICHIRO SHINKAI
Merck Sharpe & Dohme Research Laboratories, Rahway, NJ, USA

5.5.1 Thiazoles

The first synthesis of micrococcinic acid was reported. Micrococcinic acid is a degradation product of the micrococcins that embodies the unprecedented five-ring heterocyclic core of the intact antibiotics. This synthesis utilizes palladium-catalyzed biaryl coupling reactions and provides the first entry into the structurally complex heterocyclic chromophores that distinguish micrococcin and related sulfur-rich cyclic peptide antibiotics <91TL4263>. Interesting naturally occuring cyclic peptides have been isolated from marine organisms. Many of these peptides display useful antimicrobial, neurophysiological and/or anticancer properties. The synthesis of homochiral thiazoline (**1**) and thiazole containing amino acid (**2**) has appeared <91T8267>.

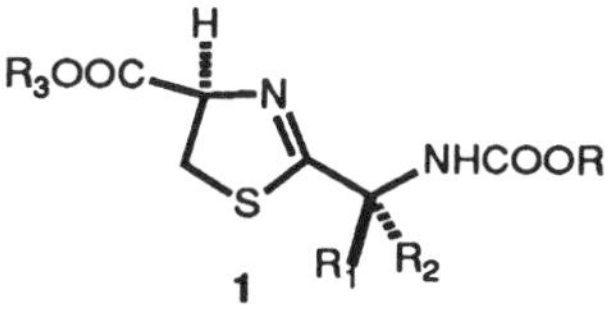

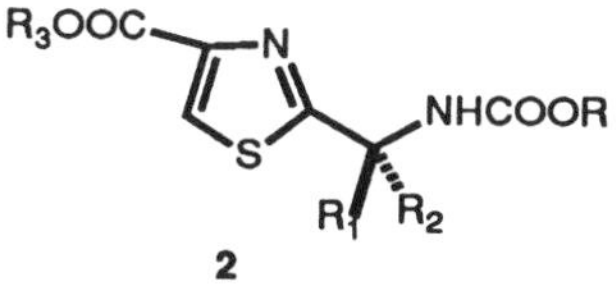

The thiazole containing amino acids (**2**) were shown to be useful precursors of a range of small chiral peptides. The condensation reactions between cysteine derivatives and imino ethers derived from amino acid provide a convenient approach towards the synthesis of (**1**) and (**2**). The 1,3-dipolar cycloaddition of 4,4-dimethyl-1,3-thiazole-5(4H)-thione (**3**) with the nitrilium betaines (**4, 5** and **6**) gave the spiro adducts (**7, 8** and **9**) in a regiospecific manner. The reaction of (**3**) with azides in toluene at 90°C gave the imine (**10**) with the elimination of elemental sulfur and nitrogen <91PSS281, 90HCA2287>.

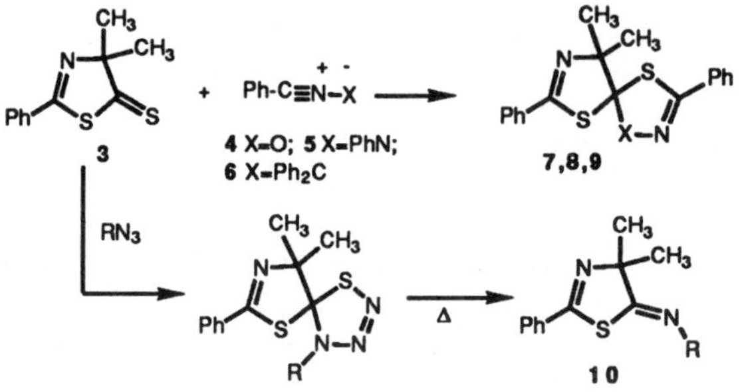

3-(4-Thiazolyl)-L-alanine (**11**) can be considered as an annular equivalent of L-histidine based on the concept of isosterism in medicinal chemistry. Both subtilisin and chymotrypsin were found effective in resolving the racemic ethyl ester of N-acetyl-3-(4-thiazolyl)alanine (**12**) <90S3507>.

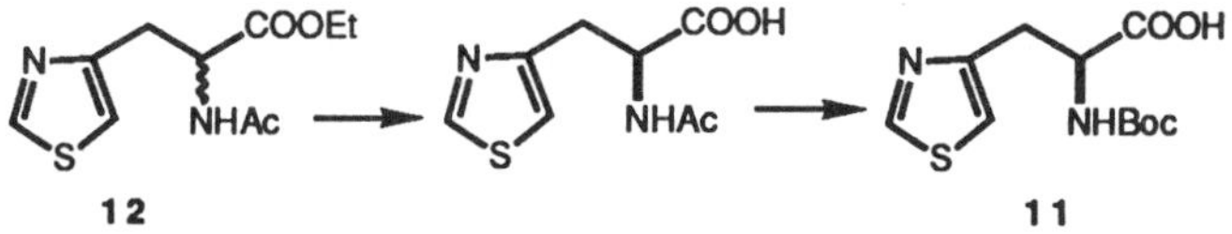

The thiazolyl-hydroxymethyl phosphonate (**13**) was readily converted to the corresponding O,O-thiocarbonate (**14**) on treatment with p-tolyloxythiocarbonyl chloride in CH_3CN/DMAP, while reaction in pyridine/CH_2Cl_2 afforded the isomeric O,S-thiocarbonate (**15**). Homolytic cleavage of (**14**) and (**15**) proceeded similarly, using tributyltinhydride/AIBN to give the new heteroarylmethylphosphonate (**16**). Conversion of the O,S-thiocarbonate to α-methylthio substituted derivative (**17**) was also reported <91S362>.

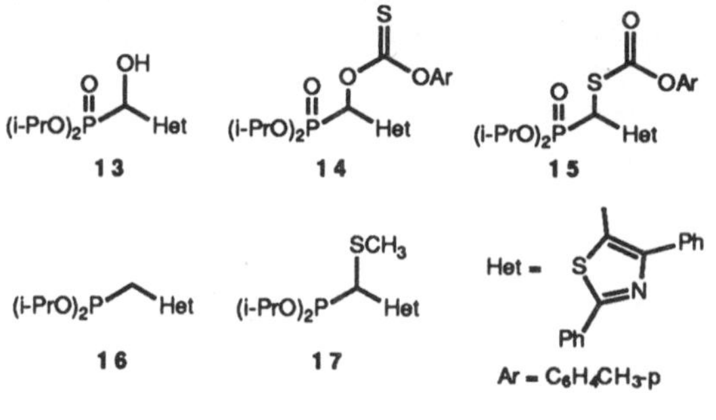

Intramolecular ring closure between sulfur and a reactive carbon atom bearing a good leaving group is a well-known route to the 4,5-dihydrothiazole ring system <91S494>. The reaction of 1-benzothiazolyl-2-aza-pentadienyl anions with aldehydes gave α and/or γ-regioisomeric adducts depending upon the experimental conditions. On the other hand, the addition to α,β–unsaturated carbonyl compounds was highly 1,4-regioselective <91T4465>. Functionalized thiazoles were prepared by a one-pot reaction <91HCA531>. A modified Hantzsch reaction was reported for the synthesis of optically pure thiazole amino acid derivatives <90SC2235>. The treatment of aminopropanedinitrile tosylate with isothiocyanates gave 4-cyanothiazole derivatives, which reacted with amidines to give thiazolo[5,4-d]pyrimidines <91JOC4645>. The chemical behavior of the thiazolium ylide under non-aqueous, aprotic conditions was investigated. It was concluded that the reactivity of the C-2 ylide derived from N-alkylthiazolium ions can be rationalized according to an ionic addition reaction rather than the insertion of the carbene intermediate <91JOC5029>. Treatment of the allylic or benzylic benzothiazole sulfones with LDA gave the corresponding α-lithio derivatives, which upon reaction with ketones afforded olefins <91TL1175>.

When the diazoacetylthiazolidine (**18**) was treated with methanolic sodium methoxide, the derived diazoamide (**19**) was isolated in 36% yield and the enantiomerically pure cyclized product (**20**) in 55% yield. The absolute configuration of (**20**) was established by X-ray crystallography. The conditions required for the cyclization suggest that an ester enolate intermediate intervenes. Classically, such species are planar and, in the absence of other chiral features, they react with electrophiles to give racemic product. However, the enolate derived from (**18**), required for the cyclization reaction, possesses axial chirality, and a sizeable energy barrier would be expected to separate it from its equatorial enantiomer. Therefore, intramolecular trapping of the enolate by the diazo group occurs more rapidly than racemization <91JCS(C)924>.

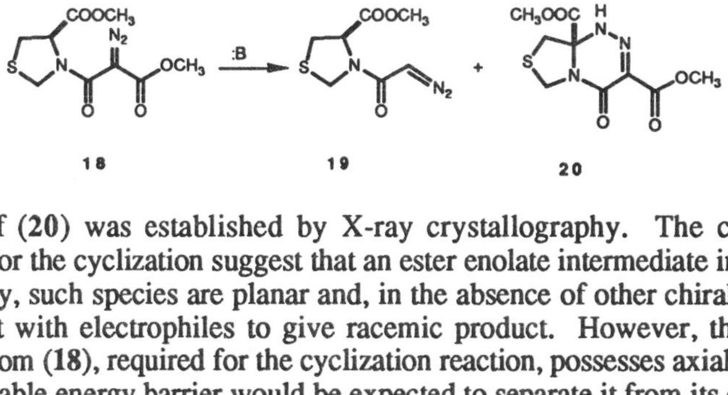

The synthesis and stereochemistry of dimethyl-4-thiazolidinyl-phosphonates were reported <91LA305>. The 1,3-dipolar cycloaddition reaction of 5-azido-4-trifluoromethylthiazole (**21**) with 2,3-dimethyl-buta-1,3-diene gave 5,6-dihydro-2H-pyrane (**22**) predominantly <91ZN1695>.

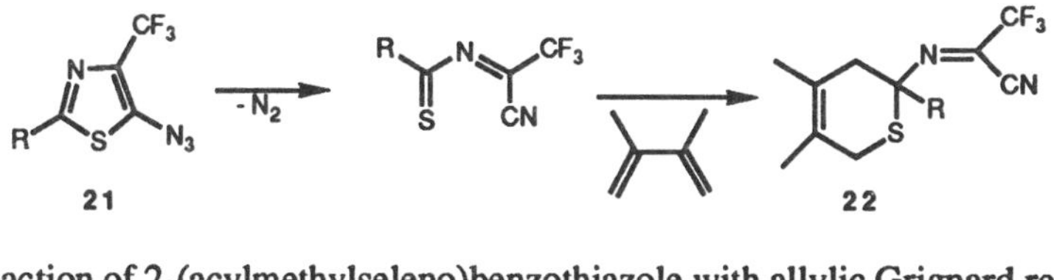

The reaction of 2-(acylmethylseleno)benzothiazole with allylic Grignard reagents gave the corresponding alcohol adducts, which on treatment with NaH/Ph3P, gave 1,4-dienes in excellent yields <91CL661>.

Treatment of 2-benzo-thiazolylbenzyl lithium with cyclohexene oxide gave the γ-hydroxyalkylbenzothiazole with a pronounced syn-diastereoselectivity (syn/anti=95/5). The transition state leading to the syn-diastereomer (re-si face matching) is thermodynamically favored over the transition state leading to the antidiastereomer (re-re face matching), which experiences a large steric compression <90T8169>.

The use of thiazoles as the synthetic equivalent of a formyl group has been well-demonstrated. Thus, 2-acetylthiazole, after selective metalation can be reacted with aldehydes to give the corresponding aldols. The condensation with chiral aldehydes gave adducts derived from the Felkin-Ahn model for asymmetric induction. This methodology was applied to the total synthesis of the octulosonic acid KDO from D-arabinose <91JOC5294, 91TL3247>. A stereoselective route to syn and anti α-hydroxy-β-amino aldehydes employing 2-trimethylsilylthiazole as a marked formyl anion equivalent has appeared. This method was applied to the homologation of L-threonine <91CC1313>.

The dianion of 2-trifluoroacetamido-4-(trifluoromethyl)thiazole was reacted preferentially at the 5-position with a variety of electrophiles <91JHC1017>. The 4-trifluoromethyl group of 5-aminothiazole derivatives showed an unusual labilization <91JHC1013>. Electrophilic substitution of imidazo[2,3-b]thiazoles at position 5 also appeared <91JCS(P1)855>. 3-Hydroxy-4-methylthiazole-2(3H)-thione carbamate (TTOC carbamate) can be used as a precursor for the formation of monoalkylaminium cation radicals, which are synthetically useful intermediates <91JOC1309>. Methylation of 2-acetylbenzothiazole (**23**) with methyl trifluoromethanesulfonate gave the N-methyl-2-acetylbenzothiazolium cation, which underwent rearrangement with ring expansion to give 2-hydroxy-2,4-dimethyl-2H-1,4-benzothiazin-3(4H)-one (**24**) in aqueous solution <90CC362>.

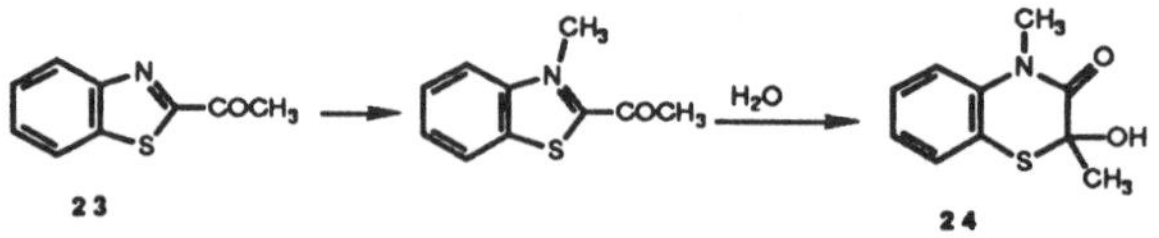

Enzymatic reduction of prochiral 2-acylthiazole has been studied <91MI55001>. A large rate-accelerating metal ion effect on the decarboxylation of α-keto acids by employing the thiazolium ion bearing a bipyridine moiety was reported <91CC373>. The 1,4-dipolar cycloaddition of thiazolium betaine with phenacyl bromide gave a new heterocyclic system, thiazino-imidazo-triazines <91H253>. Simple efficient procedures for preparing 2-(alkylideneamino)-thiazole derivatives were developed <91S621>. The chemistry of biologically interesting thiazole C-nucleosides was reported <91T5539, 91T5549>.

Several new thiazolo[2,3-c]*as*-triazines and triazolo[2,3-c]1,2,4-triazole derivatives were synthesized utilizing 4-methyl-5-ethoxy carbonyl-thiazol-2-diazonium sulfate and active methylene reagents <91PSS119>. When the lithiated unsubstituted triazolothiazole mixture (3- and 6-) was treated with trimethylsilyl chloride only the 6-trimethylsilyl derivative was isolated in 52% yield. On the other hand, the lithiation and substitution of 3-substituted lithio derivatives to give 3,6-disubstituted triazolothiazoles, which could be converted into 2,4-disubstituted thiazoles, were more satisfactory <91T2851, 91T2861>.

3-Hydroxy-4-methylthiazole-2(3H)-thione derivatives were utilized as precursors to synthetically useful vinyl bromides <91JOC6199>. Similarly, thermal decomposition of thiohydroxamic esters of 6-phenylthio-4-hexanoic acids in boiling toluene gave 2-vinylcyclopropane derivatives in moderate yields. On the other hand, 3-pentenyl type radicals could be intercepted by an intermolecular addition to the radicophilic olefins, such as acrylonitrile to give the corresponding cyclopentanecarbonitrile <91TL6085>. Thiazole derivatives were recognized as antitumor agents <91JMC1975>, anthelmintic agents <91JMC1630>, platelet aggregation inhibitors <91CPB651> and aldose reductase inhibitors <91JMC108>.

5.5.2 Isothiazoles

Several syntheses of benzo[d,d']diisothiazoles by cyclization of appropriate o-alkylthioaryl ketoximes were reported. 3,5-Dimethylbenzo[1,2-d:5,4-d']diisothiazole (**25**) was prepared from 4,6-dichloro-1,3-dimethylbenzene. The chloro groups were displaced by methanethiolate ion and the methyl groups were converted to alkyl ketoximes. Cyclization with acetic anhydride and pyridine gave the appropriate diisothiazoles. The other systems, benzo [1,2-d:4,5-d'] (**26**), benzo [1,2-d:3,4-d']-(**27**) and benzo[1,2-d:4,3-d']diisothiazoles (**28**), were synthesized in a similar approach <91JHC445>.

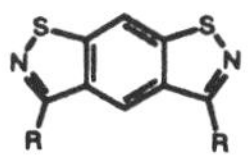

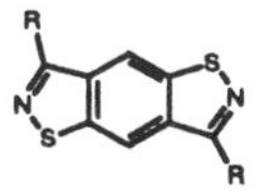

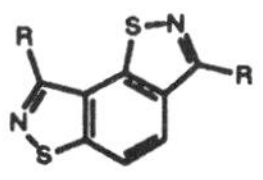

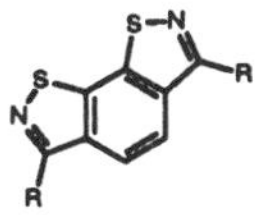

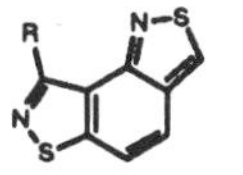

Benzo[c,d']diisothiazoles were obtained by reaction of aminomethyl-1,2-benzisothiazoles with N-sulfinylmethanesulfonamide. Nitration of 3,5-dimethyl-1,2-benzisothiazole gave only the 4-nitro isomer, which was reduced to the amine. Cyclization with N-sulfinylmethanesufonamide afforded benzo[1,2-c:6,5-d']diisothiazole system (**29**). The benzo[1,2-c:5,6-d'] system (**30**) was constructed similarly <91JHC347>.

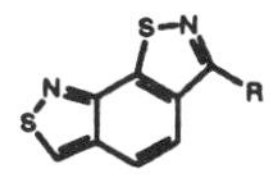

Diaminosulfoxonium salts (**31**) gave dihydro-2,1-benzisothiazole derivatives (**32**), and the corresponding sulfonimidamides (**33**) upon treatment with base. Intramolecular [2,3]sigmatropic rearrangement of the ylides led to ortho substitution *via* intermediate cyclohexadienimines. Hydrogen transfer, accompanying rearomatization, and the subsequent action of a base gave the isothiazole derivatives in moderate yields <90BCJ3223>.

Oxidation of isothiazoles with H_2O_2 in AcOH gave 1,2-thiazol-3(2H)-one-1,1-dioxides (**34**). On the other hand, oxidation of cyclohepta[c]isothiazole gave the tricyclic oxaziridine (**35**) <91HCA1059>.

Catalytic transfer hydrogenation (Pd/C--sodium phosphinate) of a pseudosaccharyl ether of phenols prepared from an inexpensive pseudosaccharyl chloride gave arenes in good yields <90TL5789>. For the study of the rational design and synthesis of new asymmetric oxidizing reagents, 3-substituted-1,2-benzisothiazole 1,1-dioxide oxides were demonstrated. These reagents, however, were less effective in symmetric oxidations than camphor-based oxaziridines <91JOC809>. The camphor based oxaziridines were conveniently used for the asymmetric syntheses of chiral phosporus compounds <91CB1627>. A chiral acryloylsultam was applied in the key synthetic step of the DNA-reactive alkaloid quinocarcin. This auxiliary-controlled dipolar cycloaddition assembled the core of quinocarcin, 6-exo-substituted 3,8-diazabicyclo[3,2,1]octane <91JOC5893>. Several isothiazoles showed biological activities as 5-lipoxygenase and cycloxygenase inhibitors <91JMC518>, and as 5-HT_2 antagonists <91JMC2477>.

5.5.3 Thiadiazoles

5.5.3.1 1,2,3-Thiadiazoles

Ring transformations of 5-chloro-1,2,3-thiadiazole-4-carbaldehyde (**36**) with alkyl- and aryl-amines gave 1,2,3-triazole-4-thiocarboxamides (**37**). In constrast, treatment with phenylhydrazine and hydroxylamine afforded unrearranged hydrazone and oxime (**38**), respectively. These rearrangements can be explained to occur *via* the diazo intermediate by ring opening. Cyclization of the diazo compound to triazole-thiocarboxylic acid chloride, followed by trapping with the excess amine would afford the rearrangement products. The similar 5-chloro-1,2,3-thiadiazoles bearing a hydrogen, phenyl or ester function at the 4-position reacted with hydrazine to afford the Dimroth rearrangement products<91JCS(P1)607>.

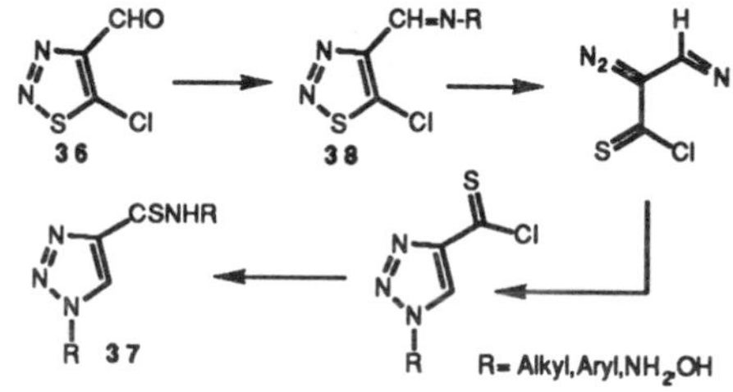

A report of the preparation of β-hydroxysulfides from 4-aryl-1,2,3-thiadiazoles appeared. Base decomposition of 1,2,3-thiadiazoles with n-butyl lithium or methyl lithium resulted in 2-(4-substituted-phenyl)ethynylthiolate anions, which were immediately reacted with α-bromoketones to give thioketones. Reaction of the α-thioketones with phenylmagnesium bromide afforded the corresponding β-hydroxysulfides by nucleophilic attack at the carbonyl group. In contrast, a similar reaction of the seleno-compounds gave the selenoethers <90JHC2037>.

The substituent at the 4-position of 1,2,3-thiadiazoles controlled the regioselectivity of methylation. The N-phenylhydrazone derived from 4-methoxycarbonyl-1,2,3-thiadiazole-5-carbaldehyde (**39**) was methylated at N-3 with methyl fluorosulfonate or Meerwein's reagents to give the mesoionic compound (**40**). On the other hand, methylation of the 4-phenyl isomer occurred at N-2 to give the thiapentalene ring systems <91JHC477>.

5.5.3.2 1,2,4-Thiadiazoles

A ring transformation of thioacylamidines (**41**) by reaction with 3,3-pentamethyleneoxaziridine, a versatile aminating agent, afforded novel ω-aminoalkyl-1,2,4-thiadiazoles(**42**) in good yield. The formation can be explained by primary S-amination and subsequent nucleophilic attack of the S-amino group at the amidine C-atom, giving a ring transformation *via* spiro intermediates <90S1071>.

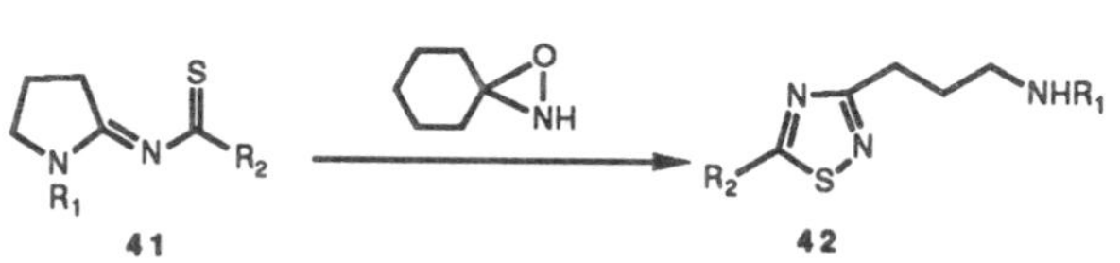

The oxidative cyclization of thioisobiurets using diethyl azodicarboxylate (DEAD) gave the corresponding 1,2,4-thiadiazoles under mild conditions <90S1020>.

5.5.3.3 1,2,5-Thiadiazoles

Reaction of tetrasulfur tetranitride with naphthalenoles and some related hydroxy-substituted benzoheterocycles in refluxing toluene afforded mono-, bis- and tris-(1,2,5)-thiadiazole-fused skeletons depending on the position of hydroxy group. The product selectivity seems to be correlated with the difference in the mode of keto-enol tautomerism of hydroxy group <91BCJ68>. The effect of the cyclophane ring on the reduction of the 1,2,5-thiadiazole was investigated <91JHC55>.

ESR studies on the reaction of radicals with benzo-2,1,3-thiadiazole derivatives were carried out. Substituents exerting only steric effects (Me,Cl) direct the addition to the less hindered nitrogen, while substituents like MeO and CN favor radical attack on the more hindered nitrogen <91JOM13>.

Bromination of naphtho[1,2-c][1,2,5]thiadiazole behaved much like phenanthrene, affording either a 4,5-addition product or a 5,6-substitution product depending on reaction conditions. Chlorination gave the corresponding addition product or 5-chloronaphtho[1,2-c][1,2,5]thiadiazole. Nitration produced a mixture of 6- and 9-isomers, since the protonated heterocyclic ring behaved as a meta-directing group <91JHC813>.

CsF mediated cyclization of 3-trimethylsilyl-1,3-diaza-2-thiaallene derivatives gave trifluoro substituted 2,1,3-benzothiadiazoles <90MI55001>. A new major metabolite of sirdaluol (5-chloro-4-(4,5-dihydroimidazol-2-ylamino)-2,1,3-benzothiadiazole hydrochloride), a novel myotonolytic agent, was dihydro-2,1,3-benzothiadiazole 2,2-dioxide, which suggests the existence of a new metabolic pathway <90JCS(P2)1705>. The 1,2,5-thiadiazole system has been used for organic semiconductors and as a ligand of metal complexes <91CL1213, 90CC1593>.

5.5.3.4 1,3,4-Thiadiazoles

Alkylidenethiocarbohydrazides were reacted with cyanogen bromide and acetyl chloride to give cyclization products, 4,5-dihydro-1,3,4-thiadiazoles. A similar reaction with α-bromo-γ-butyrolactone afforded the N-amino thiazolidine <90H1129>.

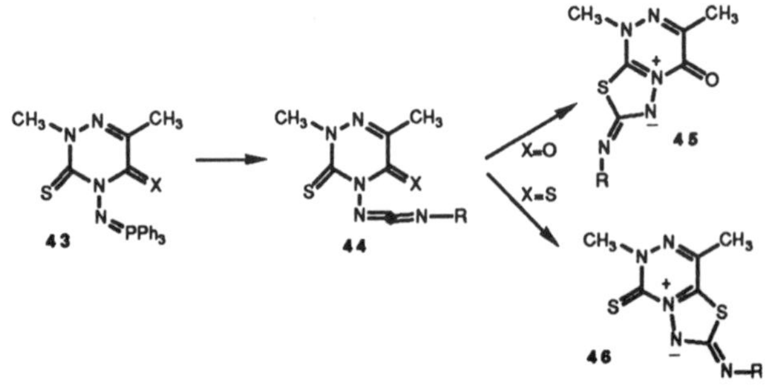

Aza Wittig-type reaction of iminophosphorane (**43**), available from 4-amino-2,6-dimethyl-5-oxo(thioxo)-3-thioxo-2,3,4,5-tetrahydro-1,2,4-triazine and triphenylphosphine dibromide, with iso (thio)cyanates leads to the mesoionic or zwitterionic fused thiadiazolo-triazine derivatives *via* postulated carbodiimide intermediates (**44**). The mode of cyclization depends on the exocyclic heteroatom in position 5. Thus, for the 5-oxo derivative cyclization occurred through the thiocarbonyl at position 3 to give [1,3,4]thiadiazolo-[2,3-c][1,2,4]triazines (**45**). The 5-thioxo derivative afforded [1,3,4]thiadiazolo[3,2-d][1,2,4]triazine derivatives (**46**) by nucleophilic attack of the sulfur atom at position 5 <91T6747>.

Cyclization of thiohydrazide derivatives with chloroacetyl chloride afforded 1,3,4-thiadiazoles <90ZC283>. Activation of thioamides with t-butyl hypochlorite gave thioamide-S-oxide and α-iminosulfenic acid equilibrium mixtures. Subsequent condensation with arenethiohydrazides gave 2,5-disubstituted 1,3,4-thiadiazoles in moderate yield <91PSS533>.

The regiospecific synthesis of novel N-sulfonyl derivatives of 2,5-diamino-1,3,4-thiadiazole was first reported. Condensation of S,S-dimethyl N-(arylsulfonyl)carbondithioimidates (**47**) with aminoguanidine bicarbonate and hydrazine carbothioamide led to N-sulfonyl derivatives of 3,5-diamino-1H-1,2,4-triazole (**48**) and 2,5-diamino-1,3,4-thiadiazole (**49**), respectively <91S220>.

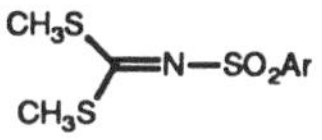

47 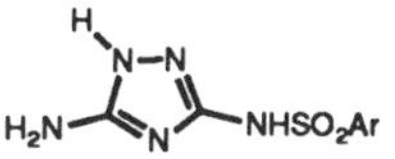 **48** 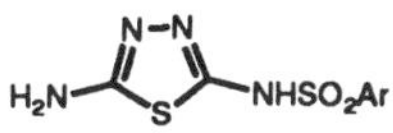 **49**

2-Aza-3-methylthiopropeniminium salts were reacted with the dithiocarbazic acid ester to give 5-benzylthio-2-(4-methoxyphenyl)-1,3,4-thiadiazole in moderate yield under mild conditions. In this case these salts behaved as activated carboxylic acid derivatives. The reaction seems to occur by the usual primary replacement of the methylthio group, followed by nucleophilic attack at the amidine carbon <91JPR149>.

Methyl 2-methyldithiocarbazate is useful for the preparation of novel bis-(1,3,4-thiadiazolium) salts and of 2,5-disubstituted 1,3,4-thiadiazoles and macrocycles containing 1,3,4-thiadiazoles subunits. Cyclization of the iminophosphorane, readily available from the carbazate and carbon disulfide, gave the mesoionic 4-methyl-5-methylthio-1,3,4-thiadiazolium-2-thiolate. On the contrary, sequential treatment of the carbazate with dicarboxylic acid chlorides and perchloric acid led to bis-(1,3,4-thiazolium) salts. These salts can be converted to compounds containing the 1,3,4-thiadiazole <91JCS(P1)1159>. Several fused 1,3,4-thiadiazole derivatives were prepared <91JHC489, 91IJC(B)435, 91G113>. Several 1,3,4-thiadiazoles, such as 3-acyl-5-methyl-1,3,4-thiadiazole-2(3H)-thiones (**50**) and 2-acylthio-5-methyl-1,3,4-thiadiazoles (**51**), were reported as useful acylating reagents similar to or superior to 3-acylthiazolidine-2-thiones (**52**).

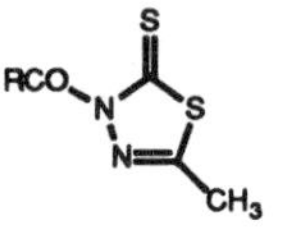

50 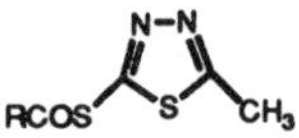 **51** 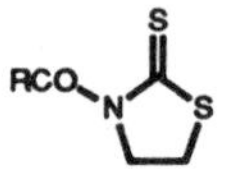 **52**

Acylation of free glycosylamine using (**50**) or (**51**) provided N-acylglycosylamines in high yield. A new selective acylation of free sucrose with (**50**) and DABCO afforded 6-O-acylsucroses. 6-O-acyl-D-glycopyranoses were synthesized from the corresponding sugar without protecting groups <91T3817, 91TL3495, 91TL1557>. Optically active 3-substituted-2,3-dihydro-1,3,4-thiadiazole derivatives were prepared <91CPB777>. The synthesis, crystal and molecular structures and magnetic properties of complexes containing 2,5-diamino-1,3,4-thiadiazole with Ni were reported <91JCS(D)2331>. Many pharmaceutically interesting compounds possessing the 1,3,4-thiadiazole nucleus were synthesized <91JMC1594, 90AJC1767, 91JMC439>.

5.5.4 Selenazoles and Selenadiazoles

Treatment of various primary aromatic or aliphatic selenoamides, which were prepared from the corresponding nitriles and $(Me_3Si)_2Se$, with N-bromosuccinimides as an oxidizing reagent provided a convenient preparation of 3,5-disubstituted 1,2,4-selenadiazoles <91BCJ1037>. Reaction of 1-acyl-4-(*p*-tolyl)-selenosemicarbazides with chloroacetone or ω-bromoacetophenone yielded selenazoles, 1,3,4-oxadiazoles, or 1,3,4-selenadiazines, depending on the solvent and acyl group <89MI55001, 89MI55002>. The synthesis of benzo[1,2-d:5,4-d']diisoselenazole, starting from 4,6-dimethylselenoisophthalaldehyde, was reported <91BSB1>. The regioselectivity in the cyclization of N-benzyl-4-homopiperidinone showed 1,2,3-selenadiazoles unusual medium effects to semicarbazone. Comprehensive mechanistic studies were presented <91JOC5203>.

The 1,2,3-selenadiazole ring system (**53**) is interesting for the facile formation of reactive intermediates such as selenirene (**55**), diradical (**54**), and selenoketene (**56**). Monocyclic 1,2,3-selenadiazoles are known to have a tendency to decompose to selenium and alkynes. Thermolysis or photolysis (λ =365nm) of electronically activated and strained olefins afforded regiospecifically 1,3-cycloadduct (**57**) among the expected products. The dimeric 1,4-diselenin was formed in the absence of any trapping reagents. In no case were selenoketene (**56**) and cyclopentyne (**58**) found as intermediates <88MI55001>.

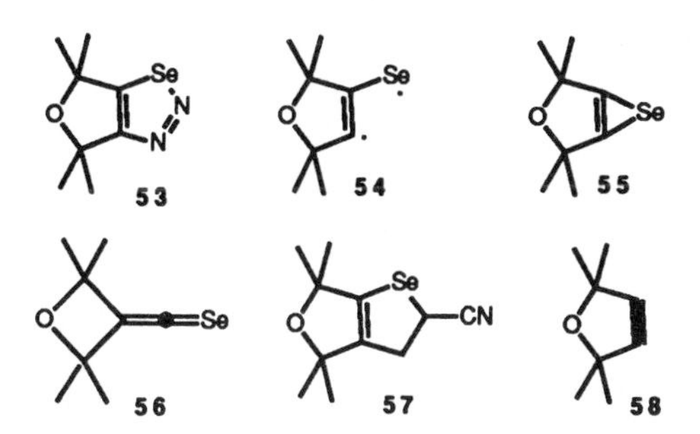

Analogously, photolysis or thermolysis of 1,2,3-selenadiazoles with kinetically stabilized phosphaalkynes afforded rare 1,3-selenaphospholes, presumably *via* sextet dipole or diradical intermediates <91MI55002>. Thermolysis of the 1,2,3-selenadiazole ring was useful for the synthesis of medium-sized carbocyclic diacetylenenes <91JA9258>. Upon pyrolysis, extremely sterically hindered 1,3,4-selenadiazole derivatives gave no extrusion of nitrogen <91JOC3178>. Electrophilic and nucleophilic substitution reactions on 4-bromo-6-methyl-2,1,3-benzoselenadiazole were reported <90IJC(B)781>. 1,4-Bis(1,2,3-selenadiazolo-4-yl)benzene and 1,2,3-selenazole derivatives were prepared from the corresponding semicarbazones and selenium dioxide <90IJC757, 90ACH433>. Thermal and photochemical reactions of fused 1,2,3-selenadiazoles were reviewed <90MI55002>.

REFERENCES

<88MI55001>	W. Ando, Y. Kumamoto, and T., N. Tokitoh; *J. Phys. Org. Chem.* , 1988, **6**, 317.
<89MI55001>	S. Bilinski, L. Bielak, J.B. Chmielewski, B. Marcewicz-Rojewska, and I. Musik, *Acta Pol. Pharm.* , 1989, **46**, 219.
<89MI55002>	S. Bilinski, L. Bielak, J. Chmielewski, B. Marcewicz-Rojewska, and I. Musik, *Acta Pol. Pharm.*, 1989, **46**, 343.
<90ACH433>	S. El-Bahaie and B. Bayoumy, *Acta. Chim. Hung.*, 1990, **127**, 433.
<90AJC1767>	R.D. Allan, C. Apostopoulos and J.A. Richardson, *Aust. J. Chem.*, 1990, **43**, 1767.
<90BCJ3223>	K. Okuma, N. Higuchi, S. Kaji, K. Takeuchi, H. Ohta, H. Matsuyama, N. Kamigata and M. Kobayashi, *Bull. Chem. Soc. Jpn.*, 1990, **63**, 3223.
<90CC362>	T.C. Owen and G.J.S. Doad, *J. Chem. Research*, 1990, 362.
<90CC1593>	I. Hawkins and A.E. Underhill, *J. Chem. Soc.,Chem. Commun.*, 1990, 1593.
<90H1129>	S. Kabashima, T. Okawara, T. Yamasaki and M. Furukawa, *Heterocycles*, 1990, **31**, 1129.

<90HCA2287> P.-C. Tromm and H. Heimgartner, *Helvetica Chimica Acta*, 1990, **73**, 2287.

<90IJC(B)781> S. Kumari, K.S. Sharma and S.P. Singh, *Indian J. Chem., Sect. B.*, 1990, **29B**, 781.

<90JHC2037> I. Ganjian, *J. Heterocycl. Chem.*, 1990, **27**, 2037.

<90IJC757> S. El-Bahaie, M.G. Assy and M.M. Hassanien, *J. Indian Chem. Soc.*, 1990, **67**, 757.

<90JCS(P2)1705> P. Koch, J.J. Boelsterli, D.R. Hirst, and M.D. Walkinshaw, *J. Chem. Soc., Perkin 2*, 1991, 1705.

<90MI55001> A.V. Zibarev and A.O. Miller, *J. Fluorine Chem.*, 1990, **50**, 359.

<90MI55002> W. Ando and N. Tokitoh, *Heteroat. Chem.*, 1990, 1.

<90S1020> Y. Kihara, S. Kabashima, K. Uno, T. Okawara, T. Yamasaki and M. Furukawa, *Synthesis*, 1990, 1020.

<90S1071> M. Paetzel, J. Liebscher, S. Andreae and E. Schmitz, *Synthesis*, 1990, 1071.

<90SC2235> M.W. Bredenkamp, C.W. Holzapfel and W.J. van Zyl, *Syn. Comm.*, 1990, **20**, 2235.

<90SC3507> C.-N. Hsiao, M.R. Leanna, L. Bhagavatula, E. De Lara, T.M. Zydowsky, B.W. Horrom and H.E. Morton, *Syn. Comm.*, 1990, **20**, 3507.

<90T8169> E. Epifani, S. Florio and G. Gasparri Fava, *Tetrahedron*. 1990, **46**, 8169.

<90TL5789> A.F. Brigas and R.A.W. Johnstone, *Tetrahedron Lett.*, 1990, **31**, 5789.

<90ZC283> W. Thiel, *Z. Chem.*, 1990, **30**, 283.

<91BCJ68> S. Mataka, K. Takahashi, Y. Ikezaki, T. Hatta, T. Akiyoshi and M. Tashiro, *Bull. Chem. Soc. Jpn.*, 1991, **64**, 68.

<91BCJ1037> K. Shimada, Y. Matsuda, S. Hikage, Y. Takeishi and Y. Takikawa, *Bull. Chem. Soc. Jpn.*, 1991, **64**, 1037.

<91BSB1> A. Jakobs, L. Christiaens and M. Renson, *Bull. Soc. Chim. Belg.*, 1991, **100**, 1.

<91CB1627> U. Verfuerth and I. Ugi, *Chem. Ber.*, 1991, **124**, 1627.

<91CC373> T. Nabesha, K. Moriyama and Y. Yano, *J. Chem. Soc., Chem. Commun.*, 1991, 373.

<91CC924> B. Beagley, M.J. Betts, R.G. Pritchard, A. Schofield, R. J. Stoodley and S. Vohra, *J. Chem. Soc., Chem. Commun.*, 1991, 924.

<91CC1313> A. Dondoni, D. Perrone and P. Merino, *J. Chem. Soc., Chem. Commun.*. 1991, 1313.

<91CL661> K. Shibata, H. Shiono, and O. Mitsunobu, *Chem. Lett.*, 1991, 661.

<91CL1213> Y. Yamashita, S. Tanaka, K. Imaeda and H. Inokuchi, *Chem. Lett.*, 1991, 1213.

<91CPB651> N. Seko, K. Yoshino, K. Yokota and G. Tsukamoto, *Chem. Pharm. Bull.*, 1991, **39**, 651.

<91CPB777> K. Toyooka , Y. Takeuchi, H. Ohto, M. Shibuya and S. Kubota, *Chem. Pharm. Bull.*, 1991, **39**, 777.

<91G113> M.B. El-Ashmawy, I.A. Shehata, H.I. El-Subbagh and A.A. El-Enam, *Gazz, Chim. Ital.*, 1991, **121**, 113.

<91H253> K.H. Yoo, D.J. Kim, D.C. Him, and S.W. Park, *Heterocycles*. 1991, **32**, 253.

<91HCA531> A. Pascual, *Helv. Chim. Acta.*, 1991, **74**. 531.

<91HCA1059> B. Schulze, G. Kirsten, S. Kirrbach, A. Rahm and H. Heimgartner, *Helv. Chim. Acta.*, 1991, **74**, 1059.

<91IJC(B)435> V.S. R. Prasad and K.K. Reddy, *Indian, J. Chem., Sect. B.*, 1991, **30B**, 435.

<91JA9258> R. Gleiter, D. Kratz, W. Schafer and V. Scheblmann, *J. Am. Chem. Soc.*, 1991, **113**, 9258.

<91JCS(C)924> B. Beagley, M.J. Betts, R.G. Pritchard, A. Schofield, R.J. Stoodley and S. Vohra, *J. Chem. Soc., Chem. Commun.*, 1991, 924.

<91JCS(D)2331> A.C. Fabretta, W. Malvasi, D. Gatteschi and R. Sessoli, *J. Chem. Soc., Dalton Trans.*, 1991, **9**, 2331.

<91JCS(P1)607> G. L'abbe, E. Vanderstede, W. Dehaen, P. Delbeke and S. Toppert, *J. Chem. Soc., Perkin Trans.1*, 1991, 607.

<91JCS(P1)855> M.A. O'Daly, C.P. Hopkinson, G.D. Meakins, and A.J. Raybould, *J. Chem. Soc., Perkin Trans. 1*, 1991, 855.

<91JCS(P1)1159> P. Molina, A. Espinosa, A. Tarraga, F. Hernandez Cano, M.C. Foces-Foces, *J. Chem. Soc., Perkin Trans. 1*, 1991, 1149.

<91JHC55> T. Hatta, S. Mataka, M. Tasho, K. Numano, K. Suzuki and A. Torii, *J. Heterocycl. Chem.*, 1991, **28**, 55.

<91JHC347> D.M. McKinnon and A. Abouzeld, *J. Heterocycl. Chem.*, 1991, **28**, 347.

<91JHC477> G. L'abbe, A. Frederix, S. Toppet and J.P. Declercq, *J. Heterocycl. Chem.*, 1991, **28**, 477.

<91JHC489> T. Tsuji, *J. Heterocycl. Chem.*, 1991, **28**, 489.

<91JHC813> W.T. Smith, Jr., J.M. Patterson and A.C. Kovelesky, *J. Heterocycl. Chem.*, 1991, **28**, 813.

<91JHC1013> M.S. South, *J. Heterocyclic Chem.*, 1991, **28**, 1013.

<91JHC1017> M.S. South and K.A. Van Sant, *J. Heterocyclic Chem.*, 1991, **28**, 1017.

<91JMC108> B.L. Mylari, E.R. Larson, T.A. Beyer, W.J. Zembrowski, C.E. Aldinger, M.F. Dee, T.W. Siegel, and D.H. Singleton, *J. Med. Chem.*, 1991, **34**. 108.

<91JMC439> C. Bennion, R.C. Brown, A.R. Cook, C.N. Manners, D.W. Payling, and D.H. Robinson, *J. Med. Chem.*, 1991, **34**, 439.

<91JMC518> D.L. Flynn, T.R. Belliotti, A.M. Boctor, D.T. Connor, C.R. Kostian, D.E. Nies, D.F. Ortwine, D.J. Schrier and J.C. Sircar, *J. Med. Chem.*, 1991, **34**, 518.

<91JMC1630 R.J. Weikert, S. Bingham, Jr., M.A. Emanuel, E.B. Fraser-smith, D.G. Loughhead, P.H. Nelson, and A.L. Poulton, *J. Med. Chem.*, 1991, **34**, 1630.

<91JMC1975> R.C. Schnur, R.J. Gallaschun, D.H. Singleton, M. Grissom, D.E. Sloan, P. Goodwin, P.A. McNiff, A.F.J. Fliri, F.M. Mangano, T.H. Olson, and V.A. Pollack, *J. Med. Chem.*, 1991, **34**, 1975.

<91JMC2477> J.L. Malleron, M.T. Comte, C. Gueremy, J.F. Peyronel, A. Truchon, J.C. Blanchard, A. Doble, O. Piot, J.L. Zundel, et al., *J. Med. Chem.*, 1991, **34**, 2477.

<91JOC809> F.A. Davis, R.T. Reddy, J.P. McCauley, Jr., R.M. Przeslawski, M.E. Harakal and P.J. Carroll, *J. Org. Chem.*, 1991, **56**, 809.

<91JOC1309> M. Newcomb and K.A. Weber, *J. Org. Chem.*, 1991, **56**, 1309.

<91JOC3178> F.S. Guziec, Jr. and L.J. SanFilippo, *J. Org. Chem.*, 1991, **56**, 3178.

<91JOC4645> F. Freeman and D.S.H.L. Kim, *J. Org. Chem.*, 1991, **56**, 4645.

<91JOC5203> B.E. Maryanoff and M.C. Rebarchak, *J. Org. Chem.*, 1991, **45**, 5203.

<91JOC5029> Y-T. Chen and F. Jordon, *J. Org. Chem.*, 1991, **56**, 5029.

<91JOC5294> A. Dondoni and P. Merino, *J. Org. Chem.*, 1991, **56**, 5294.

<91JOC5893> P. Garner, W.B. Ho, K. Sunitha, W.J. Youngs and V.O. Kennedy, *J. Org. Chem.*, 1991, **56**, 5893.

<91JOC6199> L.A. Paquette, K. Dahnke, J. Doyon, W. He, K. Wyant, and D. Friedrich, *J. Org. Chem.*, 1991, **56**, 6199.

<91JOM13> A. Alberti, M. Benaglia, D. Fiorentini dal Monte, M. Lucarini, and G.F. Pedulli, *J. Organomet. Chem.*, 1991, **401**, 1.

<91JPR149> M. Paetzel and J. Liebscher, *J. Prakt. Chem.*, 1991, **333**, 149.

<91LA305> J. Martens, J. Kintscher, K. Lindner, S. Pohl, W. Saak, and D. Haase, *Liebigs Ann. Chem.*, 1991, 305.

<MI55001> F. Gardini and M. E. Guerzoni, *Tetrahedron Asymmetry*, **2**, 243.

<91MI55002> B. Burkhart, S. Krill, Y. Okano, W. Ando and M. Regitz, *Synlett*, 1991, 356.

<91PSS119> M.K.A. Ibrahim and A.H. Elghandour, *Phosphorus Sulfur Silicon*, 1991, **60**, 119.

<91PSS281> H. Heimgartner, *Phosphorus Sulfur Silicon*, 1991, **58**, 281.

<91PSS533> G.I. Nikish, E.I. Troyanski and D.V. Demchuk, *Phosphorus Sulfur Silicon*, 1991, **59**, 533.

<91S220> S.P. Maybhate, P.P. Rajamohanan and S. Rajappa, *Synthesis*, 1991, 220.

<91S362> M. Drescher, E. Ohler, and E. Zbiral, *Synthesis*, 1991, 362.

<91S494> K.-J. Lee, J.U. Jeong, D.O. Choi, S.H. Kim and K. Park, *Synthesis*, 1991, 494.

<91S621> C. Hopkinson, G.D. Meakins, and R.J. Purcell, *Synthesis*, 1991, 621.

<91T2851> G. Jones and H. Ollivierre, *Tetrahedron.*, 1991, **47**, 2851.

<91T2861> G. Jones and H. Ollivierre, *Tetrahedron*, 1991, **47**, 2861.

<91T3817> K. Baczko and D. Plusquellec, *Tetrahedron*. 1991, **47**, 3817.

<91T4465> E. Epifani, S. Florio, D. Perrone, and G. Valle, *Tetrahedron*, 1991, **47**, 4465.

<91T5539> L. Kovacs, P. Herczegh, G. Batta, and I. Farkas, *Tetrahedron*, 1991, **47**, 5539.

<91T5549> L. Kovacs, P. Herczegh, G. Batta, and I. Farkas, *Tetrahedron*. 1991, **47**, 5549.

<91T6747> P. Molina, M. Alajarin, and A. Lopez-Lazaro, *Tetrahedron*, 1991, **47**, 6747.

<91T8267> M. North and G. Pattenden, *Tetrahedron*. 1991, **46**, 8267.

<91TL1175> J.B. Baudin, G. Hareua, S.A. Julia, and O. Ruel, *Tetrahedron Lett*., 1991, **32**, 1175.

<91TL1557> P. Leon-Raud, M. Allainmat and D. Plusquellec, *Tetrahedron Lett*.. 1991, **32**, 1557.

<91TL3247> A. Dondoni, P. Merino, and J. Orduna, *Tetrahedron Lett*., 1991, **32**, 3247.

<91TL3495> C. Chauvin and D. Plusquellec, *Tetrahedron Lett*.. 1991, **32**, 3495.

<91TL4263> T.R. Kelly, C.T. Jagoe, and Z. Gu, *Tetrahedron Lett*., 1991, **32**, 4263.

<91TL6085> Z. Cekovic and R. Saicic, *Tetrahedron Lett*., 1991, **31**, 6085.

<91ZN1695> K. Burger, E. HoB, N.Sewald, and K. Geith, *Z. Naturforsch. B*, 1991, **45**, 1695.

CHAPTER 5.6

Five-Membered Ring Systems: With O & S (Se,Te) Atoms

R. ALAN AITKEN
University of St. Andrews, UK

5.6.1 1,3-Dioxoles and Dioxolanes

A comprehensive review of O,O-acetals including a detailed survey of dioxolane chemistry has appeared [91MI1]. New catalysts reported for the addition of epoxides to aldehydes and ketones to form dioxolanes include zeolites [91MI211], N-benzylpyridinium salts [90CL2019], and HBF_4 [91KGS33]. A range of catechols are conveniently methylenated to give benzo-1,3-dioxoles using CH_2BrCl and Cs_2CO_3 in DMF or MeCN [91TL2461]. The efficient conversion of 2,2-dichlorobenzo-1,3-dioxole to the 2,2-difluoro-analogue has been described [91EP423009].

The 4-methoxy-1,3-dioxolanes (**1**), selectively protected 1,3-diketones, are formed by reaction of $RC(=O)OCH_2CH(OMe)_2$ with enamines [90SC2197]. Treatment of compounds (**2**) with base results in cyclisation of the resulting anions to afford 2-methylenedioxolanes (**3**) for X = CO_2Me [90IZV2662] and CF_3 [91DOK(319)619] and these are claimed to be the first examples of O-alkylation of a malonate anion. A new mild procedure for the conversion of α-hydroxy acids to dioxolan-4-ones (**4**) gives improved *cis/trans* ratios in the products [91TL1441]. The combination of $LiAlH_4$ with $ZrCl_4$ gives a useful

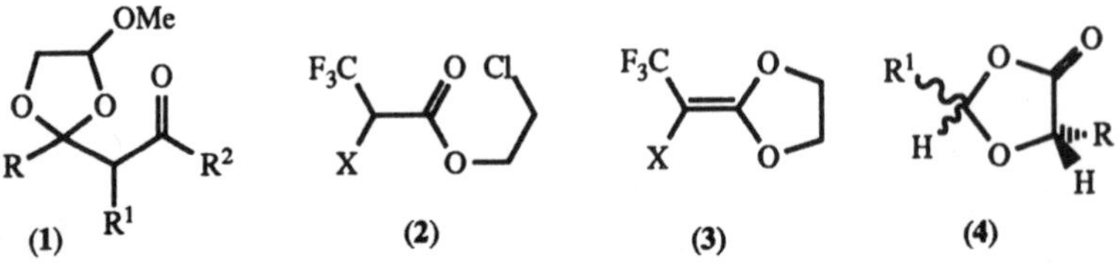

(**1**) (**2**) (**3**) (**4**)

reagent for reductive cleavage of dioxolanes [91OPP427] and the products are solvent dependant. Thus, for example, 2-phenyl-1,3-dioxolane gives $PhCH_2OCH_2CH_2OH$ in ether but toluene and benzyl alcohol in benzene. A convenient large scale preparation of chiral dioxolanediols (**5**) has been reported [91MI238] and (**6**) has been prepared as a new chiral ligand with C_2 symmetry [90G731]. Procedures for enantiomeric enrichment of chiral dioxolane carboxylic acids (**7**) by recrystallisation have been described [91EP426253].

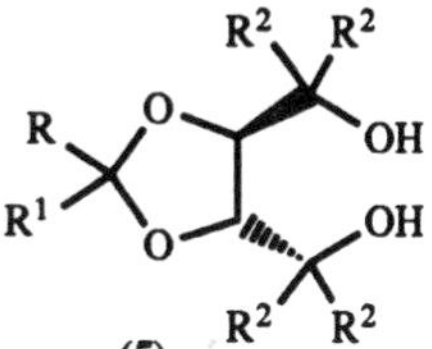

(**5**) 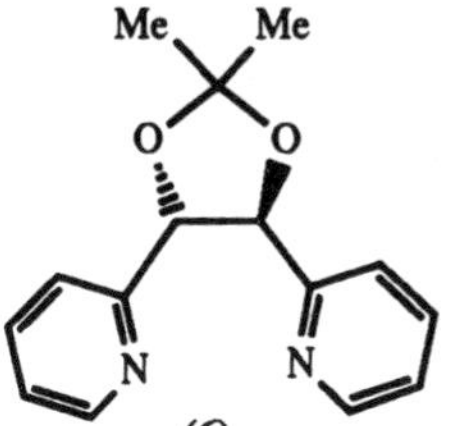 (**6**) 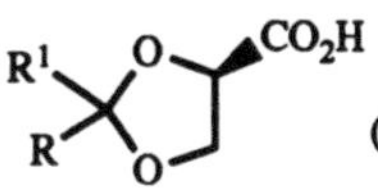 (**7**) 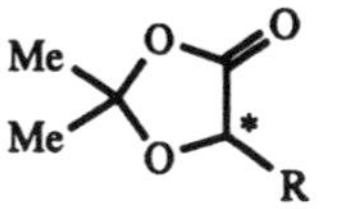 (**8**)

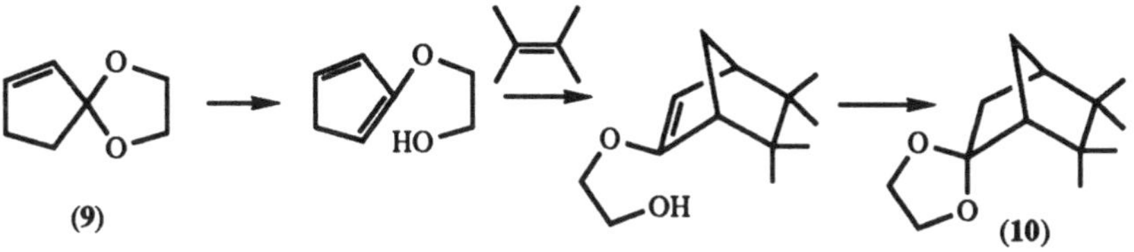

(**9**) (**10**)

Asymmetric protonation of the lithium enolates of dioxolanones (**8**) with chiral tricyclic lactams gives (**8**) in up to 91% enantiomeric excess [90AG(E)1420]. Cyclopentenone acetal (**9**) undergoes Diels Alder cycloaddition with a wide variety of dienophiles *via* its enol ether form to give norbornan-2-one acetals (**10**) as shown [91CC37]. Details of the preparation and reactivity of chiral and achiral ketene acetals (**11**) have appeared [91T975] [91TL465] and the related acetals (**12**) undergo highly diastereoselective Diels Alder reactions [91T4519]. The asymmetric hetero Diels Alder reaction of (**13**) with, for example, acrolein

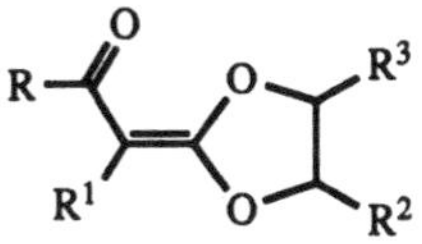

(**11**) 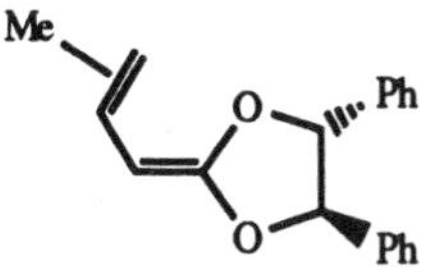 (**12**) 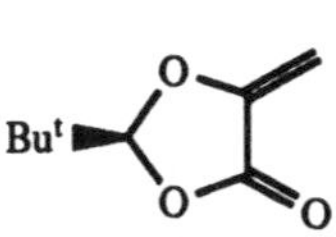 (**13**) 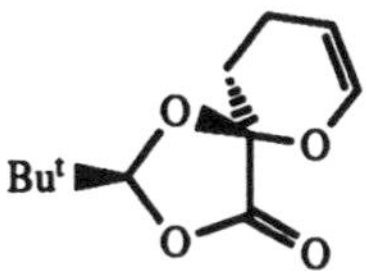 (**14**)

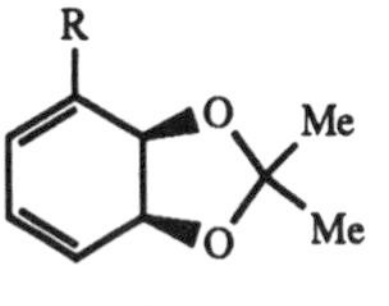

(**15**) 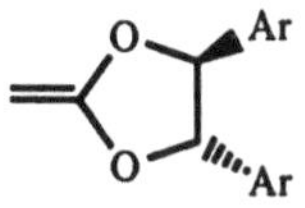 (**16**) 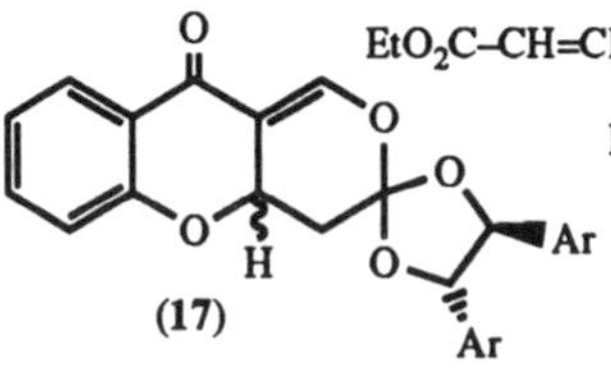 (**17**) 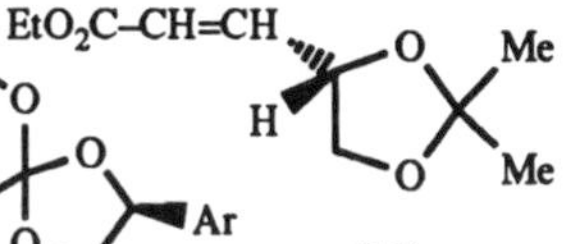 (**18**)

giving exclusively the isomer (**14**) has been reported [90SYL145], as have a wide variety of cycloadditions of (**15**) [91JCS(P1)1255]. Cycloaddition of methylenedioxolanes (**16**) to 3-formylchromone to give (**17**) is the key step in an elegant asymmetric synthesis of 2-substituted chromanones [91CC1707]. Good diastereoselectivity is observed in the addition of silicon centred radicals to the double bond of *E* or *Z*-(**18**) [91TL3683].

5.6.2 1,3-Dithioles and Dithiolanes

Activity in the field of tetrathiafulvalenes has continued at a high level. A new direct preparation for substituted TTF's involves desulphurisation of 1,3-dithiole-2-thiones with $(MeO)_3P$ at high pressure [90JCS(P1)3358]. Convenient preparations of mono- and dibromo-TTF [91CC92], mono-, di-, and tetrachloro-TTF [91S263], and tetrakis(trifluoromethyl)-TTF [91LA395] [91SM(42)2381] have been described. Further new substituted TTF's include (**19**) [91MI107], (**20**) [90MI535], and a range of examples such as (**21**) bearing double bonded groups suitable for polymerisation [91CC1638]. The complexation of TTF itself with a paraquat-related cyclophane host molecule has been observed in solution and the solid state [91CC1584]. Dimeric TTF

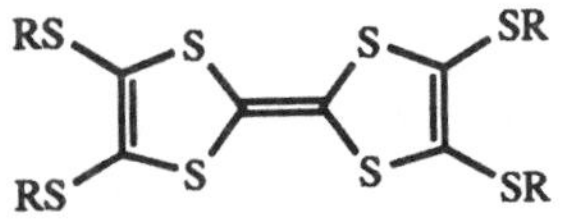

(**19**) R = CH_2CH_2SMe
(**20**) R = CH_2CH_2CN
(**21**) R = $CH{=}CH_2$

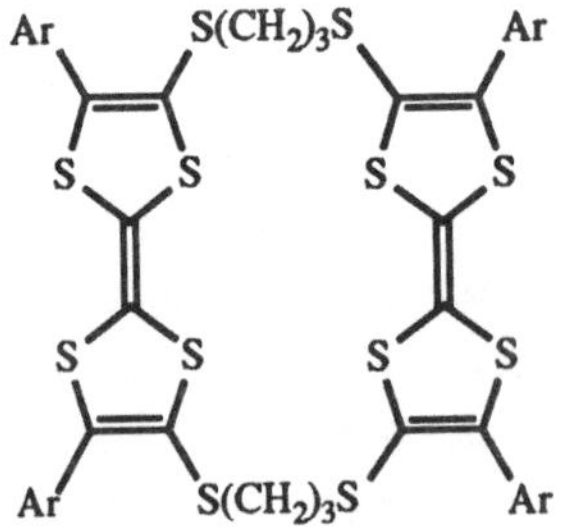

(**23**)

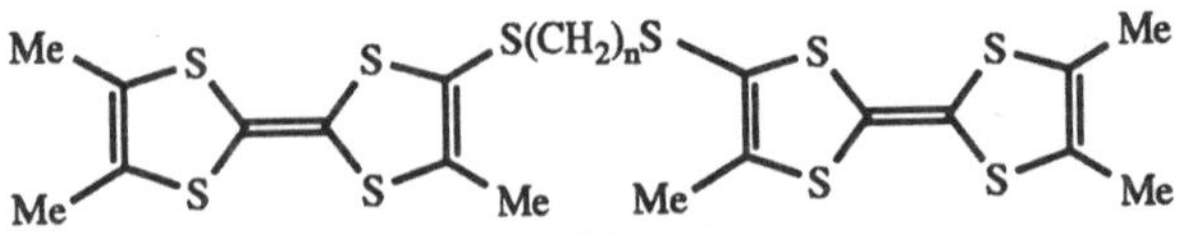

(**22**)

derivatives including (**22**) [91JOC5684] [91SM(42)2561] and (**23**) [91CC843] have been prepared. The synthesis and X-ray structure of tris(tetrathiafulvalenyl)-phosphine have been reported [91CC1370].

New bis-annulated TTF derivatives include (**24**) [91SM(42)2381], (**25**) [91KGS848], (**26**) and (**27**) [91CC952], and (**28**), (**29**), and (**30**) [91TL479]. The unsymmetrical derivatives (**31**) [90CL2129] and (**32**) [91TL2737] [91TL2741] have also been prepared.

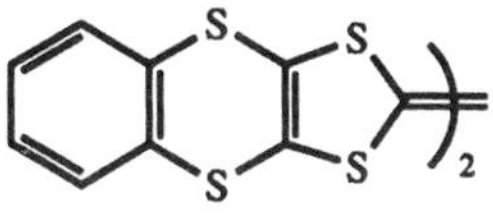

(24) 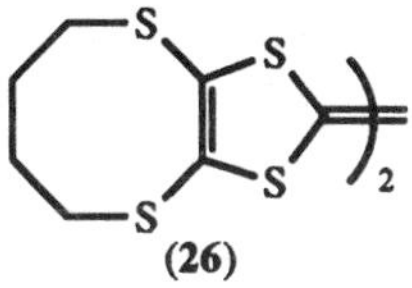 (26) 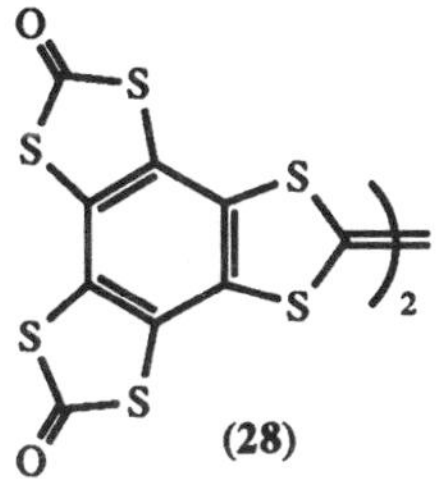 (28) 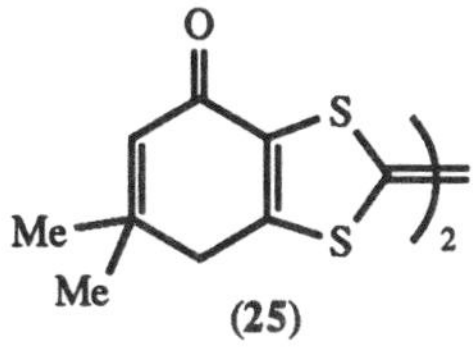 (25) 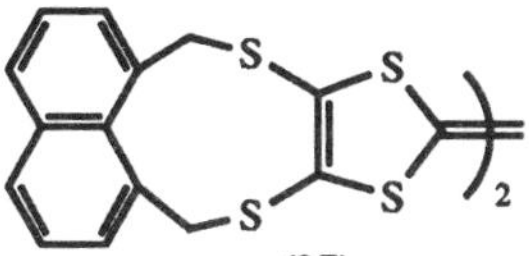 (27) 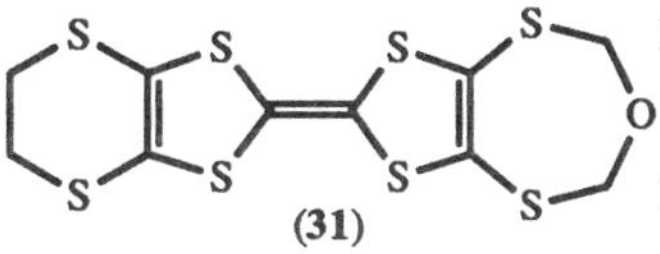 (31) 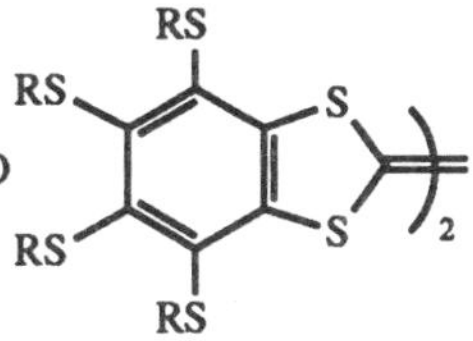 (30) 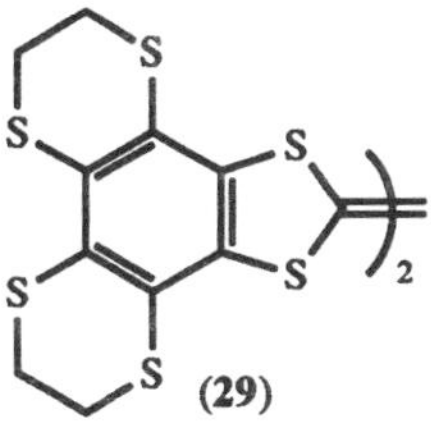 (29) 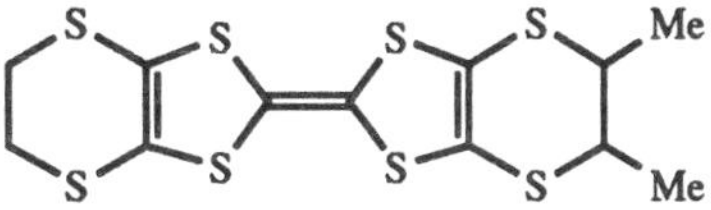 (32)

Extended TTF analogues have also been reported as exemplified by **(33)** [91CC320], **(34)** [91S26], **(35)**, **(36)** and **(37)** [91JCS(P1)157], **(38)**

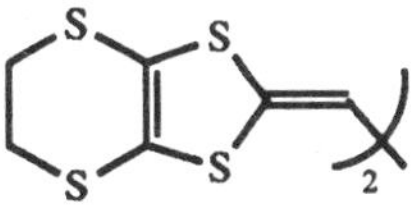

(33) 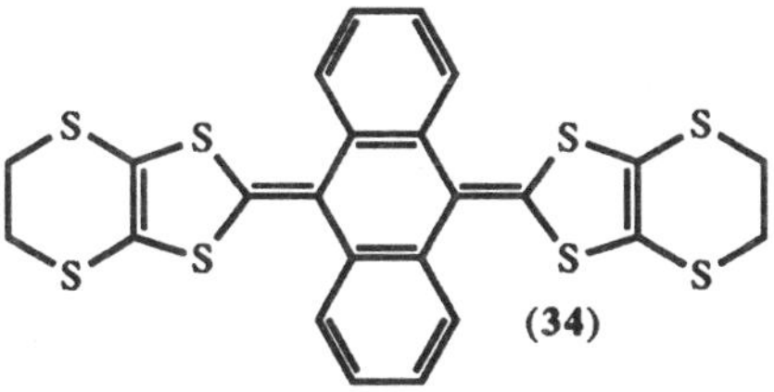 (34) 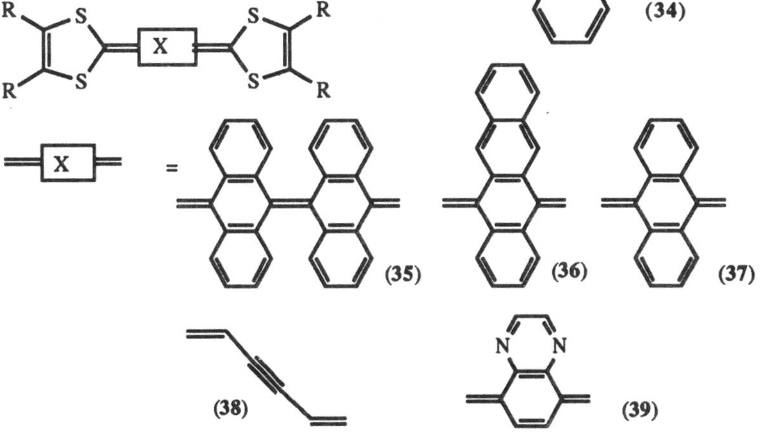 (35) (36) (37) (38) (39)

[90TL7307] [90TL7311] [91SM(42)2327] and **(39)** [91CC1132]. More complex systems include **(40)** [91TL2897] and **(41)** [91SM(42)2323]. Full details of the formation of (**42**), (**44**) and related compounds from $C_6(SH)_6$ have appeared [90S1149] and (**43**) and (**45**) have been prepared [91TL4897].

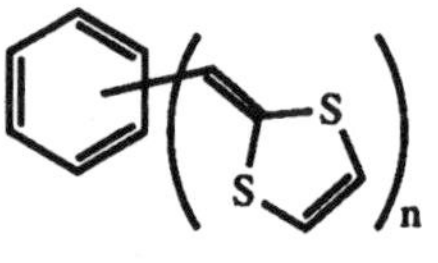

(**40**) n = 2 (1,2- and 1,4-)
n = 4 (1,2,4,5-)

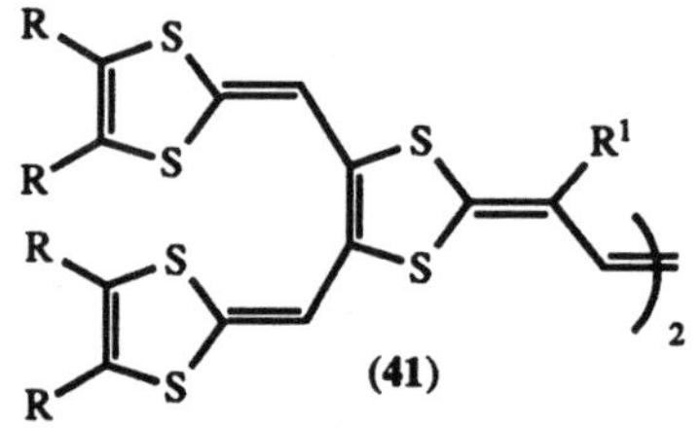

(**41**)

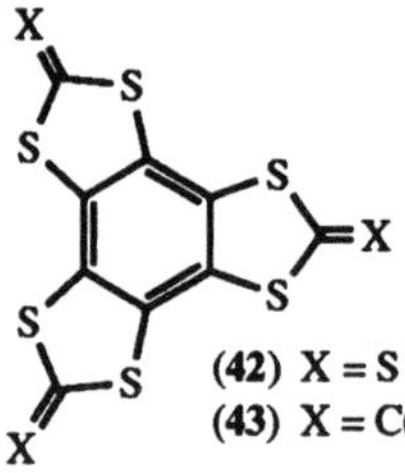

(**42**) X = S
(**43**) X = $C(SMe)_2$

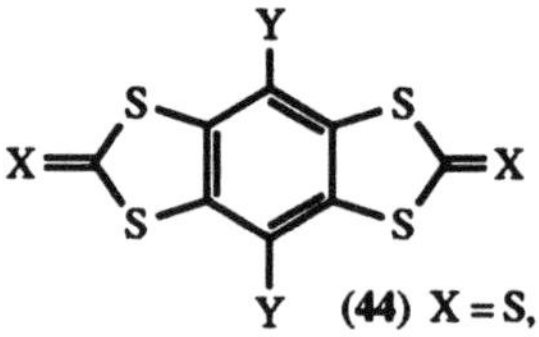

(**44**) X = S, Y = SR
(**45**) X = $C(SMe)_2$ Y = H

Reduction of the bis methyl iodide adduct of (**46**) to give a diradical has been reported [91AG(E)871]. A detailed study of the chemistry of compounds (**47**) includes preparation of spiro derivatives such as (**48**) [91CB2025]. Reaction of $Bu^n_3P.CS_2$ with DMAD under neutral conditions gives the 1 : 2 adduct (**49**) *via* ring opening of intermediate (**50**) [91CC812]. New methods for the formation of 2-imino-1,3-dithioles include reaction of dicyano epoxides with dithiocarbamates to afford (**51**) [91CC906] and benzopentathiepine with RNCS or RNCSe to give (**52**) [91CL355].

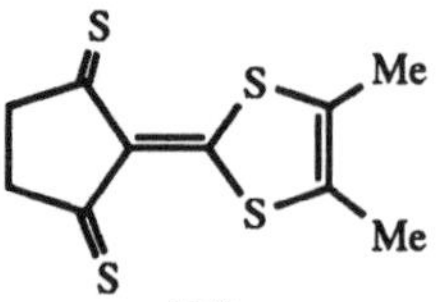

(**46**)

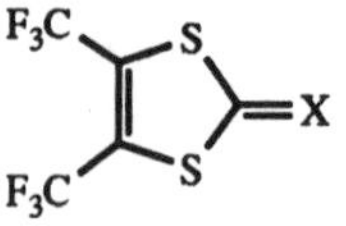

(**47**) X = O, S, Cl_2

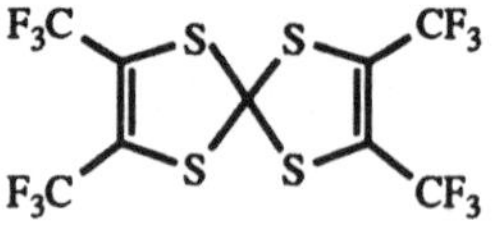

(**48**)

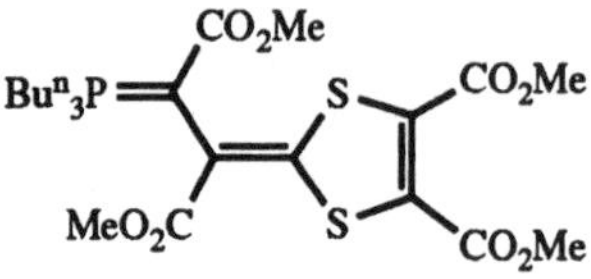

(**51**)

(**49**)

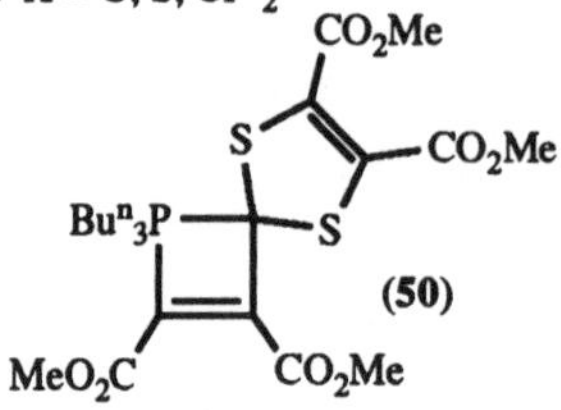

(**50**)

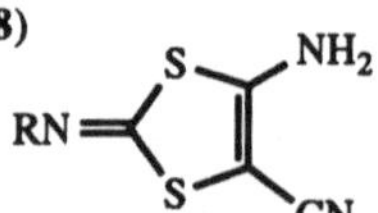

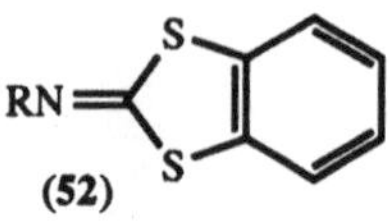

(**52**)

A convenient one pot conversion of active methylene compounds to 2-alkylidene 1,3-dithiolanes has been reported [91S301]. Insertion of two molecules of C≡S into the S–Cl bonds of $ClSCH_2CH_2SCl$ is accompanied by extrusion of $CSCl_2$ to give 1,3-dithiolane-2-thione [91SL143]. Anhydrous

$FeCl_3$ on silica is an efficient catalyst for conversion of aldehydes and ketones to 2-substituted-1,3-dithiolanes [91TL2259] and the reverse reaction can be achieved by trans-thioacetalisation with *p*-nitrobenzaldehyde catalysed by trialkylsilyl triflates [91CC1750]. The monosulphone of 1,3-dithiolane has been reported for the first time [91PS(58)207]. Reaction of 2-aryl-1,3-dithiolanes with MeMgI in the presence of a nickel catalyst gives the corresponding styrenes [91JOC4035].

Intramolecular addition of the 1,3-dithiolan-2-yl radical derived from (**53**) by photolysis gives (**54**) [90TL7035]. Ring expansion of 2-hydroxyalkyl-1,3-dithiolanes (**55**) to give dithiins (**56**) and methylenedithianes (**57**) has been reported [91S575]. The generation of 1,3-dithiolanylideneketene (**58**) by FVP of Meldrum's acid or isoxazolone precursors has been described and it decomposes

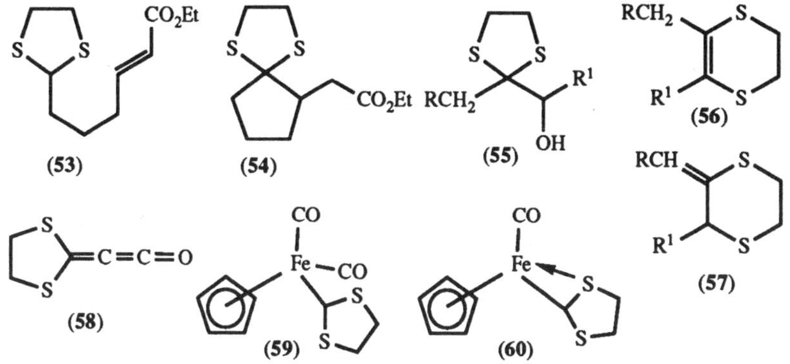

(**53**) (**54**) (**55**) (**56**) (**57**) (**58**) (**59**) (**60**)

on further pyrolysis to CO, ethene and S=C=C=S [91JA3130]. Photochemical elimination of CO from the dithiolanyl iron complex (**59**) is accompanied by a change in bonding to give (**60**) [91MVC99].

5.6.3 1,3-Oxathioles and Oxathiolanes

A comprehensive review of O,S-acetals including a detailed survey of oxathiolane chemistry has appeared [91MI1]. Rhodium catalysed decomposition of α-diazoketones and esters (**61**) in the presence of CS_2 gives oxathiole-2-thiones (**62**) [90BCJ3096]. For the ester-containing diazoketones (**63**) simple oxathiole-2-thione formation is only a minor process and the main product is the spiro compound (**64**), the 2 : 1 CS_2 adduct of the carbonyl ylide formed by interaction of the carbene with the ester [91BCJ771]. Further mechanistic studies on the reaction of $HOCMe_2$–C≡C–CN with metal thiocyanates to give oxathiolanes have appeared [91SL63]. Enantiomerically pure oxathiolane-2-thiones (**65**) derived from 2,3-epoxyalcohols have been used in the synthesis of thionucleosides [91CC1421]. Pyrido-oxathiolone (**66**) has been prepared and used as a reagent for conversion of primary amines into symmetrical and unsymmetrical ureas [91JOC891]. Asymmetric Pummerer reaction of the

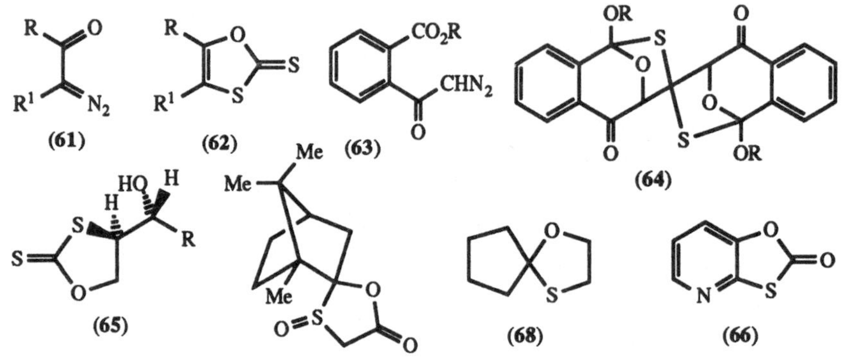

camphor-derived oxathiolanone S-oxide (**67**) has been described [90TAS143]. Radical attack of $(Me_3Si)_3Si\cdot$ results in C-2–S cleavage of oxathiolanes such as (**68**) [91TL2853].

5.6.4 1,2-Dioxoles and Dioxolanes

Treatment of unsaturated hydroperoxides such as (**69**) with Bu^tOCl causes cyclisation to the chlorodioxolanes (**70**) [90TL7077]. In an unusual reaction, photo-oxygenation of 2-alkoxyfurans results in the formation of *3H*-1,2-dioxoles (**71**) [91JCS(P1)1479].

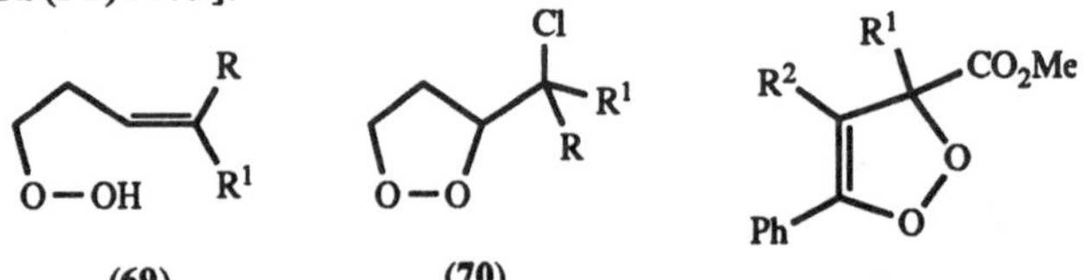

5.6.5 1,2-Dithioles and Dithiolanes

Oxidative dealkylation occurs upon treatment of compounds (**72**) with NBS to give benzo-1,2-dithioles (**73**) [91CC540]. Reaction of (**74**) with thiourea in the presence of base gives a mixture of (**75**) and (**76**), while either (**74**) or (**77**) can be converted to (**75**) with Na_2S [91BCJ699]. The condensation of two molecules of $ArCOCH_2SCN$ with Ar^1CHO under basic conditions to afford (**78**) has been described [90EGP278340]. Pyrolysis of dibutyl tetrasulphide, $Bu^nSSSSBu^n$, above 350°C gives dithiolethione (**79**) in 5% yield [90ZOR1225]. A series of dithiolethiones (**80**) have been prepared and converted to the corresponding sulphines [91JCS(P2)913]. The product of the Ogston colour test for chloral hydrate has been identified as (**81**) [91AP131].

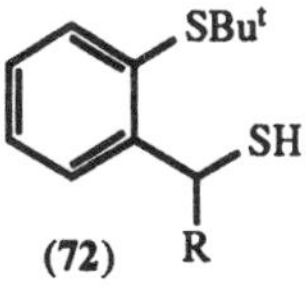

(72) 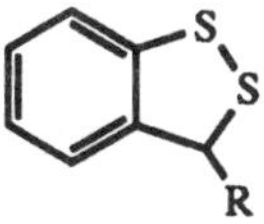 (73) 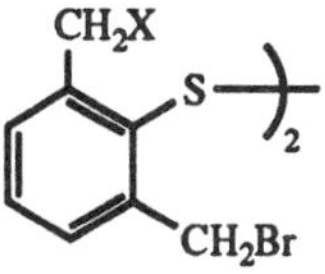 (74) X = Br (77) X = H (75) Y = H (76) Y = SH

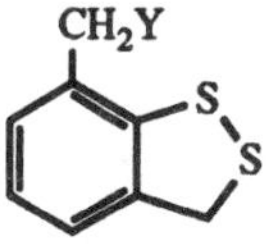

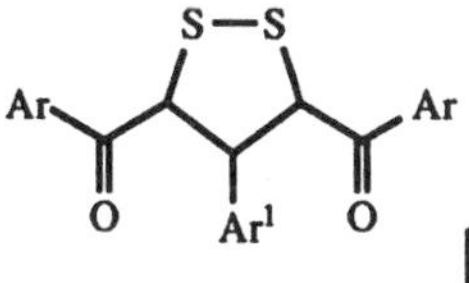

(78) 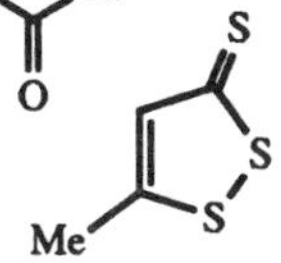 (79) 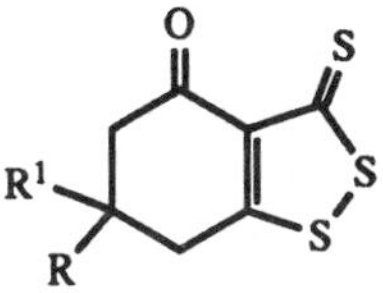 (80) 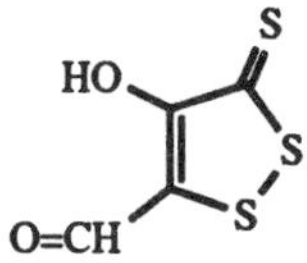 (81)

The new tetraceno bis(dithiole) (**82**) has been prepared as a potential electron donor [91CMA630]. The molybdenum and iron complexes of 1,2-dithiole S-oxides (**83**) are formed by cycloaddition of S_2O to metal alkynyl complexes [91OM2936]. The first bridged bicyclic thiosulphinate ester (**84**) has been prepared by oxidation of the corresponding dithiolane and its X-ray structure determined [91JOC904]. A convenient preparation of (**85**) and its conversion to (**86**) have been reported [91MON71] and the reaction of (**85**) with primary

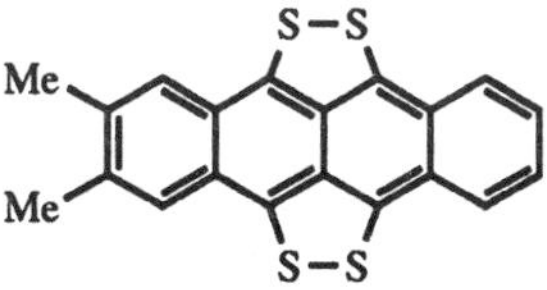

(82) 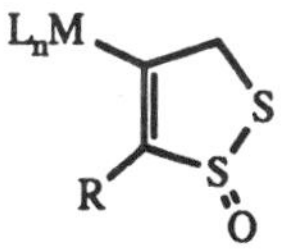 (83) 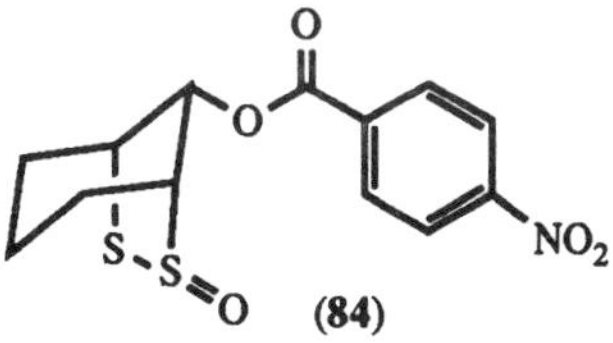 (84)

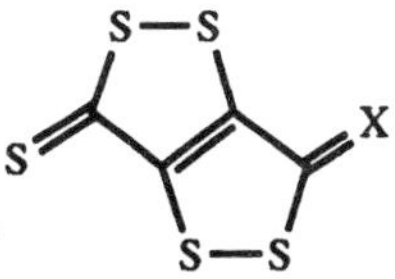

(85) X = S (86) X = O (87)

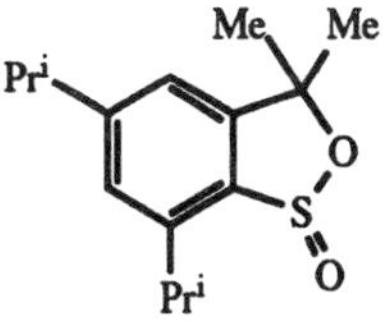

amines has been examined [91JPR19]. A review lecture dealing with the evidence from various techniques on the structure of trithiapentalenes has been reported [91PS(58)17] and the chemical and biological properties of naturally occurring 1,2-dithiolanes have also been reviewed [90SR257].

5.6.6 1,2-Oxathioles

A by-product formed in Friedel Crafts reactions using 2,4,6-tri-isopropylbenzenesulphonyl chloride has been identified as the benzo-1,2-oxathiole S-oxide (**87**) [90SC2999].

5.6.7 Three Heteroatoms

The ^{17}O NMR spectra of 1,2,4-trioxolanes such as (**88**) show a wide separation between the peaks for the ether and peroxide oxygens (δ_O +130 *vs* +295 respectively) [91CC816]. A new approach to asymmetric synthesis of sulphoxides relies upon sequential attack by Grignard reagents on cyclic sulphite (**89**) derived from *S*-lactic acid [91PS(58)89]. Reaction of diazomalonates with CS_2 produces 1,2,4-trithiolanes (**90**) in addition to the 1,3-oxathiol-2-ones (**62**) mentioned earlier [90BCJ3096].

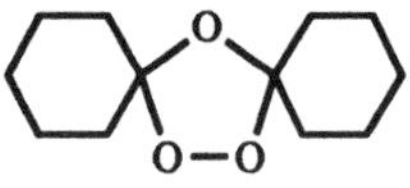

(**88**) 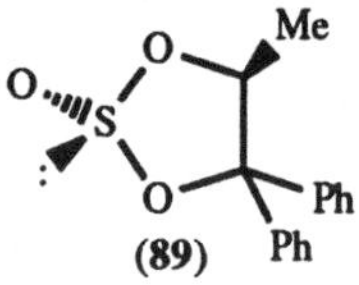 (**89**) 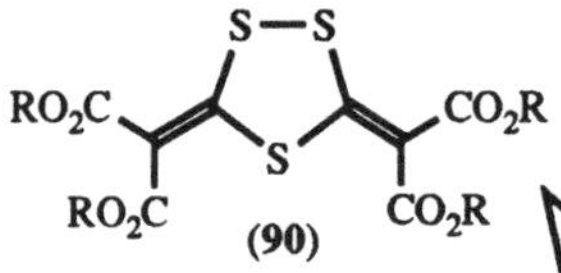 (**90**)

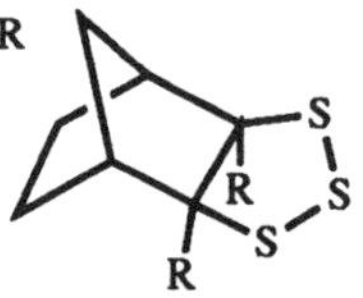

(**91**) R = H
(**92**) R = Ph

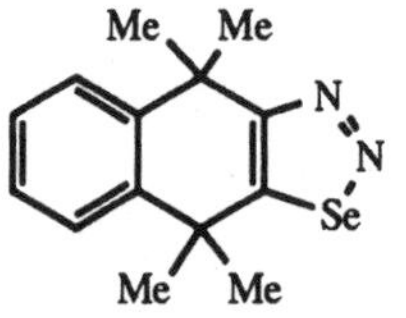

(**93**) 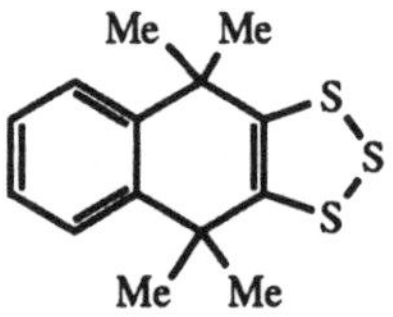 (**94**) 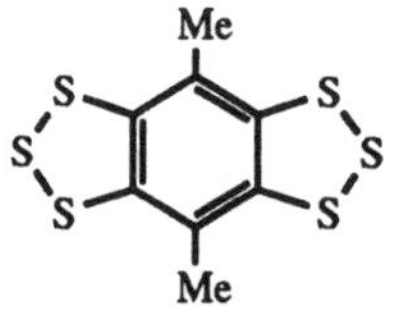 (**95**)

A number of new results in 1,2,3-trithiole and trithiolane chemistry have been described, including the formation of (**91**) by heating norbornene with sulphur in DMF, and its ready transfer of the S_3 unit to 1,2-diphenylnorbornene to give (**92**) [91PS(58)179]. Reaction of 1,2,3-selenadiazoles with sulphur can give the 1,2,3-trithioles as illustrated by conversion of (**93**) to (**94**) [91PS(58)179]. Finally, the bis-trithiole (**95**) is readily formed from tetrabromo-*p*-xylene and its conversion to the 1- and 2-oxides has been examined [91PS(59)137].

5.6.8 References

90AG(E)1420 D. Potin, K. Williams, and J. Rebek, *Angew. Chem., Int. Ed. Engl.*, 1990, **29**, 1420.

90BCJ3096 T. Ibata and H. Nakano, *Bull. Chem. Soc. Jpn.*, 1990, **63**, 3096.

90CL2019 S. B. Lee, T. Takata, and T. Endo, *Chem. Lett.*, 1990, 2019.

90CL2129 H. Nakano, K. Yamada, T. Nogami, Y. Shirota, A. Miramoto, and H. Kobayashi, *Chem. Lett.*, 1990, 2129.

90EGP278340 J. Teller, M. Kleist, B. Olk, and H. Dehne, *E. Ger. Pat.*, 278 340 (1990) [*Chem. Abstr.*, 1991, **114**, 122350].

90G731 G. Chelucci, M. Falorni, and G. Giacomelli, *Gazz. Chim. Ital.*, 1990, **120**, 731.

90IZV2662 V. G. Andreev, A. F. Kolomiets, and G. A. Sokol'skii, *Izv. Akad. Nauk SSSR, Ser. Khim.*, 1990, 2662.

90JCS(P1)3358 Y. Yamashita, M. Tomura, and S. Tanaka, *J. Chem. Soc., Perkin Trans. 1*, 1990, 3358.

90MI535 Y. Liu and Z. Yao, *Youji Huaxue*, 1990, **10**, 535 [*Chem. Abstr.*, 1991, **114**, 143190].

90S1149 A. M. Richter, N. Beye, and E. Fanghänel, *Synthesis*, 1990, 1149.

90SC2197 K. Saigo, S. Machida, K. Kudo, Y. Saito, Y. Hashimoto, and M. Hasegawa, *Synth. Commun.*, 1990, **20**, 2197.

90SC2999 D. E. Bugner, *Synth. Commun.*, 1990, **20**, 2999.

90SR257 L. Teuber, *Sulfur Reports*, 1990, **9**, 257.

90SYL145 J. Mattay, G. Kneer, and J. Mertes, *Synlett.*, 1990, 145.

90TAS143 S. Y. Po, H. H. Liu, and B. J. Uang, *Tetrahedron: Asymmetry*, 1990, **1**, 143.

90TL7035 A. Nishida, M. Nishida, and O. Yonemitsu, *Tetrahedron Lett.*, 1990, **31**, 7035.

90TL7077 A. J. Bloodworth and N. A. Tallant, *Tetrahedron Lett.*, 1990, **31**, 7077.

90TL7307 A. Khanous, A. Gorgues, and F. Texier, *Tetrahedron Lett.*, 1990, **31**, 7307.

90TL7311 A. Khanous, A. Gorgues, and M. Jubault, *Tetrahedron Lett.*, 1990, **31**, 7311.

90ZOR1225 E. N. Sukhomazova, L. P. Turchaninova, N. A. Korchevin, E. N. Deryagina, and M. G. Voronkov, *Zh. Org. Khim.*, 1990, **26**, 1225.

91AG(E)871 M. R. Bryce, M. A. Coffin, M. B. Hursthouse, and M. Mazid, *Angew. Chem., Int. Ed. Engl.*, 1991, **30**, 871.

91AP131 H. D. Stachel and T. Zoukas, *Arch. Pharm. (Weinheim, Ger.)*, 1991, **324**, 131.

91BCJ699 H. Fujihara, J. J. Chiu, and N. Furukawa, *Bull. Chem. Soc. Jpn.*, 1991, **64**, 699.

91BCJ771 H. Nakano, H. Tamura, and T. Ibata, *Bull. Chem. Soc. Jpn.*, 1991, **64**, 771.

91CB2025 C. Maletzko, W. Sundermeyer, H. Pritzkow, H. Irngartinger, and U. Huber-Patz, *Chem. Ber.*, 1991, **124**, 2025.

91CC37 M. Ohkita, T. Tsuji, and S. Nishida, *J. Chem. Soc., Chem. Commun*, 1991, 37.

91CC92 J. Y. Becker, J. Bernstein, S. Bittner, L. Shahal, and S. S. Shaik, *J. Chem. Soc., Chem. Commun.*, 1991, 92.

91CC320 A. J. Moore, M. R. Bryce, D. J. Ando, and M. B. Hursthouse, *J. Chem. Soc., Chem. Commun.*, 1991, 320.

91CC540 M. F. Chan and M. E. Garst, *J. Chem. Soc., Chem. Commun.*, 1991, 540.

91CC812 R. A. Aitken, G. Ferguson, and S. V. Raut, *J. Chem. Soc., Chem. Commun.*, 1991, 812.

91CC816 J. Lauterwein, K. Griesbaum, P. Krieger-Beck, V. Ball, and K. Schlindwein, *J. Chem. Soc., Chem. Commun.*, 1991, 816.

91CC843 F. Bertho-Thoraval, A. Robert, A. Souizi, K. Boubekeur, and P. Batail, *J. Chem. Soc., Chem. Commun.*, 1991, 843.

91CC906	M. Guillemet, M. Baudy-Floc'h, and A. Robert, *J. Chem. Soc., Chem. Commun.*, 1991, 906.
91CC952	S. K. Kumar, H. B. Singh, K. Das, U. C. Sinha, and A. Mishnev, *J. Chem. Soc., Chem. Commun.*, 1991, 952.
91CC1132	Y. Yamashita, S. Tanoka, K. Imaeda, H. Inokuchi, and M. Sano, *J. Chem. Soc., Chem. Commun.*, 1991, 1132.
91CC1370	M. Fourmigné and P. Batail, *J. Chem. Soc., Chem. Commun.*, 1991, 1370.
91CC1421	J. Uenichi, M. Motoyama, Y. Nishiyama, and S. Wakabayashi, *J. Chem. Soc., Chem. Commun*, 1991, 1421.
91CC1584	D. Philp, A. M. Z. Slawin, N. Spencer,J. F. Stoddart, and D. J. Williams, *J. Chem. Soc., Chem. Commun.*, 1991, 1584.
91CC1638	A. J. Moore and M. R. Bryce, *J. Chem. Soc., Chem. Commun.*, 1991, 1638.
91CC1707	T. W. Wallace, I. Wardell, K.-D. Li, and S. R. Challand, *J. Chem. Soc., Chem. Commun*, 1991, 1707.
91CC1750	T. Ravindranathan, S. P. Chavan, R. B. Tejwani, and J. P. Varghese, *J. Chem. Soc., Chem. Commun.*, 1991, 1750.
91CL355	R. Sato and S. Yamaichi, *Chem. Lett.*, 1991, 355.
91CMA630	T. Maruo, M. T. Jones, M. Singh, and N. P. Rath, *Chem. Mater.*, 1991, **3**, 630.
91DOK(319)619	V. G. Andreev, A. F. Kolomiets, and A. V. Fokin, *Dokl. Akad. Nauk SSSR*, 1991, **319**, 619.
91EP423009	M. Casado, M. Crochemore, and B. Langlois, *Eur. Pat.*, 423 009 (1991) [*Chem. Abstr.*, 1991, **115**, 49673].
91EP426253	H. G. M. Walraven, P. B. M. Groen, and E. J. A. M. Leenderts, *Eur. Pat.*, 426 253 (1991) [*Chem. Abstr.*, 1991, **115**, 114490].
91JA3130	C. Wentrup, P. Kambouris, R. A. Evans, D. Owen, G. MacFarlane, J., Chuche, J. C. Pommelet, A. B. Cheikh, M. Plisnier, and R. Flammang, *J. Am. Chem. Soc.*, 1991, **113**, 3130.
91JCS(P1)157	A. J. Moore and M. J. Bryce, *J. Chem. Soc., Perkin Trans. 1*, 1991, 157.
91JCS(P1)1255	M. F. Mahon, K. Molloy, C. A. Pittol, R. J. Pryce, S. M. Roberts, G. Ryback, V. Sik, J. O. Williams, and J. A. Winders, *J. Chem. Soc., Perkin Trans. 1*, 1991, 1255.
91JCS(P1)1479	M. L. Graziano, M. R. Iesce, F. Cermola, G. Cimminiello, and R.Scarpati, *J. Chem. Soc., Perkin Trans. 1*, 1991, 1479.
91JCS(P2)913	A. Z. Q. Khan, J. Sandström, and S. L. Wang, *J. Chem. Soc., Perkin Trans. 2*, 1991, 913.
91JOC891	D. A. Laufer and E. Al-Farhan, *J. Org. Chem.*, 1991, **56**, 891.
91JOC904	P. L. Folkins, D. N. Harpp,and B. R. Vincent, *J. Org. Chem.*, 1991, **56**, 904.
91JOC4035	Z.-J. Ni, N.-W. Mei, X. Shi, Y.-L. Tzeng, M. C. Wang, and T.-Y. Luh, *J. Org. Chem.*, 1991, **56**, 4035.
91JOC5684	M. Joergensen, K. A. Lerstrup, and K. Bechgaard, *J. Org. Chem.*, 1991, **56**, 5684.
91JPR19	E. Fanghänel, U. Laube, B. Kordts, and A. M. Richter, *J. Prakt. Chem.*, 1991, **333**, 19.
91KGS33	A. A. Gevorkyan, P. I. Kazaryan, O. V. Avakyan, and R. A. Vardanyan, *Khim. Geterosikl. Soedin.*, 1991, 33
91KGS848	O. Neilands and B. Adamsone, *Khim. Geterosikl. Soedin.*, 1991, 848.
91LA395	H. Müller, A. Lerf, and H. P. Fritz, *Liebigs Ann. Chem.*, 1991, 395.
91MI1	H. Hagemann and D. Klamann (Eds.), *Houben-Weyl Methoden der Organischen Chemie*, 4th Ed., Vol. E14a Part 1, Thieme, Stuttgart, 1991.
91MI107	E. S. Kozlov, A. A. Yurchenko, and A. A. Tolmachev, *Ukr. Khim. Zh.*, 1991, **57**, 107 [*Chem. Abstr.*, 1991, **115**, 29170].
91MI211	L. W. Zatorski and P. T. Wierzchowski, *Catal. Lett.*, 1991, **10**, 211.
91MI238	A. K. Beck, B. Bastani, D. A. Plattner, W. Petter, D. Seebach, H. Braunschweiger, P. Gysi, and L. LaVecchia, *Chimia*, 1991, **45**, 238.

91MON71	B. Kordts, A. M. Richter, and E. Fanghänel, *Monatsh. Chem.*, 1991, **122**, 71.
91MVC99	H. Adams, N. A. Bailey, J. J. Laskowski, C. Ridgway, and M. J. Winter, *Mendeleev Commun.*, 1991, **1**, 99.
91OM2936	M. E. Raseta, R. K. Mishra, S. A. Cawood, M. E. Welker, and A. L. Rheingold, *Organometallics*, 1991, **10**, 2936.
91OPP427	W. Shaozu, R. Tianhui, C. Ning, and Z. Yulan, *Org. Prep. Proced. Int.*, 1991, **23**, 427.
91PS(58)17	C. Th. Pedersen, *Phosphorus, Sulfur and Silicon*, 1991, **58**, 17.
91PS(58)89	H. B. Kagan, F. Rebiere, and O. Samuel, *Phosphorus, Sulfur and Silicon*, 1991, **58**, 89.
91PS(58)179	W. Ando, N. Tokitoh, and Y. Kabe, *Phosphorus, Sulfur and Silicon*, 1991, **58**, 179.
91PS(58)207	K. Schank, *Phosphorus, Sulfur and Silicon*, 1991, **58**, 207.
91PS(59)137	R. Sato and S. Takahashi, *Phosphorus, Sulfur and Silicon*, 1991, **59**, 137.
91S26	A. J. Moore and M. J. Bryce, *Synthesis*, 1991, 26.
91S263	M. J. Bryce and G. Cooke, *Synthesis*, 1991, 263.
91S301	D. Villemin and A. Ben Alloum, *Synthesis*, 1991, 301.
91S575	C. A. M. Afonso, M. T. Barros, L. S. Godhino, and C. D. Maycock, *Synthesis*, 1991, 575.
91SL63	B. A. Trofimov, A. G. Mal'kina, Yu. M. Skvortsov, V. K. Bel'ski, and E. I. Moshchevitina, *Sulfur Lett.*, 1991, **13**, 63.
91SL143	J. Nielsen and A. Senning, *Sulfur Lett.*, 1991, **12**, 143.
91SM(42)2323	A. Belyasmine, A. Gorgues, M. Jubault, and G. Duguay, *Synth. Met.*, 1991, **42**, 2323.
91SM(42)2327	A. Khanous, A. Gorgues, and M. Jubault, , *Synth. Met.*, 1991, **42**, 2327.
91SM(42)2381	H. Müller, A. Lerf, H. P. Fritz, and K. Andres, , *Synth. Met.*, 1991, **42**, 2381.
91SM(42)2561	M. Joergensen, K. Lerstrup, and K. Bechgaard, *Synth. Met.*, 1991, **42**, 2561.
91T975	C. N. Eid and J. P. Konopelski, *Tetrahedron*, 1991, **47**, 975.
91T4519	M. A. Böhler and J. P. Konopelski, *Tetrahedron*, 1991, **47**, 4519.
91TL465	J. P. Konopelski, *Tetrahedron Lett.*, 1991, **32**, 465.
91TL479	N. Beye, R. Wegner, A. M. Richter, and E. Fanghänel, *Tetrahedron Lett.*, 1991, **32**, 479.
91TL1441	N. Chapel, A. Greiner, and J.-Y. Orroland, *Tetrahedron Lett.*, 1991, **32**, 1441.
91TL2259	H. K. Patney, *Tetrahedron Lett.*, 1991, **32**, 2259.
91TL2461	R. E. Zelle and W. J. McClellan, *Tetrahedron Lett.*, 1991, **32**, 2461.
91TL2737	J. S. Zambounis and C. W. Mayer, *Tetrahedron Lett.*, 1991, **32**, 2737.
91TL2741	J. S. Zambounis and C. W. Mayer, *Tetrahedron Lett.*, 1991, **32**, 2741.
91TL2853	P. Arya, M. Lesage, and D. D. M. Wayner, *Tetrahedron Lett.*, 1991, **32**, 2853.
91TL2897	M. Salle, A. Belyasmine, A. Gorgues, M. Jubault, and N. Soyer, *Tetrahedron Lett.*, 1991, **32**, 2897.
91TL3683	W. Smadja, M. Zahouily, M. Journet, and M. Malacria, *Tetrahedron Lett.*, 1991, **32**, 3683.
91TL4897	Y. Gimbert and A. Moradpour, *Tetrahedron Lett.*, 1991, **32**, 4897.

CHAPTER 5.7

Five-Membered Ring Systems: With O and N Atoms

G. V. BOYD
The Hebrew University, Jerusalem, Israel

ISOXAZOLES

Thianthrene radical perchlorate in the presence of 2,4-di-t-butyl-6-methylpyridine promotes the cyclization of oximes ArCH=CHCMe=NOH to the isoxazoles 1 (91JOC1332). Treatment of 4,5-bis(bromomethyl)-3-phenylisoxazole with sodium iodide generates the dimethyleneisoxazoline 2, which was trapped by diethyl azodicarboxylate as the Diels-Alder adduct 3 (91TL4603). The metal - catalyzed extrusion of nitrogen from the diazo compound 4 gave 5, the first furo[3,4-d]isoxazole (91CB2481). The analogue 6 reacts with dimethyl acetylenedicarboxylate to form the oxepino[2,3-d]isoxazole 7 (91TL1161).

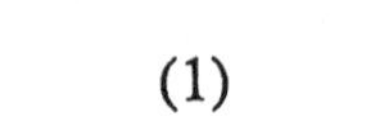

(1) (2) (3)

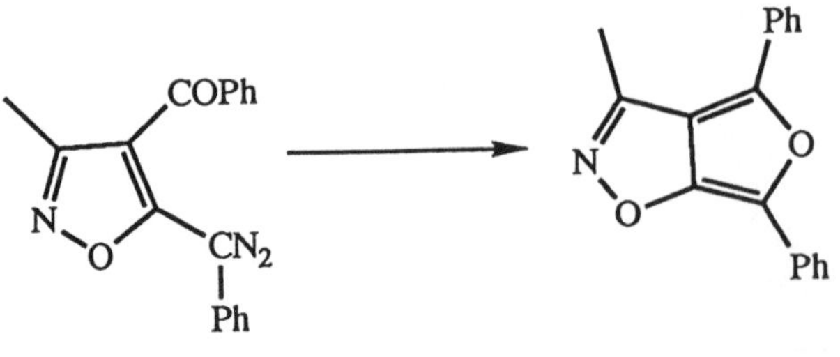

(4) (5)

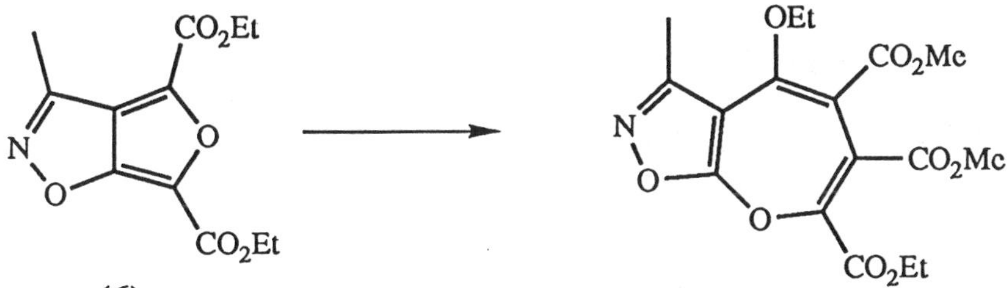

(6) (7)

Isoxazolines

As in previous years, there have been several reports on 1,3-dipolar cycloaddition reactions of nitrile oxides. 3-Phenyl-2-isoxazoline (8) is obtained by the reaction of the hydroximoyl chloride PhCCl=NOH with tributyl(vinyl)tin (91CC884). Benzonitrile oxide is generated by thermolysis of the oxime ester PhCCl=NOCOOEt; in the presence of tetradec-1-ene the isoxazoline 9 is produced (91BCJ318). N-Methylanilinoallene reacts with 3,5-dichloromesitonitrile oxide to form the isoxazoline 10, which can add a second molecule of the nitrile oxide at the exocyclic double bond (91JCS(P1)1843).

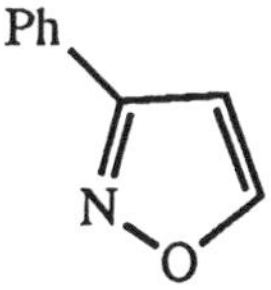

(8) 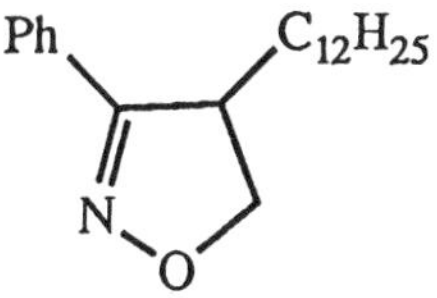 (9) 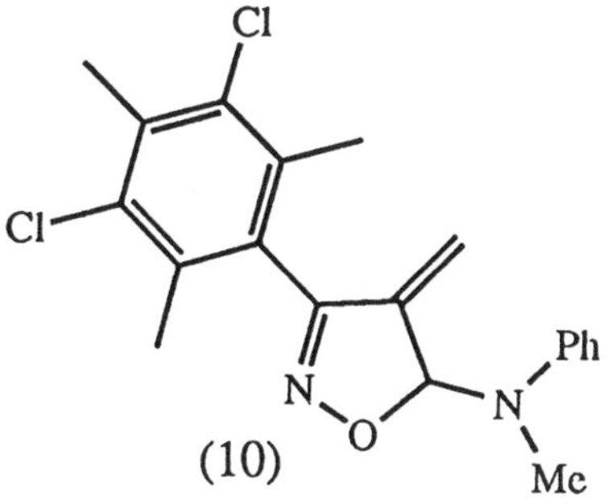 (10)

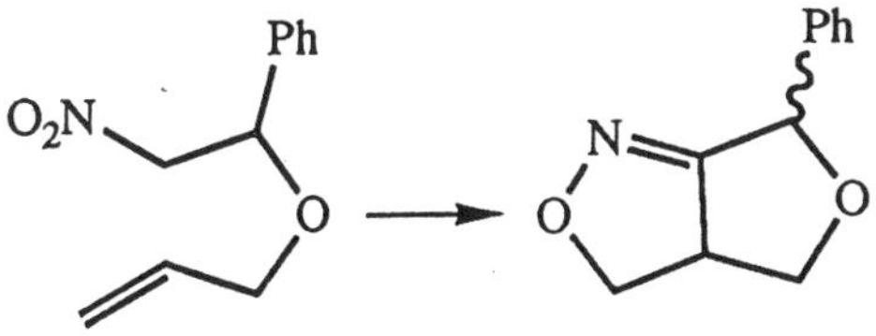

(11) (12) 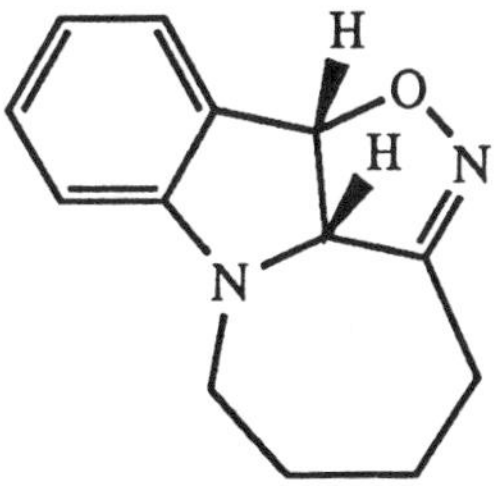 (13)

The following illustrate intramolecular cycloadditions of nitrile oxides; in each case the nitrile oxide was generated by the action of phenyl isocyanate/triethylamine on the appropriate nitro compound. The furoisoxazoline 12 was produced as a 9:1 mixture of cis- and trans-isomers from the nitro olefin 11 (91TL4259, 91CB1181), and 1-(5-nitropentyl)indole gave the fused isoxazoline 13 (91JOC896).

Benzyl bromide condenses with ethyl nitroacetate in the presence of aluminum oxide to afford mainly the trans-hydroxyisoxazoline N-oxide 14 (91JOC6258). The nitrone 15 was unexpectedly transformed into the benzisoxazolinone 16 on heating (91CL1333). 3-Amino-2-isoxazoline 17 forms stable diazonium salts (90KGS1110).

Lewis-acid catalyzed additions to the aldehyde 18 proceed with high stereoselectivity. The syn-aldol product 19 of acetophenone is produced in the presence of titanium tetrachloride, while boron trifluoride etherate causes the formation of the anti-isomer (90TL5053). Similarly, the ene reaction of 18 with 2-phenylpropene, promoted by a titanium catalyst, gives a mixture of diastereomers containing 90% of the syn-adduct 20 (90CL 1991).

The thermal retro-Diels-Alder reaction of 21 results in a mixture of 3-phenylisoxazole and the norbornene adduct 22 (90MI1023).

Isoxazolidines

C,N-Diphenylnitrone reacts with the lithium enolate of acetaldehyde to give the hydroxyisoxazolidine 23 as a mixture of cis- and trans-isomers (91T421). Fused isoxazolidines, e.g. 24, are obtained from C,N-diarylnitrones and 2,5- or 2,6-dialkyl-p-benzoquinones (91BCJ2388). The deuteriated nitrone PhCD=NMeO adds to styrene to yield the isoxazolidine 25 as a mixture of diastereomers (90MI1277). The diastereospecific cycloaddition of C,N-diphenylnitrone to di-l-menthyl benzylidenemalonate produces solely compound 26 (R = l-menthyl) (90TL4633).

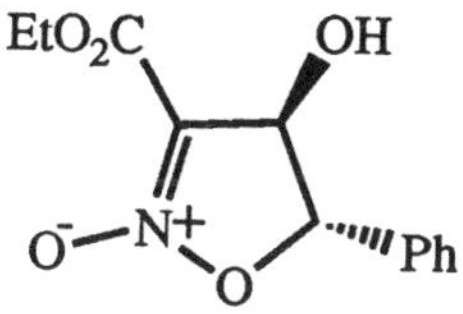

(14) 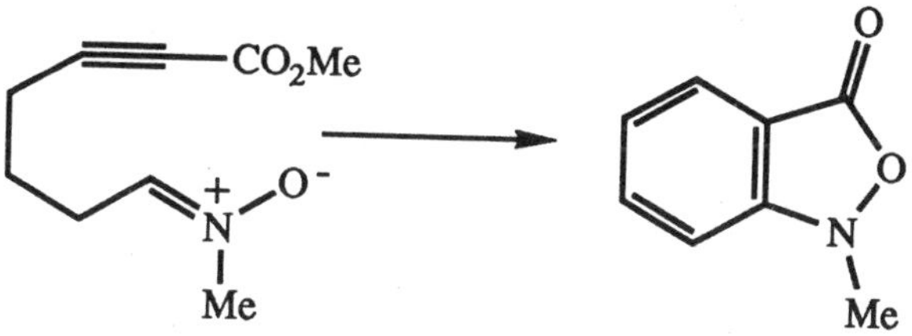 (15) (16)

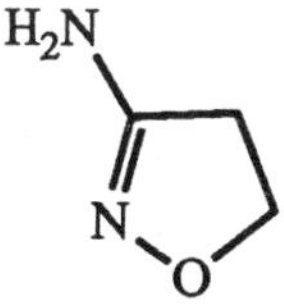

(17) 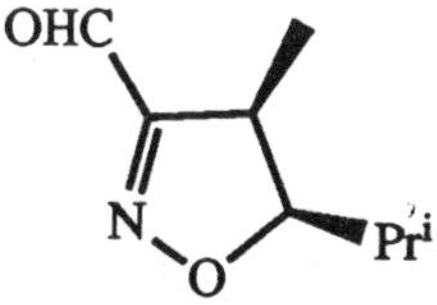 (18)

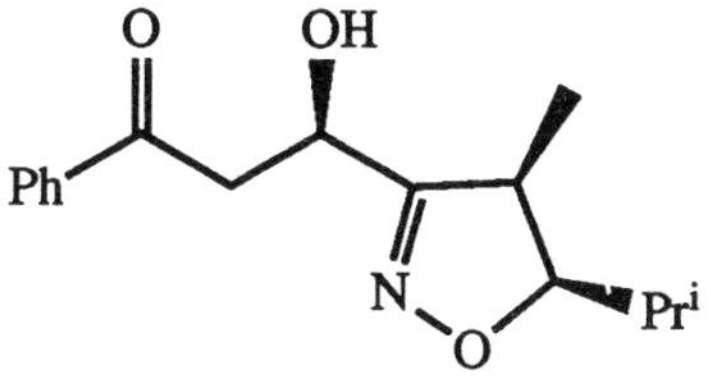

(19) 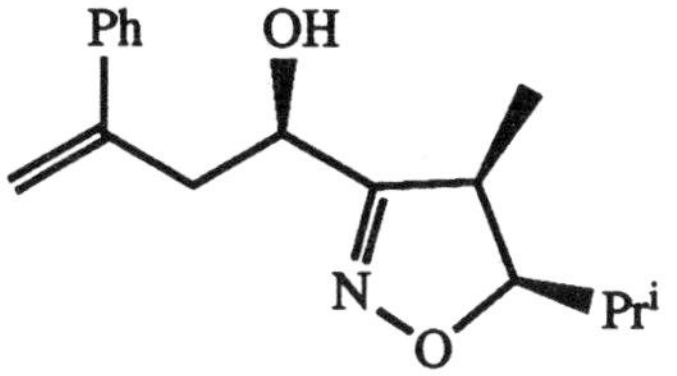 (20)

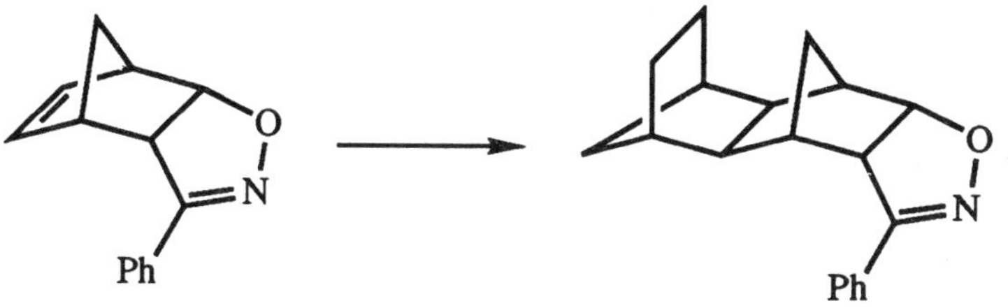

(21) (22)

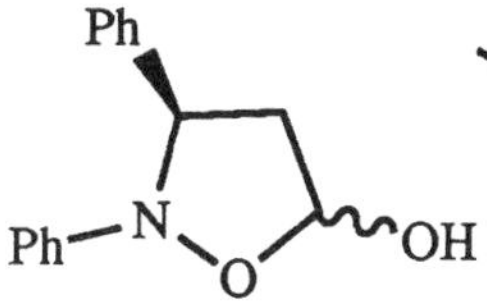

(23) 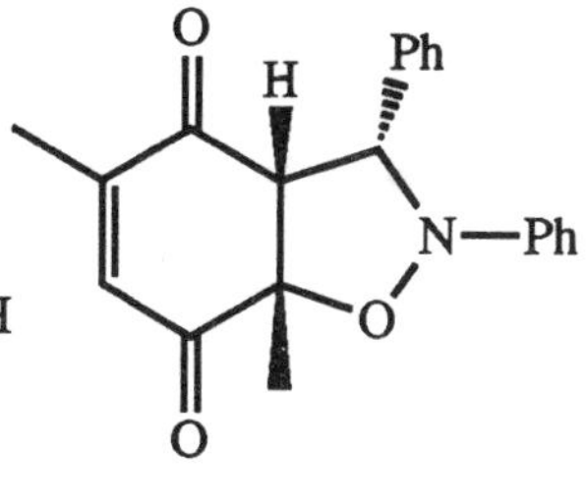 (24) 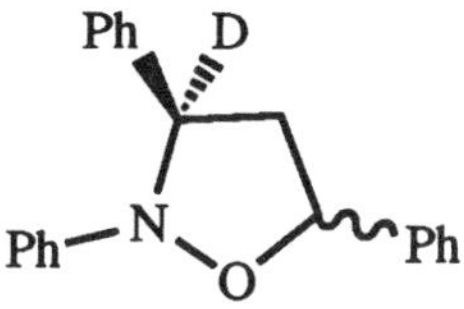 (25)

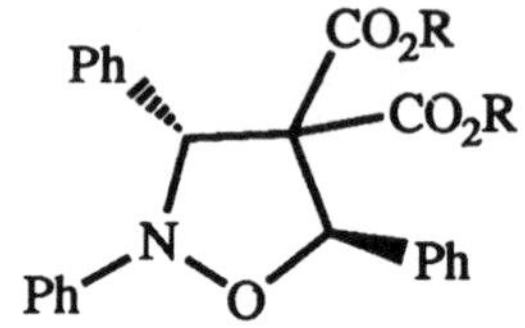

(26)

Numerous intramolecular nitrone cycloadditions have been described. Compounds 28 are formed diastereoselectively from the sulphides 27 (90MI251) and treatment of oximes HON=CRR' (R = Me or Ph; R' = H or Me) with 2,3-bis(phenylsulphonyl)buta-1,3-diene yields the bridged isoxazolidines 29 by way of a transient nitrone, followed by 1,3-dipolar cycloaddition (91JOC2154). Heating the oxime 30 results in the fused isoxazolidine 31 via the nitrone tautomer of the oxime (91T4007); a similar cyclization of the pyrrolidine derivative 32 yields 33 with the stereoselective introduction of three chiral centres (91JOC2775). Michael addition of cyclopentanone oxime to phenylsulfonylethene, followed by intramolecular 1,3-dipolar cycloaddition, affords the spiro compound 34 (91T4477); a similar reaction of phenylsulfonylethene with the oxime 35 gives the fused isoxazolidine 36 (91T4495). A reversible palladium-catalyzed [2,3]-sigmatropic shift converts benzaldoxime O-allyl ether into the nitrone 37, which was trapped by N-methylmaleimide as a mixture of endo- and exo-adducts 38 (91TL279).

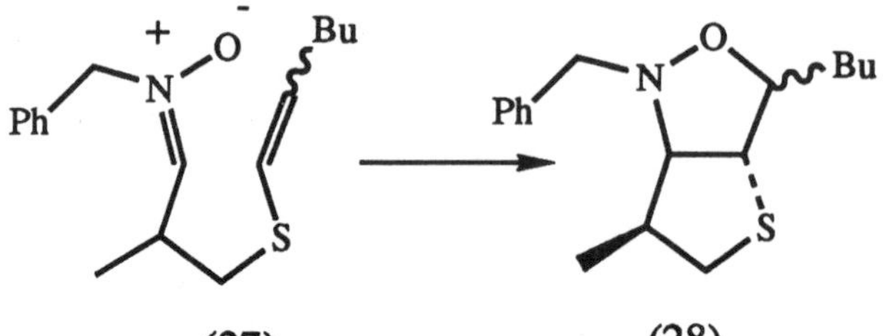

(27) (28) 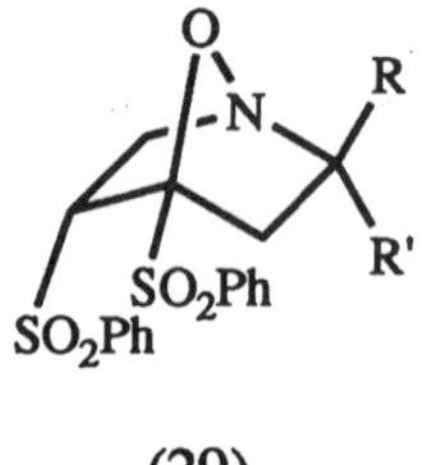 (29)

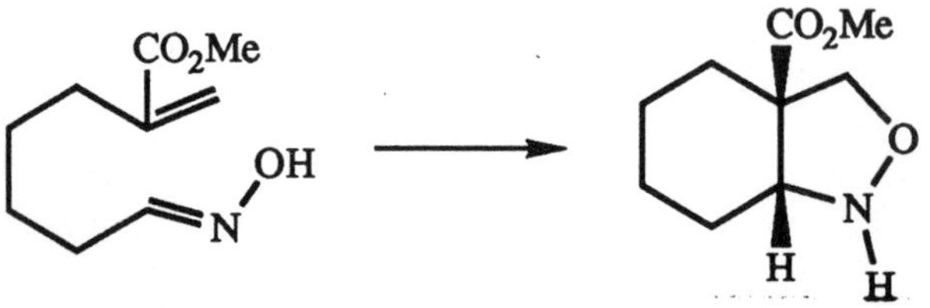

(30) (31)

(32) (33)

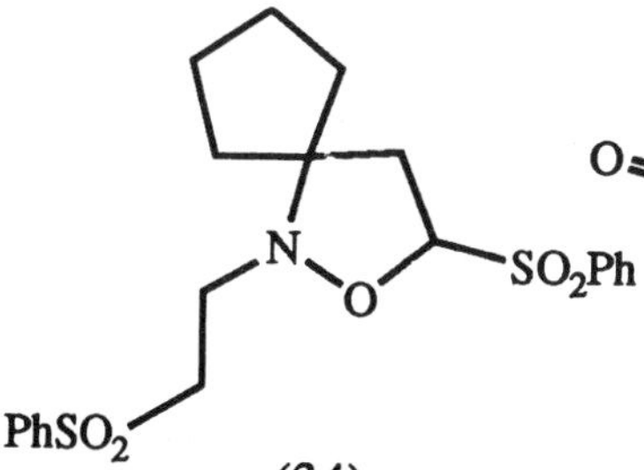

(34) 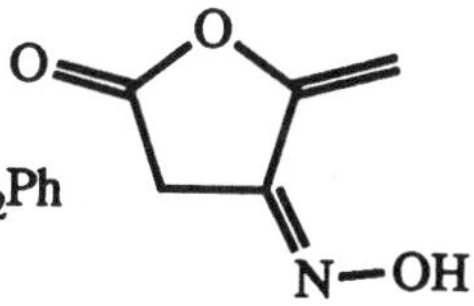 (35) 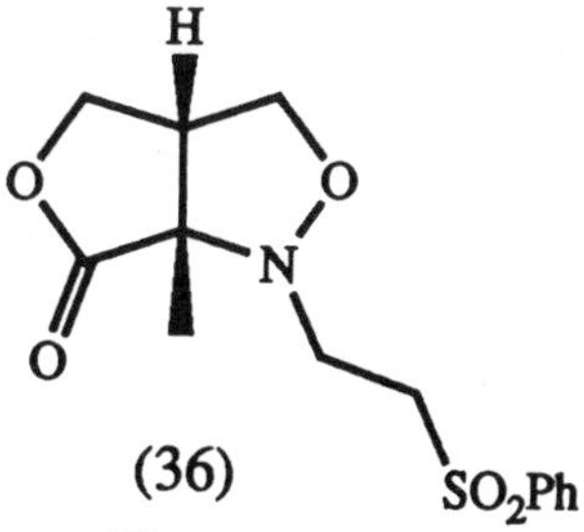 (36)

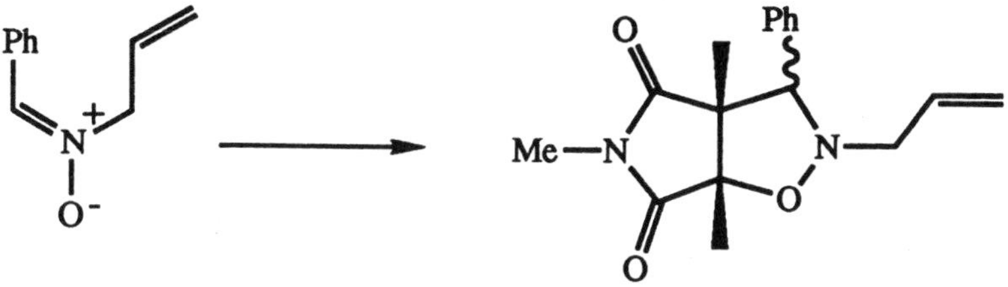

(37) (38)

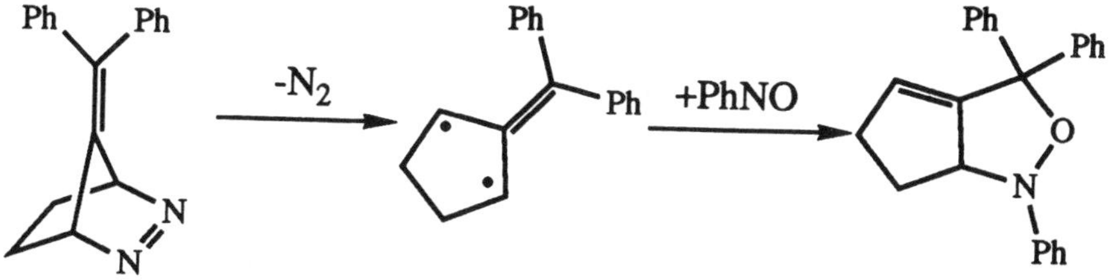

(39) (40) (41)

The trimethylenemethane diradical 40, generated by thermolysis or photolysis of the azo compound 39, has been trapped by nitrosobenzene as the isoxazolidine 41 (91TL4283).

OXAZOLES

Zinc, together with ultrasound, converts N-benzoyl-hexafluoroisopropylideneimine into the oxazole 42 (91CZ253). The oxazole 44 is formed when the 3-pyridone 43 is irradiated (91CC1200).

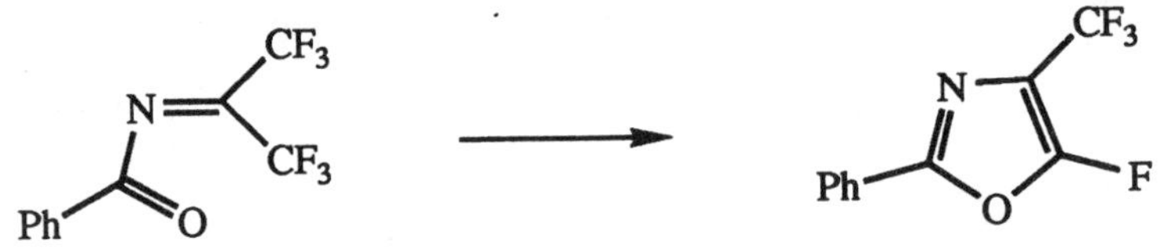

(42)

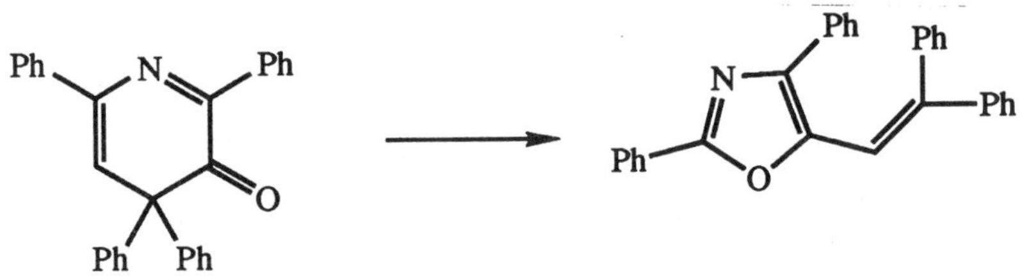

(43) (44)

Lithiation of oxazole gives the 2-lithio derivative, which is in equilibrium with the isocyanide 45. Treatment of the lithium compound with aldehydes yields 4-substituted oxazoles by reaction with the enolate 45, followed by cyclization (91JOC449). 2,4- And 2,5-diphenyloxazole are attacked by s-butyllithium at the unsubstituted carbon atom of the oxazole ring (91JOC3058). The oxazole 46 undergoes an intramolecular Diels-Alder addition to the carbon-oxygen double bond of the aldehyde group (91JOC3419. 2,4,5-Trimethyloxazole is transformed into the oxadiazoline 47 by the action of nitrosobenzene (91CL1221).

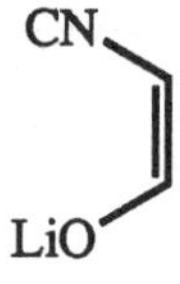

(45) 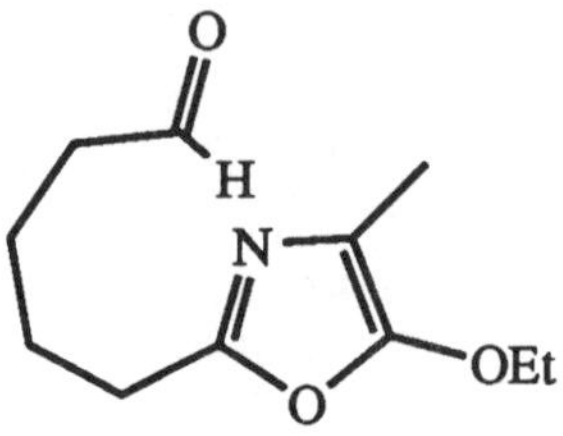 (46) 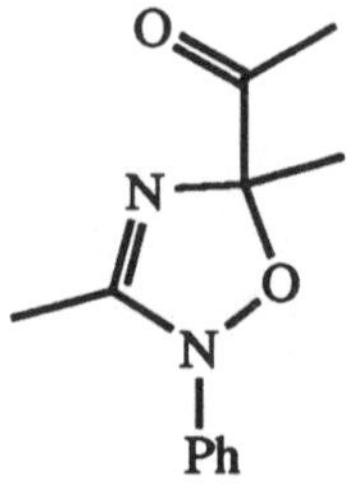 (47)

Oxazolines and Oxazolinones

Chiral 2-unsubstituted oxazolinones 48 are readily obtained from optically active β-amino alcohols and dimethylformamide dimethylacetal (91JOC1961). Aldehydes RCHO react with methyl isocyanoacetate in the presence of a chiral ferrocenylphosphine-silver complex to yield the oxazolines 49 in better than 80% enantiomeric excess (91TL2799); a similar reaction of benzaldehyde with toluene-p-sulfonylmethyl isocyanide gives the trans-oxazoline 50 in 83% ee (90JOC5935).

The dichloromethyloxazoline 51 yields chlorocarbene adducts 52 on treatment with aldehydes RCHO under phase-transfer catalysis (91TL87). Treatment of the chiral naphthyloxazoline 53 with butyllithium, followed by methyl iodide, gave the dihydronaphthalene derivative 54 in greater than 95% diastereomeric excess (91JOC 2292); a similar reaction has been reported for the naphthyl ether 55 (91TL2095).

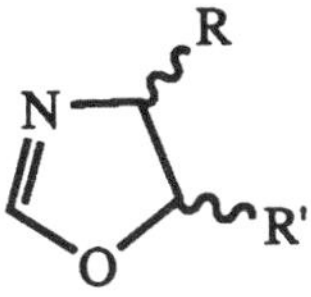

(48) 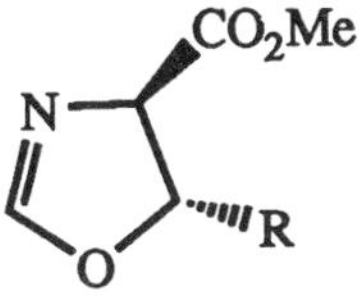 (49) 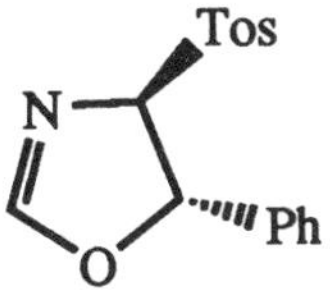 (50)

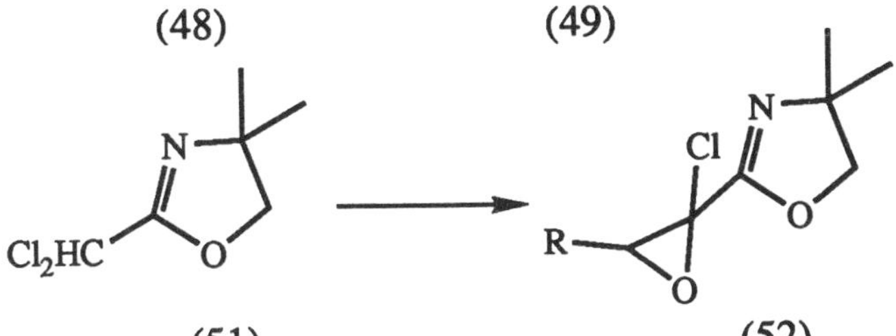

(51) (52)

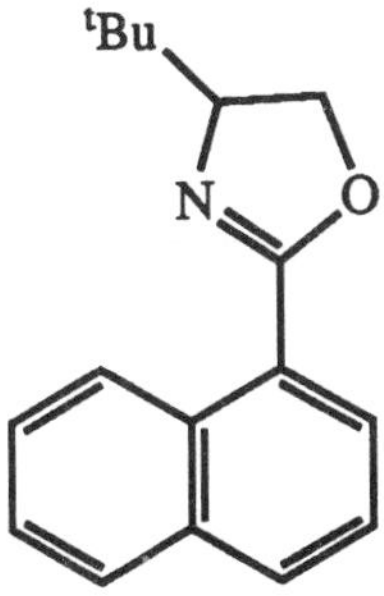

(53) 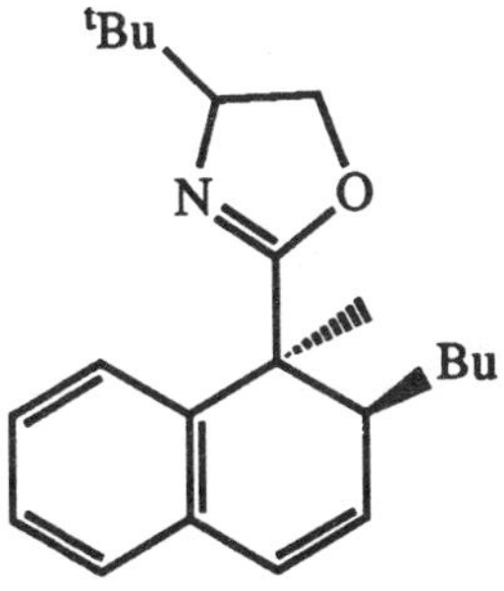 (54) 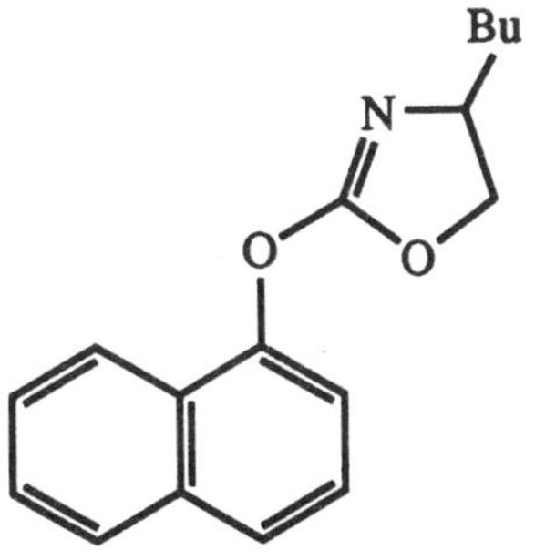 (55)

4,4-Dimethyl-2-phenyl-2-oxazoline 56 is alkylated via palladium complexes to give 2'-mono- or 2',6'-dialkyl derivatives, depending on conditions (91JOM29). Reaction of the tricarbonylchromium complex of 56 with an alkyllithium compound RLi, followed by an alkyl or aryl halide R'X, results in a cyclohexadiene 57 (91JA9676). Aziridinooxazolidines are formed from 4-toluene-p-sulfonyloxymethyl-2-oxazolines and activated Grignard reagents, e.g. 58 from 4-methyl-2-phenyl-4-toluene-p-sulfonyloxymethyl-2-oxazoline and allylmagnesium chloride (90JOC6216).

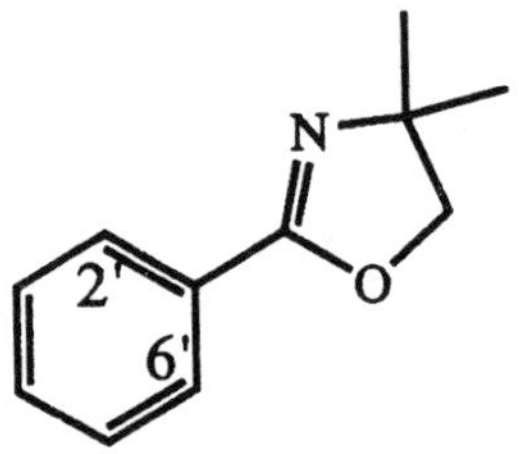

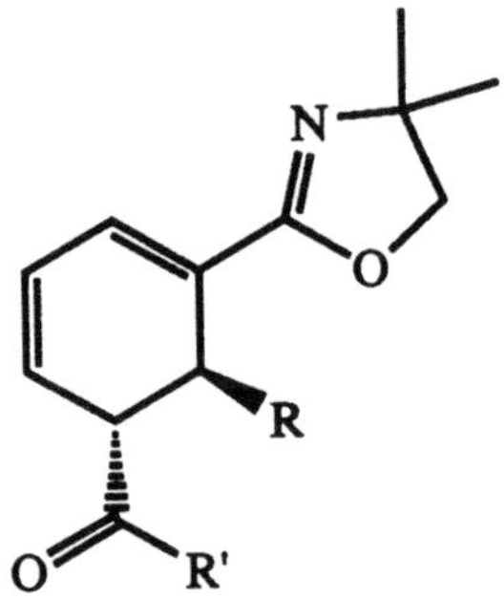

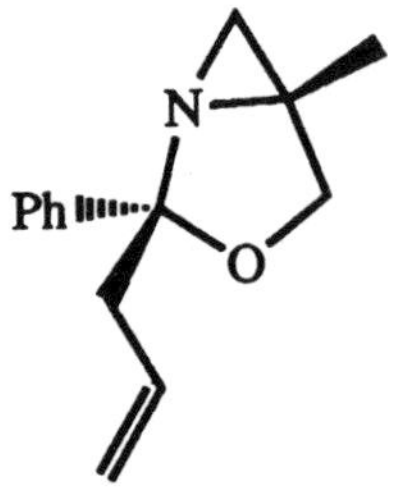

(56) (57) (58)

Oxazolidines and Oxazolidinones

A variety of oxazolidin-2-ones 59 has been obtained by the action of carbon dioxide on ethanolamines in the presence of the iron sulphide cyclopentadienyl cluster (91MI217). Such oxazolidin-2-ones are also formed by flash-vacuum pyrolysis of 1-ethoxycarbonylaziridines 60 (91JCS(P1)961). The allene 61 undergoes diastereospecific iodocyclization to the oxazolidinone 62 (91JOC4888). The stereoselective formation of the bicyclic oxazolidinone 64 by the combined action of tetrabutylammonium fluoride and tetrakis(triphenylphosphine)palladium on the carbamate 63 has been reported (90TL5339). Radical-mediated cyclization of the iodo amide 65 gives the fused oxazolidine 66 stereoselectively (91MI793).

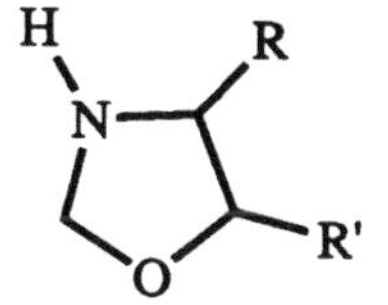

(59) 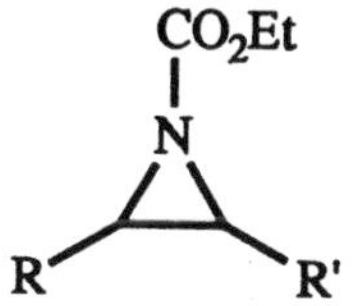 (60)

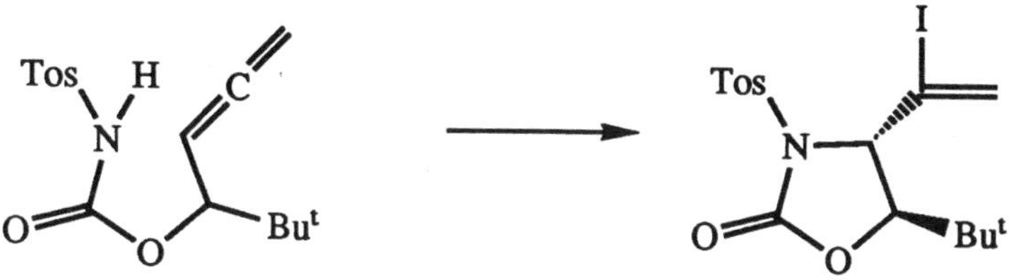

(61) (62)

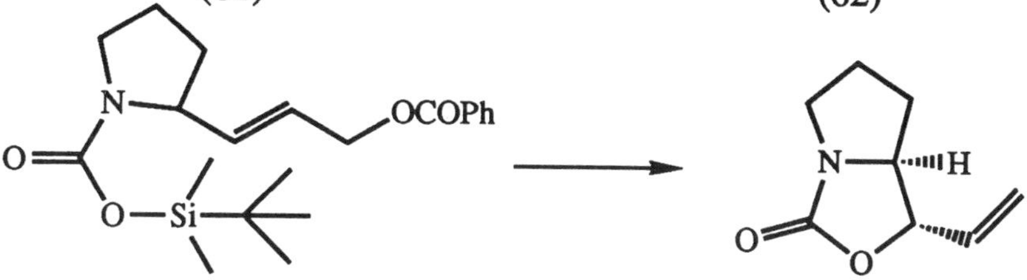

(63) (64)

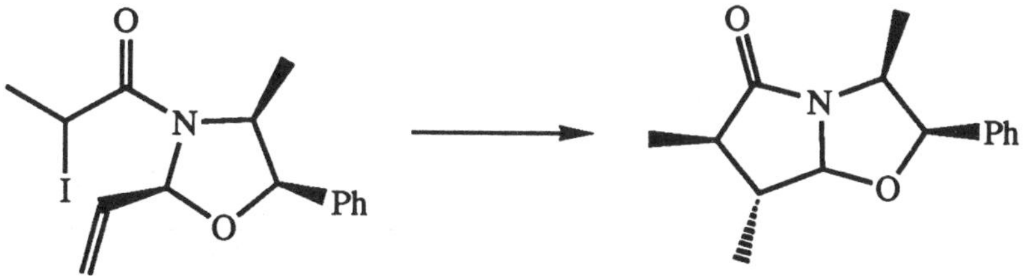

(65) (66)

(E)-5-Allyl-4-methoxyoxazolidin-2-one (67; R = OMe) reacts with alkyl or aryl cuprates in the presence of boron trifluoride to yield substitution products 67 (R = alkyl or aryl) with retained configuration (91TL3523).

There have been numerous articles on the steric course of addition reactions of various oxazolidines and oxazolidinones. The chiral oxazolidine ketone 68, derived from (S)-prolinol, reacts with organotitanium triisopropoxides to yield (S)-tertiary alcohols, whereas organolithium compounds afford (R)-alcohols (91TL2919).

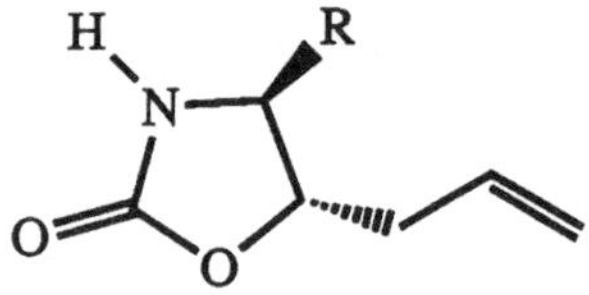

(67) 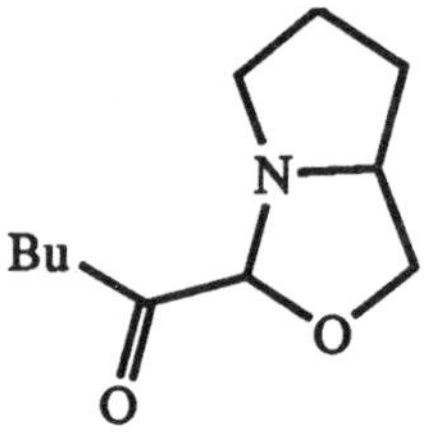 (68)

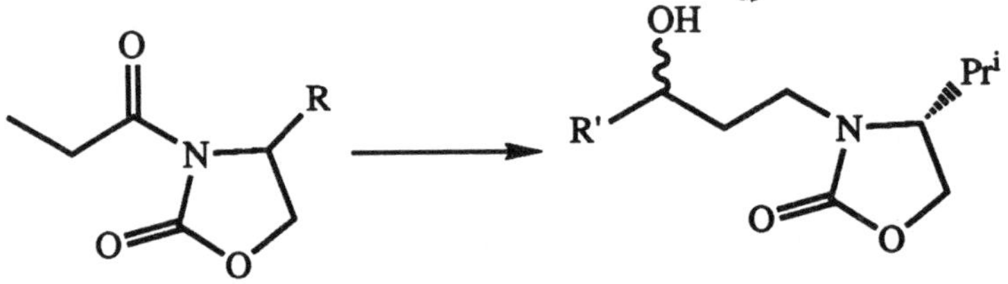

(69) (70)

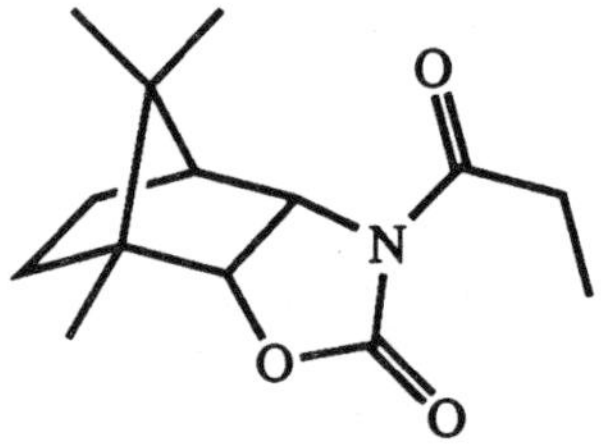

(71) 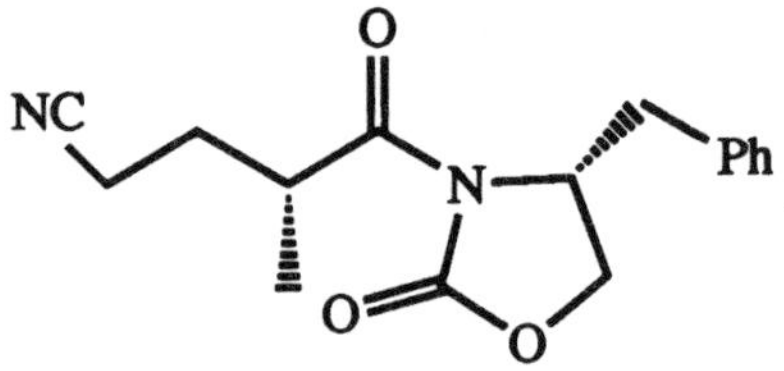 (72)

The stereochemistry of the aldol products 70 from boron enolates of the oxazolidinone 69 (R = i-Pr) and aliphatic or aromatic aldehydes R'CHO in the presence of Lewis acids (tin tetrachloride, titanium tetrachloride or diethylaluminum chloride) depends on the nature and amount of the catalyst (91JOC5747). The titanium enolate of 69 (R = i-Pr) forms syn-aldol adducts with aldehydes, while the boron enolate yields anti-compounds (91JOC2489). The lithium and titanium(IV) enolates of the oxazolidinone 71 undergo syn-selective aldol reactions with a variety of aldehydes (91JA1299). The enantioselective Michael addition of the titanium enolate of 69 (R = benzyl) to acrylonitrile yields 72 (91JOC 5750).

Stereoselective additions to the carbon-carbon double bond of 3-acryloyloxazolidin-2-ones include the reaction of 73 (R = p-chlorophenyl) with diethylaluminum chloride to give a 91:9 mixture of (S)- and (R)-74 (91AG(E)694), the dimethylaluminum chloride-catalyzed ene reaction of 73 (R = Me) with methylenecyclopentane to afford 75 (91JOC4908) and Diels-Alder syntheses with 3-acryloyloxazolidin-2-one, which adds buta-1,3-diene selectively in the presence of a chiral titanium catalyst to give 76 (91BCJ387) and which, in the presence of a chiral iron complex, affords the cyclopentadiene adduct 77 with 91% enantioselectivity and 96% endo-selectivity (91JA728).

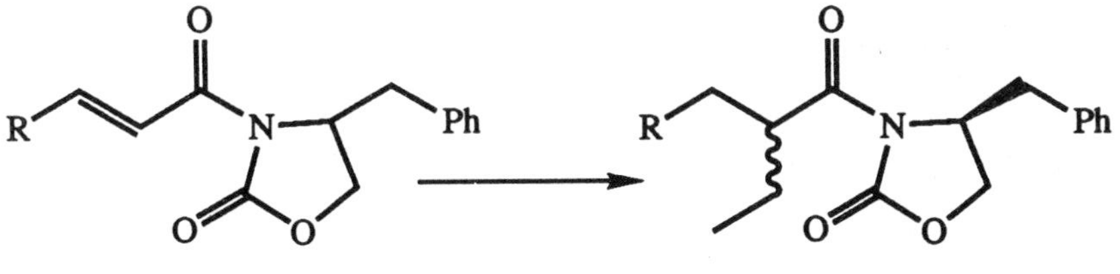

(73)

(74)

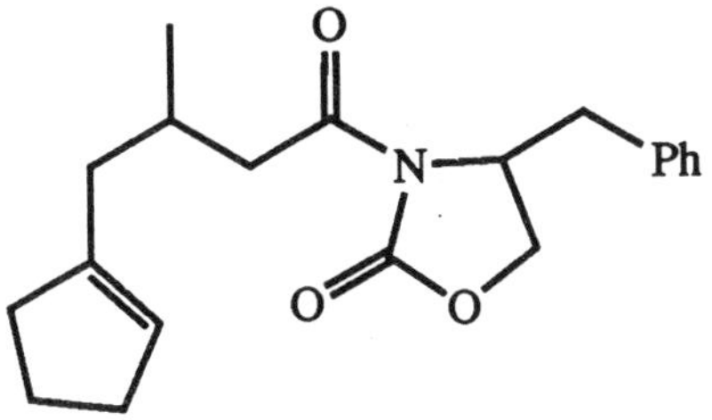

(75)

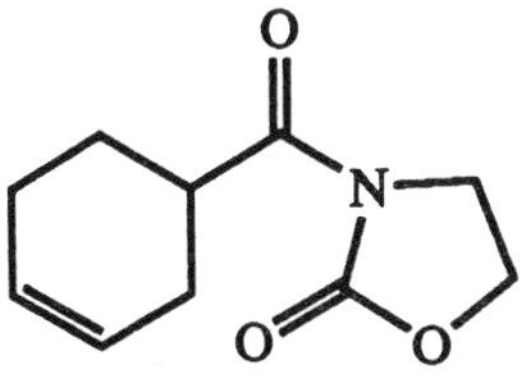

(76)

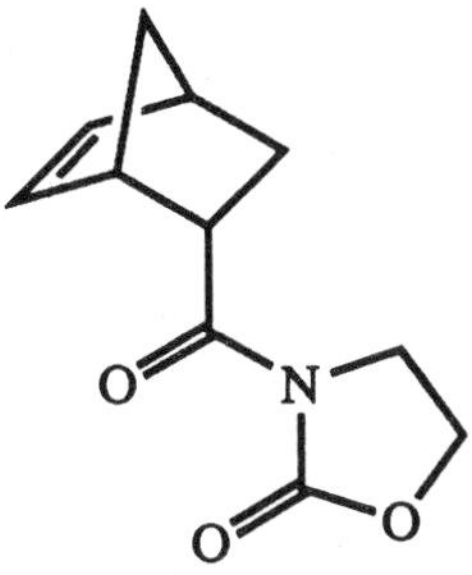

(77)

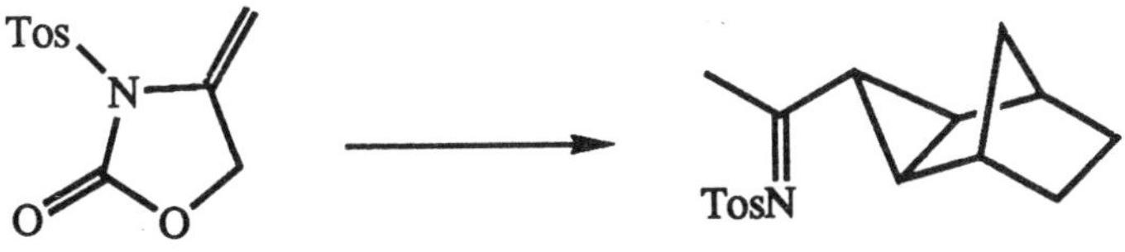

(78)

(79)

The methyleneoxazolidinone 78 reacts with norbornene in the presence of tetrakis(triphenylphosphine)palladium to give the fused cyclopropane 79 (90JA9646).

An improved synthesis of chiral oxazolidine-2-selenones, e.g. 80, as sensitive 77-Se NMR reagents for probing chirality has been described (91JOC6733). Treatment of cyanhydrins RCH(OH)CN (R = Me or Ph) with chlorosulfonyl isocyanate and subsequent acidic hydolysis results in oxazolidinediones 81 (91S697).

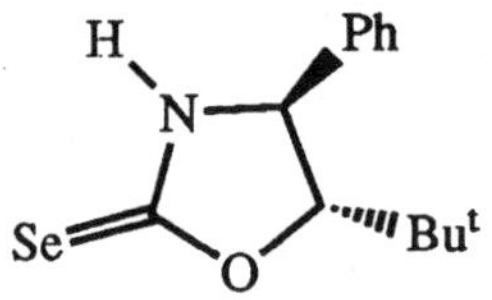

(80)

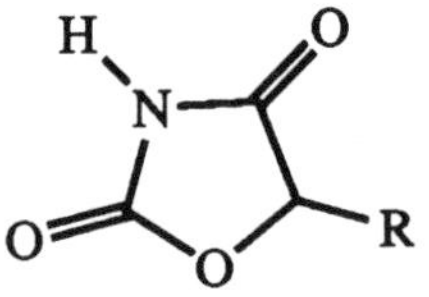

(81)

SYSTEMS WITH THREE HETEROATOMS

Benzonitrile oxide adds to aromatic aldehydes to yield 1,4,2-dioxazolines 82; an analogous reaction with Schiff's bases RCH=NAr (R = Ph, benzyl or t-Bu) gives 1,2,4-oxadiazolines 83 (91MI189). 1,4,2-Dioxazolin-5-ones 84 decompose to acylnitrenes on irradiation; in the presence of 3-methoxy-2-methylpropene, oxazolines 85 are produced by way of intermediate aziridines (90ZOR1757). Dipolar derivatives 87 of 1,2,4-oxadiazole have been made from aromatic nitrile oxides and N-cyanoimides 86 (R = Ph or OEt) (90IZV2084). Bis(5-trifluoromethyl-1,2,4-oxadiazol-3-yl) (88) yields the salt 89 by the action of sodium hydroxide (90KGS853). (E)-2-Hydroxyimino-2-phenylacetonitrile (91) is formed by photolysis of the azidofuroxan 90 (90BCJ1843). Irradiation of benzofuroxan (92) leads to a mixture of the azepinedione 93 and the tetracyclic furoxan 94 (91JHC1079).

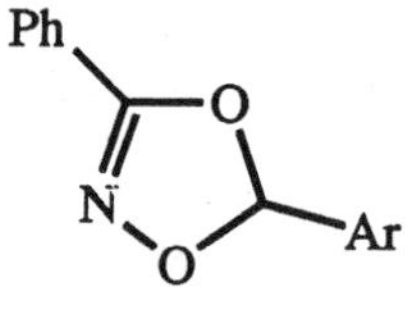

(82)

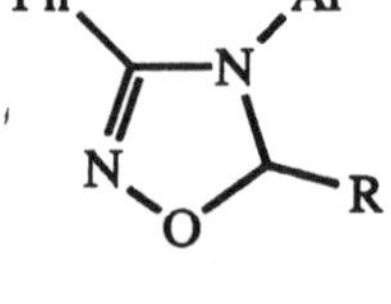

(83)

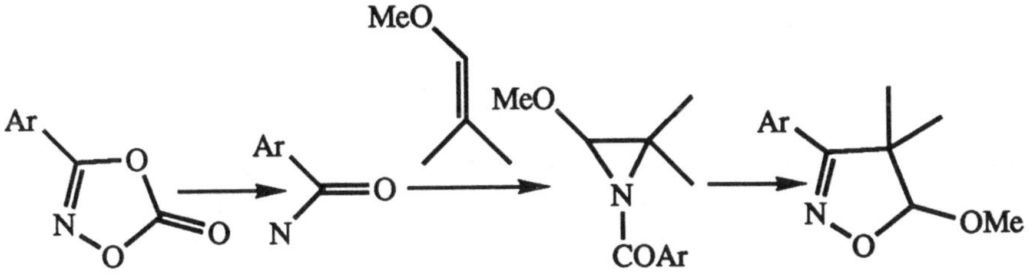

(84) (85)

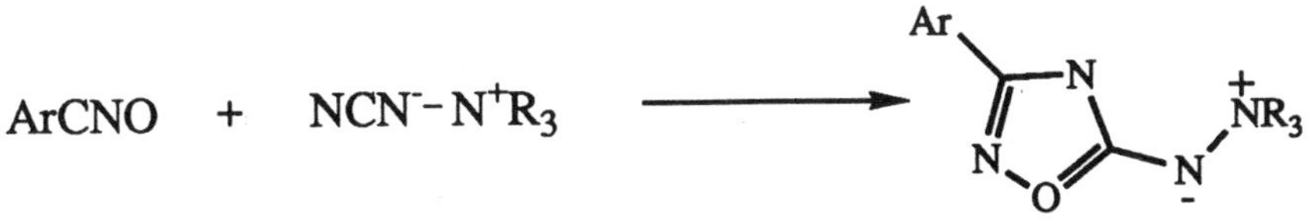

$ArCNO + NCN^{-}-N^{+}R_3$

(86) (87)

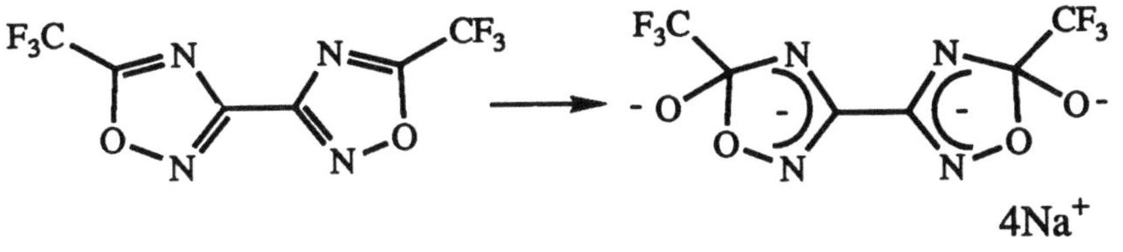

(88) (89)

$4Na^{+}$

(90) (91)

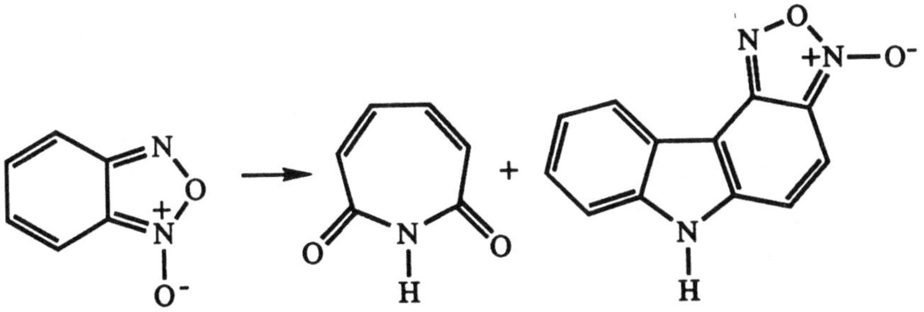

(92) (93) (94)

REFERENCES

90BCJ1843 A.Kunai, T.Doi, T.Nagaoka, H.Tagi and K. Sasaki, Bull. Chem. Soc. Jpn., 1990, 63, 1843.

90CL1991 A.Kamimura and A.Yamamoto, Chem. Lett., 1990, 1991.

90IZV2084 V.A.Ogurtsov, O.A.Rakitin, N.V.Obruchnikova, L.F.Chertanova, A.A.Gazikasheva and L.I. Khmelnitskii, Izv. Akad. Nauk SSSR, Ser. Khim., 1990, 2084.

90JA9646 K.Ohe, T.Ishihara, N.Chatani and S.Murai, J. Am. Chem. Soc., 1990, 112, 9646.

90JOC5935 M.Sawamura, H.Hamashima and Y.Ito, J. Org. Chem., 1990, 55, 5935.

90JOC6216 G.Fronza, A.Mele, G.Pedrocchi-Fantoni, D. Pizzi and S.Servi, J. Org. Chem., 1990, 55, 6216.

90KGS853 V.G.Andrianov and A.V.Eremeev, Khim. Geterotsikl. Soedin., 1990, 853.

90KGS1110 V.I.Rybinov, M.V.Gorelik, M.Ya.Mustafina and V.Ya.Gurvich, Khim. Geterotsikl. Soedin., 1990, 1110.

90MI251 M.Cinquini, F.Cozzi, P.Giaroni and L.Raimondi, Tetrahedron: Asymmetry, 1990, 1, 251.

90MI1023 P.Sohar, I.Kovesdi, S.Frimpong-Mansa, G.Sta- and G.Bernath, Magn. Reson. Chem., 1990, 28, 1023.

90MI1277 A.Liguori, P.Mascaro, G.Sindona and N.Uccella, J. Labelled Compd. Radiopharm., 1990, 28, 1277.

90TL4633 N.Katagiri, N.Watanabe, J.Sakaki, T.Kawai and C.Kaneko, Tetrahedron Lett., 1990, 31, 4633.

90TL5053 A.Kamimura and S.Marumo, Tetrahedron Lett., 1990, 31, 5053.

90TL5339 G.W.Spears, K.Nakanishi and Y.Ohfune, Tetrahedron Lett., 1990, 31, 5339.

90ZOR1757 A.N.Studenikov, V.P.Semenov, O.Yu.Ivanov and and K.A. Ogloblin, Zhur. Org. Khim., 1990, 26, 1757.

91AG(E)694 K.Rueck and H.Kunz, Angew. Chem., Int. Ed. Engl., 1991, 30, 694.

91BCJ318 T.Shimizu, Y.Hayashi, N.Furakawa and K.Taramura, Bull. Chem. Soc. Jpn., 1991, 64, 318.

91BCJ387 K.Narasaka, H.Tanaka and F.Kanai, Bull. Chem. Soc. Jpn., 1991, 64, 387.
91BCJ2388 S.Shiraishi, Y.Inoue and K.Imamura, Bull. Chem. Soc. Jpn., 1991, 64, 2388.
91CB1181 A.Hassner and W.Dehaen, Chem. Ber., 1991, 124, 1181.
91CB2481 L.Assmann, L.Palm, M.Zander and W.Friedrichsen, Chem. Ber., 1991, 124, 2481.
91CC884 O.Moriya, Y.Urata and T.Endo, J. Chem. Soc., Chem. Commun., 1991, 884.
91CC1200 Y.Mori and K.Maeda, J. Chem. Soc., Chem. Commun., 1991, 1200.
91CL1221 H.Suga and T.Ibata, Chem. Lett., 1991, 1221.
91CL1333 H.Y.Kang, Y.S.Cho, H.Y.Koh and M.H.Chang, Chem. Lett., 1991, 1333.
91CZ253 K.Burger and B.Helmreich, Chem.-Ztg., 1991, 115, 253.
91JA728 E.J.Corey, N.Imai and H.Y.Zhang, J. Am. Chem. Soc., 1991, 113, 728.
91JA1299 M.P.Bonner and E.R.Thoenton, J. Am. Chem., Soc., 1991, 113, 1299.
91JA9676 E.P.Kuendig, G.Bernardinelli, R.Liu and A. Ripa, J. Am. Chem. Soc., 1991, 113, 9676.
91JCS(P1)961 M.R.Banks, J.I.G.Cadogan, J.Gosney, P.K.G. Hodgson and D.E.Thomson, J. Chem. Soc., Perkin Trans. 1, 1991, 961.
91JCS(P1)1843 G.Broggini and G.Zecchi, J. Chem. Soc., Perkin Trans. 1, 1991, 1843.
91JCS(P1)2183 J. Cornforth and M. Du, J. Chem. Soc., Perkin Trans. 1, 1991, 2183.
91JHC1079 M.Hasegawa and T.Takabatake, J. Heterocycl. Chem., 1991, 28, 1079.
91JOC449 J.C.Hodges, W.C.Patt and C.J.Connolly, J. Org. Chem., 1991, 56, 449.
91JOC896 W.Dehaen and A.Hassner, J. Org. Chem., 1991, 56, 896.
91JOC1332 A.K.M.M.Hoque, W.K.Lee, H.J.Shine and D. C.Zhao, J. Org. Chem., 1991, 56, 1332.
91JOC1961 W.R.Leonard, J.L.Romine and A.I.Meyers, J. Org. Chem., 1991, 56, 1961.
91JOC2154 B.H.Norman, Y.Gareau and A. Padwa, J. Org. Chem., 1991, 56, 2154.
91JOC2292 D.J.Rawson and A.I.Meyers, J. Org. Chem., 1991, 56, 2292.
91JOC2489 M.Nerz-Stormes and E.R.Thornton, J. Org. Chem., 1991, 56, 2489.

91JOC2775 A.Hassner, R.Maurya, A.Padwa and W.H.Bullock, J. Org. Chem., 1991, 56, 2775.

91JOC3058 S.E.Whitney and B.Rickborn, J. Org. Chem., 1991, 56, 3058.

91JOC3419 A.Hassner and B.Fischer, J. Org. Chem., 1991, 56, 3419.

91JOC4888 R.W.Friesen and A.E.Kolaczewska, J. Org. J. Org. Chem., 1991, 56, 4888.

91JOC4908 B.B.Snider and Q.Zhang, J. Org. Chem., 1991, 56, 4908.

91JOC5747 M.A.Walker and C.H.Heathcock, J. Org. Chem., 1991, 56, 5747.

91JOC5750 D.A.Evans, M.T.Bilodeau, T.C.Somers, J. Clardy, D.Cherry and Y.Kato, J. Org. Chem., 1991, 56, 5750.

91JOC6258 G.Rosini, E.Marotta, P.Righi and T.P.Seerden, J. Org. Chem., 1991, 56, 6258.

91JOC6733 L.A.Silks, J.Peng, J.D.Odom and R.B.Dunlap, J. Org. Chem., 1991, 56, 6733.

91JOM29 J.C.Clinet and G.Balavoine, J. Organomet. Chem., 1991, 405, C29.

91MI189 R.A.Aitken and S.V.Raut, Synlett., 1991, 189.

91MI217 M.Kodaka, T.Tomohiro, A.L.Lee and H.Okuno, Chem. Express., 1991, 6, 217.

91MI793 C.Gennari, G.Poli, C.Scolastico and M. Vassalo, Tetrahedron: Asymmetry, 1991, 2, 793.

91S697 M.V.Garcia, J.C.Menendez, M.Villacampa and M. M. Sollhuber, Synthesis, 1991, 697.

91T421 L.DiNunno and A.Scilimati, Tetrahedron, 1991, 47, 421.

91T4007 R.Grigg, F.Heaney, J.Markandu, S.Surendrakumar, M.Thornton-Pett and W.J.Warnock, Tetradron, 1991, 47, 4007.

91T4477 R.Grigg, F.Heaney, S.Surendrakumar and W.J. Warnock, Tetrahedron, 1991, 47, 4477.

91T4495 P.Armstrong, R.Grigg, F.Heaney, S.Surendrakumar and W.J.Warnock, Tetrahedron, 1991, 47, 4495.

91TL87 J.C.Combret, A.Meghni, D.Postaire and J. Tekin, Tetrahedron Lett., 1991, 32, 87.

91TL279 R.Grigg and J.Markandu, Tetrahedron Lett., 1991, 32, 279.

91TL1161 L.Assmann, T.Debaerdemaeker and W.Friedrichsen, Tetrahedron Lett., 1991, 32, 1161.

91TL2095 D.J.Rawson and A.I.Meyers, Tetrahedron Lett., 1991, 32, 2095.
91TL2799 T.Hayashi, Y.Oozumi, A.Yamazaki, M.Sawamura, H.Hamashima and Y.Ito, Tetrahedron Lett., 1991, 32, 2799.
91TL2919 Y.Ukaji, K.Yamamoto, M.Fukui and T.Fujisawa, Tetrahedron Lett., 1991, 32,2919.
91TL3523 S.Ishibuchi, Y.Ikematsu, T.Ishizuka and T. Kunieda, Tetrahedron Lett., 1991, 32, 3523.
91TL4259 H.R.Kim, H.J.Kim, J.L.Duffy, M.M.Olmstead, K.Ruhlandt-Senge and M.J.Kurth, Tetrahedron Lett., 1991, 32, 4259.
91TL4283 W.Adam, S.Bottle and K.Peters, Tetrahedron Lett., 1991, 32, 4283.
91TL4603 S.Mitkidou and J.Stephanidou-Stephanatou, Tetrahedron Lett., 1991, 32, 4603.

CHAPTER 6.1

Six-Membered Ring Systems: Pyridine and Benzo Derivatives

J.G. KEAY & J.E. TOOMEY, Jr.
Reilly Industries, Inc, Indianapolis, IN, USA

6.1.1 SYNTHESIS

6.1.1.1 Pyridines

Palladium catalysis has been used for insertion and coupling of carbon monoxide with halo-olefins and acetylenes [91SL695]. This work was originally developed for making quinolones and now has been extended to pyridones (1) (Scheme 1).

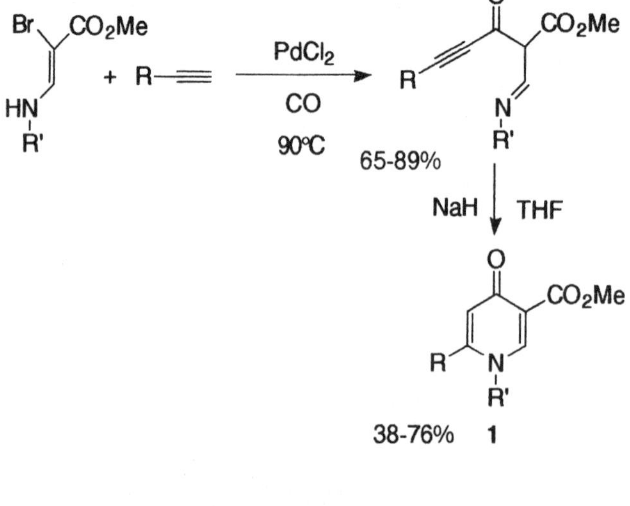

Br CO₂Me; HN R'; + R—≡; $PdCl_2$, CO, 90°C; 65-89%; NaH THF; 38-76% **1**

SCHEME 1

Carbonylation of heterodienes (2) using Lewis acid catalysis and carbonic acid derivatives to make 4-pyridones was reported last year. That reaction has been extended to make 4-chloropyridines (3) (Scheme 2) [91JOC6751], without the necessity of catalyst. In fact, Lewis acids will inhibit the thiophosgene reaction.

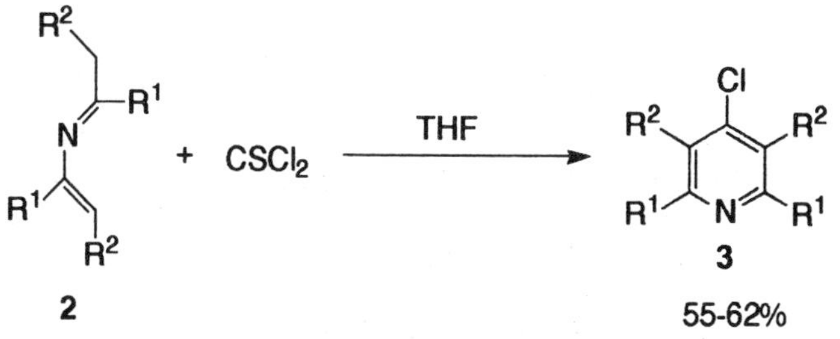

2 + $CSCl_2$ —THF→ 3 (55-62%)

SCHEME 2

Palladium (II) catalysis has also been used to cyclize alkynyl amines (4) to 2,3,4,5-tetrahydropyridine (5) [91JOC5812] (Scheme 3). Gold (III) catalysis was considered more useful for similar reactions [91S975].

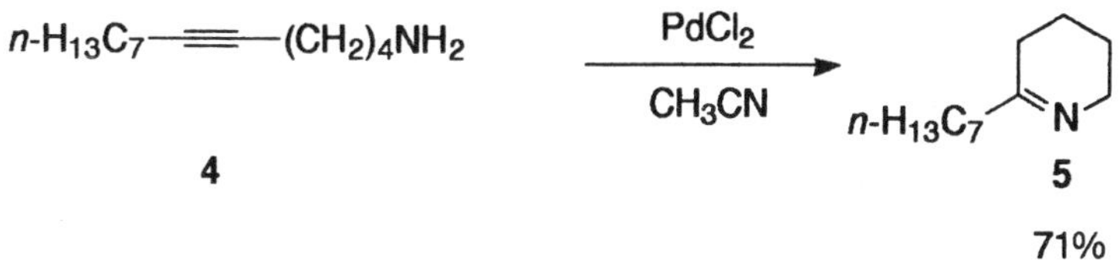

n-$H_{13}C_7$—≡—$(CH_2)_4NH_2$ (4) —$PdCl_2$, CH_3CN→ 5 (71%)

SCHEME 3

The efficient use of 1-azabutadienes in intramolecular heterodiene synthesis (Diels-Alder) is reported, key being acylation of nitrogen (Scheme 4) [91JOC6729].

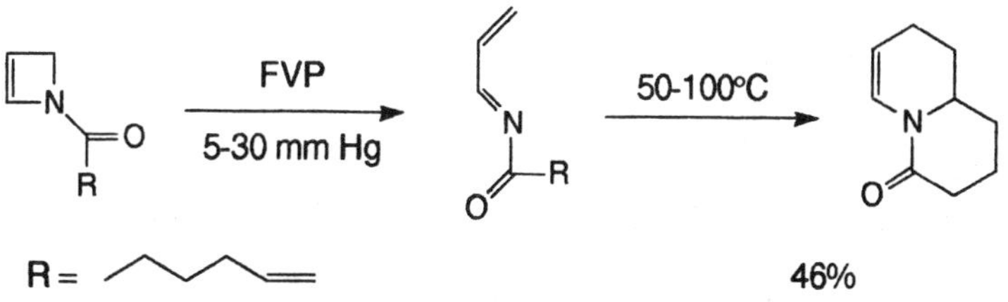

FVP, 5-30 mm Hg; 50-100°C; 46%

SCHEME 4

2-Cyano-1-azabutadienes (6) bearing an N-aryl substituent have also been used successfully for intermolecular heterodiene synthesis [91SL816] (Scheme 5).

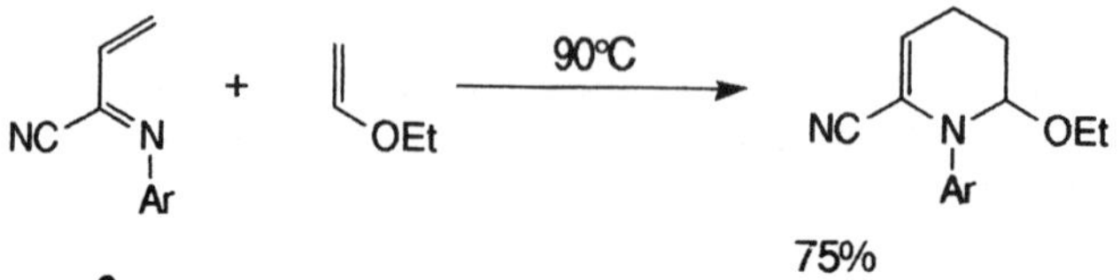

SCHEME 5

N-Vinylcarbodiimides (7) have been used to construct pyridine rings in another type of heterodiene synthesis (Scheme 6) [91BSJ1325]. Cyclic dienophiles give rise to 5,6,7,8-tetrahydroisoquinolines (8).

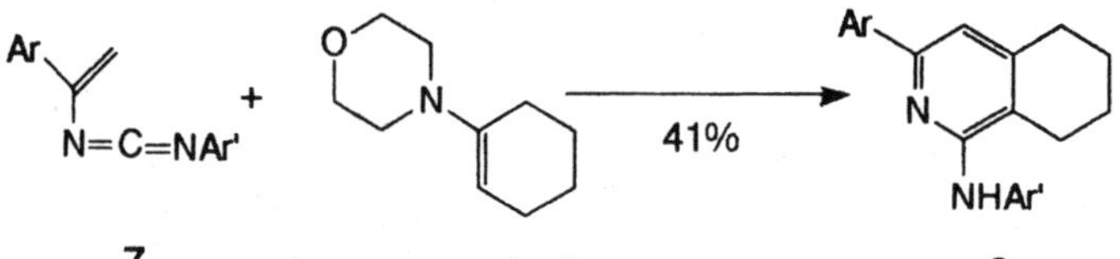

SCHEME 6

2-Furfuraldoxime (9) has been shown to be another efficient 1-aza-butadiene system that undergoes Diels-Alder reaction (Scheme 7) [91TL3199], despite prior reports that α,β-unsaturated oximes are unreactive and can undergo 1,3-dipolar cycloadditions instead.

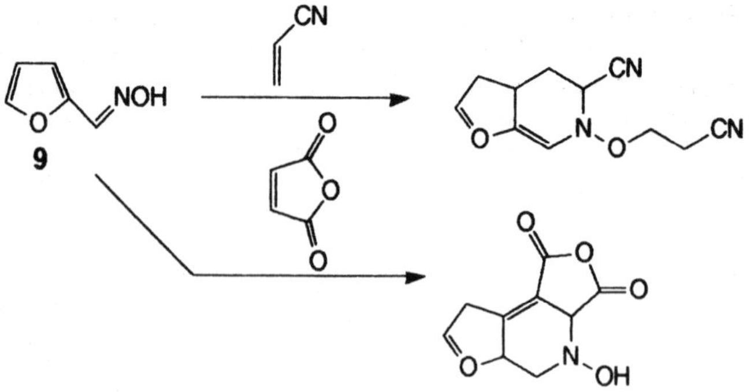

SCHEME 7

An interesting, versatile new method for making heteroaryl-substituted pyridines has been developed [91JHC161]. Two equivalents of acetonitrile condense with a heteroaryl-substituted α,β-unsaturated ketone (10) to construct the pyridine ring (Scheme 8). Sonication was used to accelerate these reactions.

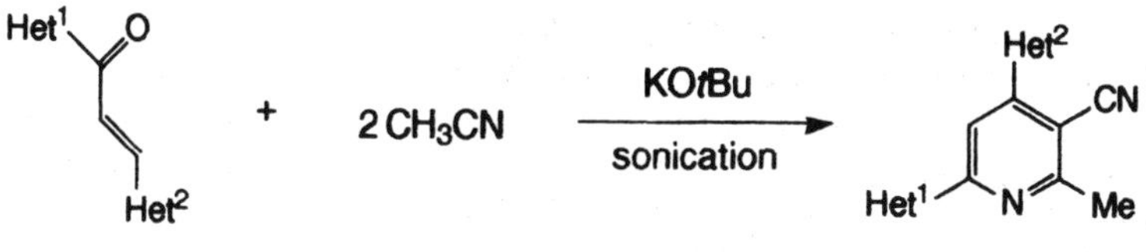

Het¹ O / Het² (**10**) + 2 CH_3CN —KO*t*Bu, sonication→ Het² / CN / Het¹ N Me (36-93%)

SCHEME 8

Photolysis of homopyrrole (**11**) gave rise to 1,2-dihydropyridines (**13**) via intermediacy of 1-azahexatriene (**12**) (Scheme 9) [91TL6895], while thermolysis of **11** has been shown to proceed by an ylide intermediate.

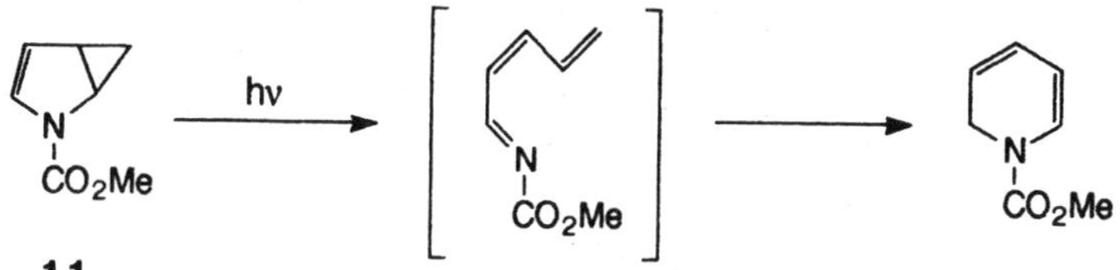

N CO_2Me (**11**) —hv→ [N CO_2Me (**12**)] → N CO_2Me (**13**)

SCHEME 9

6.1.1.2 Quinolines and Isoquinolines

Radical reaction sequencing, a new, promising tool for synthesis of a wide class of organics, has been used to create a simple route to cyclopenta-fused quinolines (**14**) (Scheme 10) [91JA2127]. This is the first example of a 4+1 radical annulation.

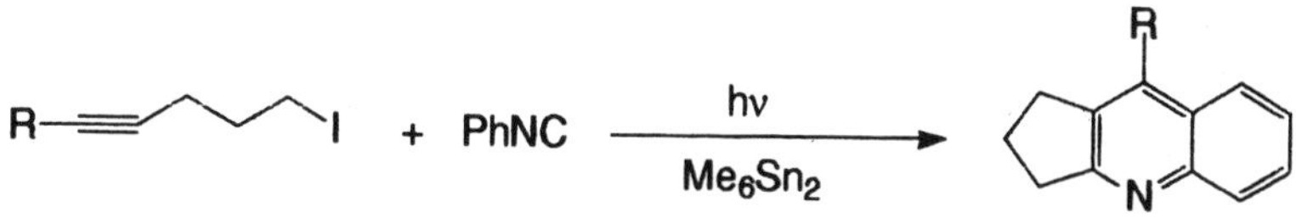

R—≡—I + PhNC —hv, Me_6Sn_2→ R / N (**14**)

SCHEME 10

Oximes have been photo-cyclized into quinolines, to provide a facile, rapid, general three-step method to make polycyclic quinolines, essentially a photo-annulation of ketones (Scheme 11) [91TL5235].

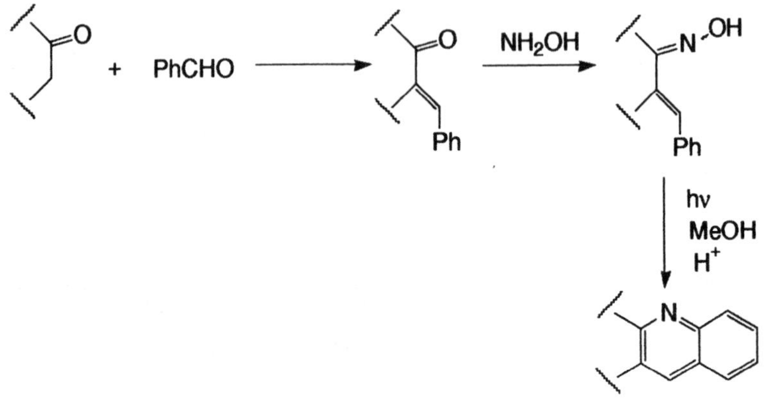

SCHEME 11

3-Arylisoquinolines can now be prepared by a modification of the usual Bischler-Napieralski scheme using oxalyl chloride. Function of oxalyl chloride is to avoid the normal nitrilium ion intermediate and subsequent loss of nitrile when the aryl moiety is inactivated [91JOC6034].

Two recent publications focus on synthesis of the 2-substituted-1,4-dihydro-4-oxoquinoline ring system (**15**). One utilizes an isoxazole moiety for reductive cyclization (**Scheme 12**) [91T5111], while the other uses a Pd-catalyzed carbonylation (**Scheme 13**) [91TL237]. Intermediates postulated for both reaction schemes are quite similar.

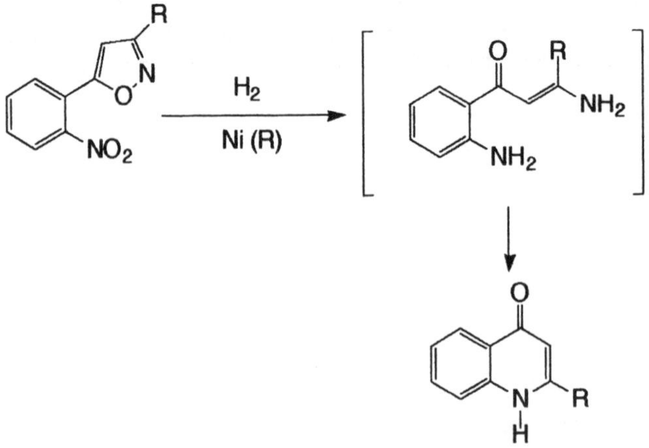

SCHEME 12

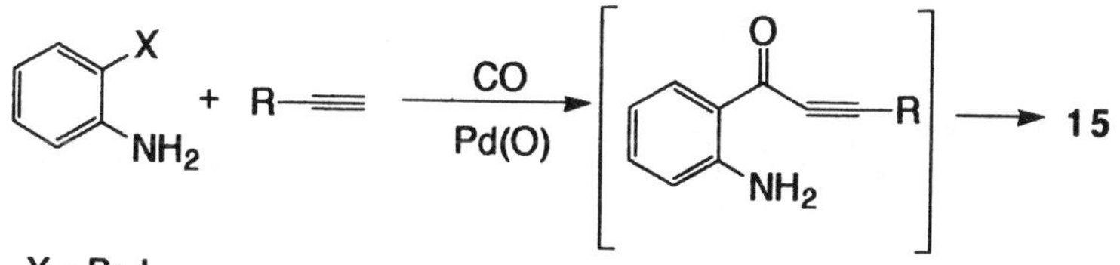

X = Br, I

SCHEME 13

The technique of Microwave-induced Organic Reaction Enhancement (MORE) has been applied to the Bischer-Napieralski reaction. 1-Phenyl-3,4-dihydroisoquinoline (**16**) could be isolated from the acylamine after microwave irradiation (MWI). The amidine was a by-product [91JOC6968].

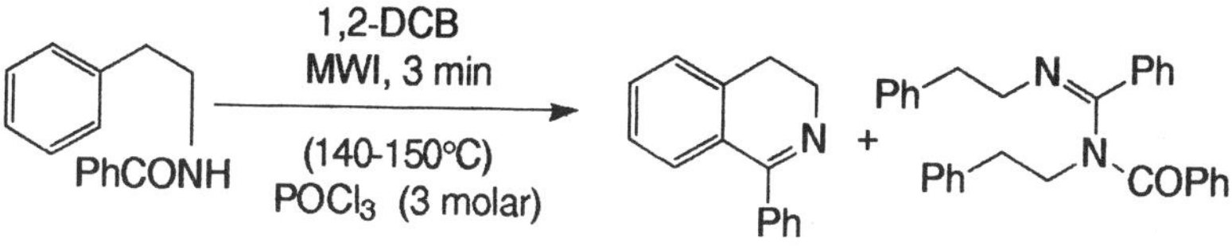

SCHEME 14

Other optically pure tetrahydroisoquinolines have been made by asymmetric Pictet-Spengler synthesis where a chiral N-acyl group was used [91TL2995]. This chiral auxiliary induced enantioselective cyclization with aldehydes (Scheme 15).

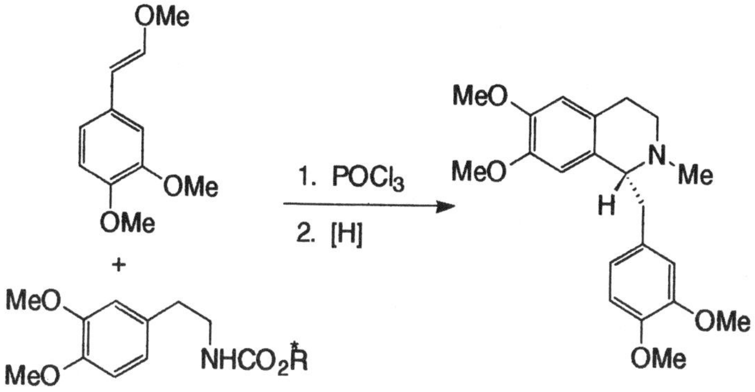

R^* = (-)-8-phenylmenthyl

SCHEME 15

6.1.1.3 Acridines and other annelated pyridines

Ullmann-Fetvadjian reaction was used to prepare benz[a]acridines (**17**) (one-step) using 2-naphthol, and to prepare benz[c]acridines (**18**) (two-step) using 1-naphthol (Scheme 16) [91JHC321].

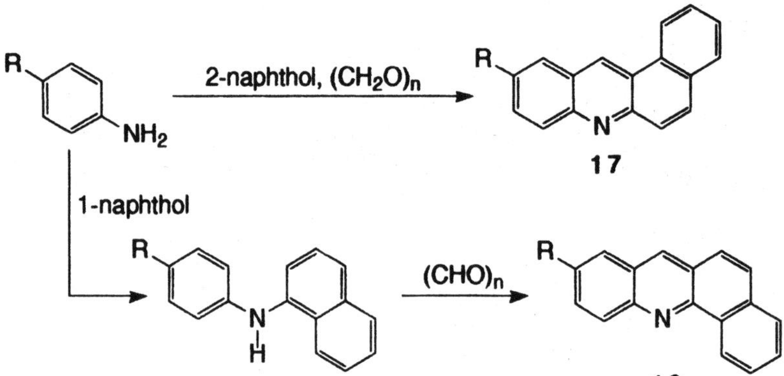

SCHEME 16

A high-yield, one-step synthesis of sulfur-substituted indolizines (**20**) has been reported using cyclopropenyl cations (**19**) as a key three carbon building block (Scheme 17) [91JHC2059].

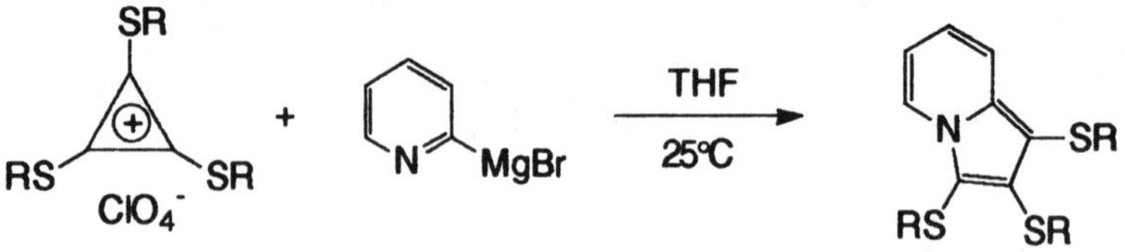

SCHEME 17

A series of cationic cyclazines (**21**) is accessible by a three-step synthesis from α-picoline N-oxide and aryl Grignard reagents (**Scheme 18**) [91JOC4858].

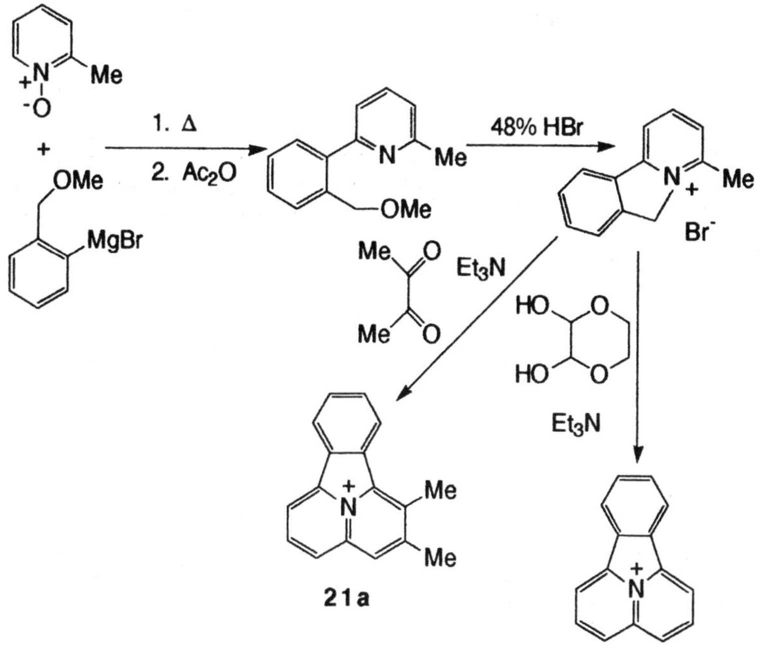

SCHEME 18

6.1.2 REACTIONS

6.1.2.1. Pyridines

Reviews have appeared on aromaticity and tautomerism [91H127,329]. Additionally, the gas and solution phase basicities for an extensive series of pyridines, quinolines and isoquinolines have been determined [91JA1770]. Lithiation of pyridines at carbon has been reviewed recently [91MI20].

The preparation of "unnatural products" - those materials synthesized for molecular properties - has been the subject of increasing focus. Much of this work utilizes the pyridine molecule for its π-electron deficient ring or chelation abilities and several communications and reviews have appeared [91T6851, 91SL677]. The self-assembly of [n]pseudorotaxanes [91CC1677,1680] and [2]catenanes [91CC634] from pyridinium species and donor molecules and their potential application in information storage and processing has aroused much interest.

Chiral tris(bipyridines) (22) have been shown to undergo helical self-organization in the presence of Cu(I) and Ag(I) ions [91HCA1843].

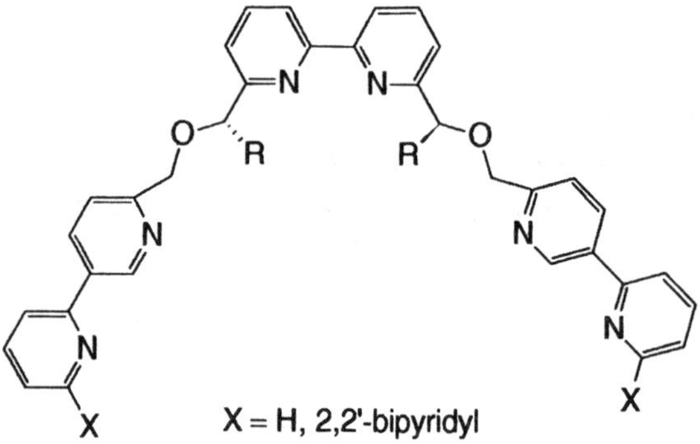

X = H, 2,2'-bipyridyl

22

The use of hydrogen bonding to control molecular aggregation [91JOC2284] or conformational changes in addition to self-assembly has also been explored [91CC1283].

Zincke-Koenig reaction is a pivotal step in making chiral piperidines (Scheme 19) [91CC570]. Oxazolidines (23) are synthetic equivalents of 3,4-dihydropyridiniums (24) having a chiral substituent directly on nitrogen. 23 is then used to control subsequent ring formation.

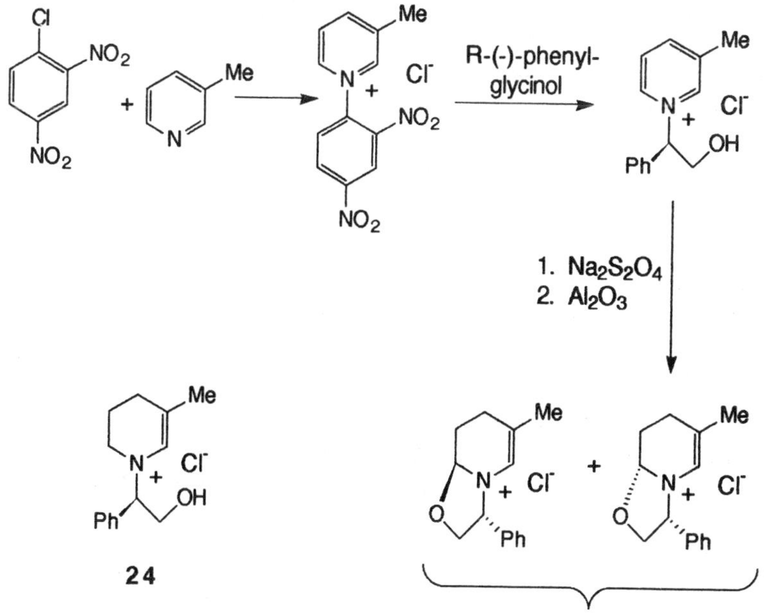

SCHEME 19

Metallation of pyridine at carbon was enhanced by use of BF_3 complex (25) (Scheme 20) [91CC570]. Direct metallation of pyridine itself ocurred with lithium amide to give bipyridyls. Ring carbon metallation was favored even when relatively acidic methyl groups were present on ring.

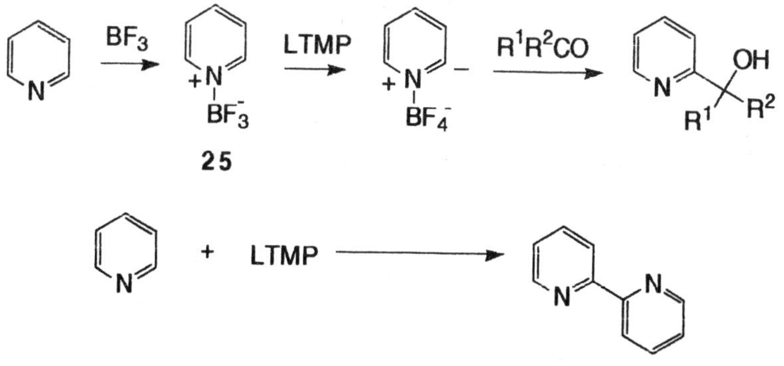

SCHEME 20

Heteroaryl lithium salts react with pyridinium salts (**26**) by addition to give 2- or 4-substituted dihydropyridines which can be oxidized by oxygen (**Scheme 21**) [91H2111] to heteroaryl substituted pyridines (**27**).

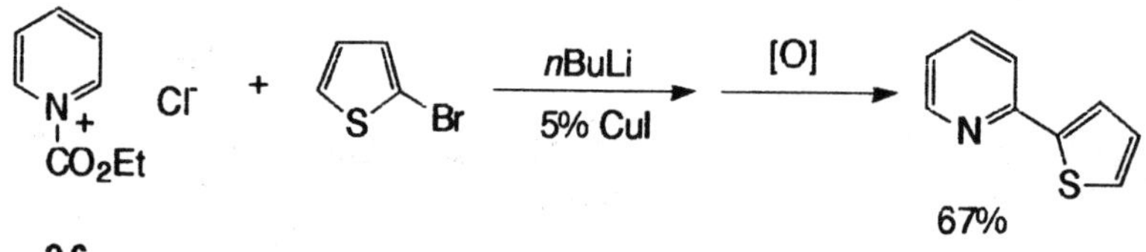

SCHEME 21

The chiral 2-cyano-6-oxazolopiperidine synthon, (**28**) (**Scheme 22**) has been further utilized for the diastereoselective synthesis of indolizidinediol alkaloids [91H879]. 1-Benzyl-2-vinylpiperidine (**29**) has been used in the synthesis of the homologous quinolizidines (**30**) [91JA6690].

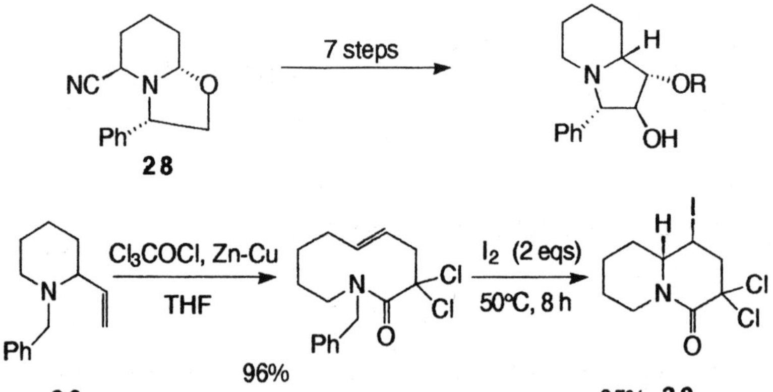

SCHEME 22

Upon alkylation, 4-*t*-butylpiperidines (**31**, **32**) will alkylate in the equatorial 2-position and subsequentially alkylate axially or equatorially in the 6-position depending upon the nature of the stabilizing group on nitrogen. Formamidines tend to give diequatorial alkylation, whereas the *t*-butoxycarbonyl moiety gives 6-axial alkylation with electrophiles (Scheme 23). A study of the controlling factors has been made [91JO2751].

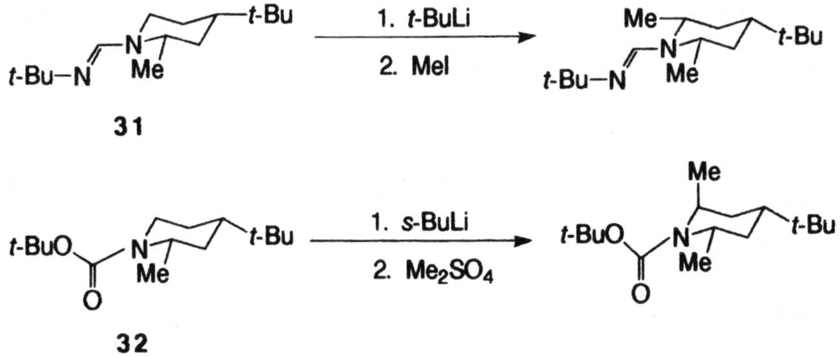

1. *t*-BuLi
2. MeI

31

1. *s*-BuLi
2. Me_2SO_4

32

SCHEME 23

Pyridine based sulfenylating agents, useful for the preparation of symmetrical and unsymmetrical disulfides (33) (Scheme 24) have been reported and make use of 2-thiopyridyl-N-oxide [91TL6127] or N-alkylpyridthione as leaving groups. These reagents react with a range of sulfur nucleophiles in addition to amines and diketones [91JOC6697].

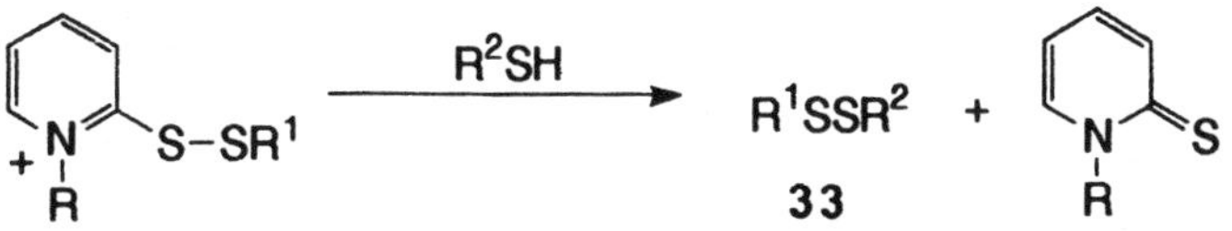

R^2SH

R^1SSR^2 +

33

R = O, alkyl

R^1 = Me, 2-pyridyl-N-oxide, cholest-5-en-3β-yl

SCHEME 24

In a continuing study, carbon radicals generated by the photolysis of N-acyloxy-2-thiopyridones have provided for the homologation of carboxylic acids [91TL3309] and conversion of carboxylic acid to nitriles [91TL3321].

α-Hydroxyacylpyridines are available from pyridine-N-oxide and α-bromoacetaldehydes via the alkoxypyridinium salts (34) (Scheme 25) and base induced rearrangement [91JHC1127].

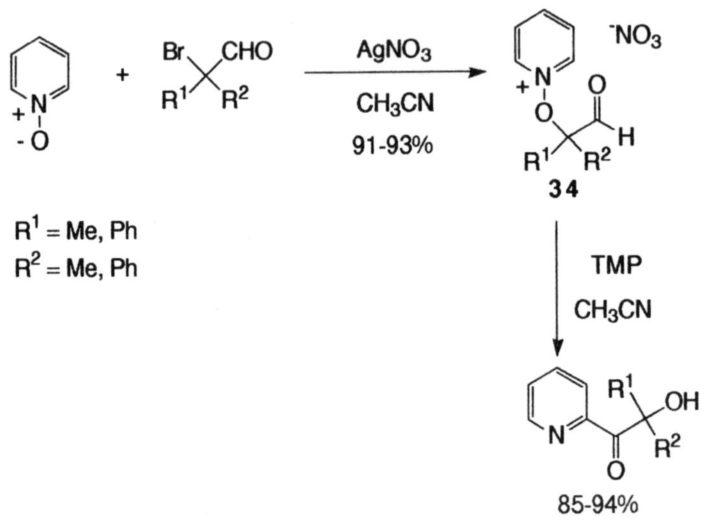
Br
CHO
R^1
R^2
AgNO3
CH3CN
91-93%
$^-NO_3$
O
H
34
R^1 = Me, Ph
R^2 = Me, Ph
TMP
CH3CN
OH
85-94%

SCHEME 25

Direct substitution of hydrogen, by nucleophiles, mediated by acetyl hypofluorite in pyridines and quinolines has been extended by Rozen. The solvent normally supplies the nucleophilic species in a mild and rapid procedure (Scheme 26) [91JO6298].

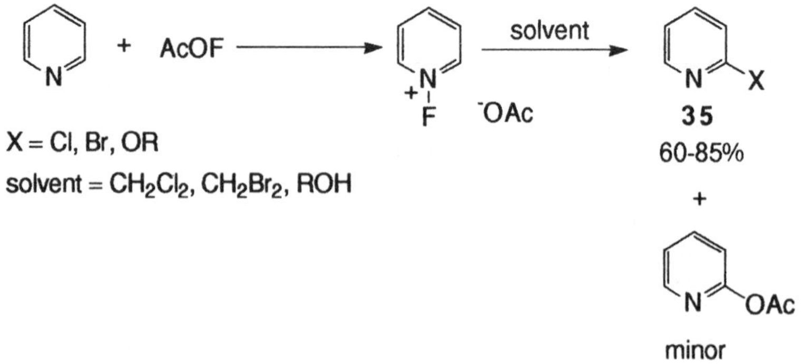
AcOF
solvent
^-OAc
X
35
60-85%
X = Cl, Br, OR
solvent = CH_2Cl_2, CH_2Br_2, ROH
OAc
minor

SCHEME 26

2-Substituted 1-acyl-1,2-dihydropyridines (37) are useful intermediates for the synthesis of a variety of natural products. The asymmetric syntheses of (37) mediated by menthyl derived chiral auxiliaries have been reported [91JOC7197] from the reaction of Grignard reagents on acylpyridinium salts.

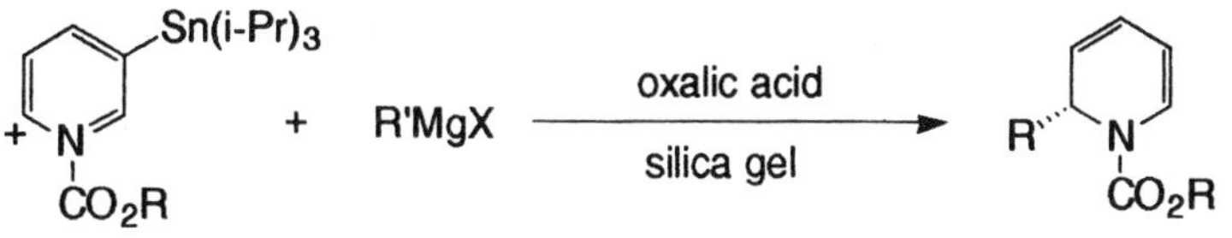

$Sn(i\text{-}Pr)_3$ + R'MgX — oxalic acid, silica gel →

36 (CO_2R) → 37 (CO_2R)

yield = 58-87%
de = 76-92%

R = (-)-8-(4-phenoxyphenyl)menthyl, (-)-8-phenylmenthyl

SCHEME 27

6.1.2.2 Quinolines

The radical alkylation of heteroaromatic bases is not new. A recent report describes the photochemical promoted alkylation of 4-methylquinoline using tetrakis(trimethylsilyl)silane (38) (Scheme 28) and an alkyl halide for the alkylation of 4-methylquinoline under non-oxidative conditions. Alkyl bromides and iodides were effective for this transformation. The problem of regiospecificity was not addressed. Bis-(tributyltin) was also investigated, but gave butylation in one case [91CL2063].

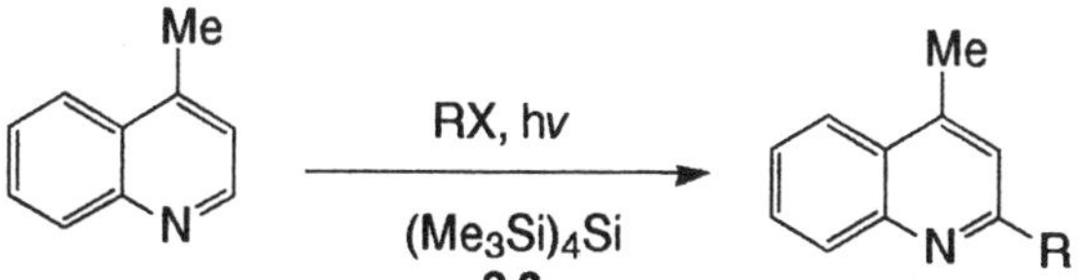

Me — RX, hv; $(Me_3Si)_4Si$ 38 → Me, R

X = Br, I

32-87%

SCHEME 28

Homolytic alkylation of quinolines using the half-esters of oxalic acid [91CL1691] or acylation using α-ketocarboxylic acids [91JOC2866] has been investigated. In the latter, a two-phase system is utilized to extract the product and combat the problem of diacylation. This system was also applied to pyridines.

The reaction of pyridylmalonates (39) (Scheme 29) with alkenes and alkynes in an oxidizing environment (Mn(III), Ce(IV), Fe(III)) leads to the formation of tetra- and dihydroquinolines and isoquinolines in good to excellent yield. In the case of 3-pyridylmalonates, both quinolines and isoquinolines were formed; the isomer ratio varied from 0.2-6.6 for the diethylmalonate and 1-octene under a variety of reaction conditions [91JOC5335].

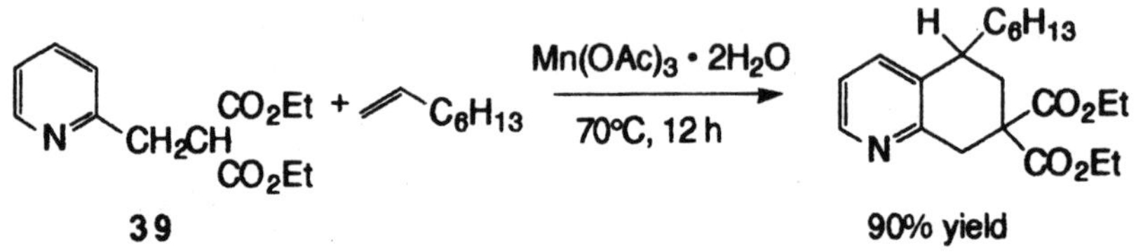

$CH_2CH(CO_2Et)_2$ + C_6H_{13} — $Mn(OAc)_3 \cdot 2H_2O$, 70°C, 12 h → H, C_6H_{13}, CO_2Et, CO_2Et

39 — 90% yield

SCHEME 29

6.1.2.3 Isoquinolines

The amide ion promotion toward substitution of 4-bromo and 4-chloro-isoquinoline by thiolate ion has been observed [91JOC2805]. In the case of 4-bromoisoquinoline, amide ion was shown to be required for reaction with thiomethoxide, 94% recovery of the quinoline was obtained in the absence of amide ion. A $S_{RN}1$ mechanism was proposed.

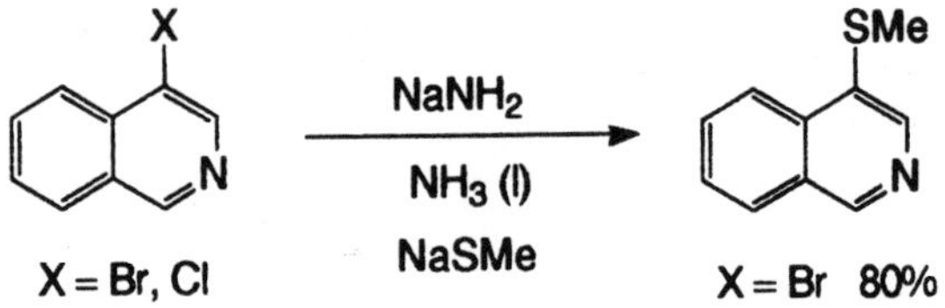

X — $NaNH_2$, NH_3 (l), NaSMe → SMe

X = Br, Cl — X = Br 80%

SCHEME 30

Diastereofacial differentiation has been achieved during addition of organometallics to N-acylisoquinolinium salts (**40**) [91H1869]. The addition product can be further reduced to form 1,2,3,4-tetrahydroisoquinolines (**41**) in an enantioselective fashion (Scheme 31). Diastereomeric ratios were from 70:30 to 80:20, at generally high yield. (+)-Carnegine was prepared 62% optically pure R-isomer.

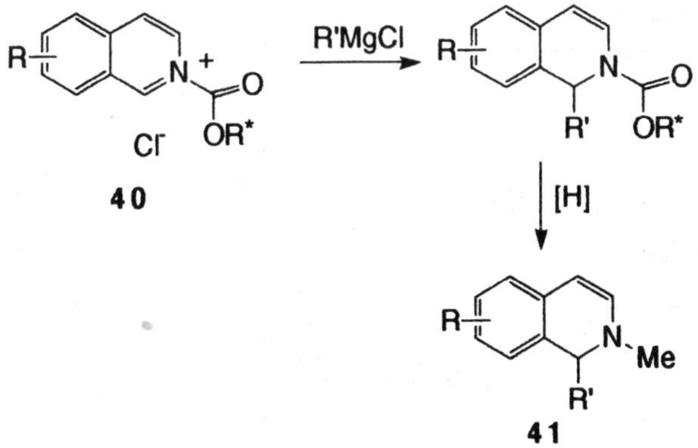

SCHEME 31

The use of the *tert*-leucine methyl ether based formamidine as a chiral auxiliary has enabled the asymmetric synthesis of 1,1-disubstituted tetrahydroisoquinolines in 76-86% ee to be achieved [91TL5501]. This auxiliary gave better results than that derived from valine (R = i-Pr). Substitution for an achiral auxiliary derived from 2-amino-2-methylpropanol led to racemization on subsequent alkylation [91TL5505]. A mechanism to explain the observed selectivity observed has been proposed [91TL5509].

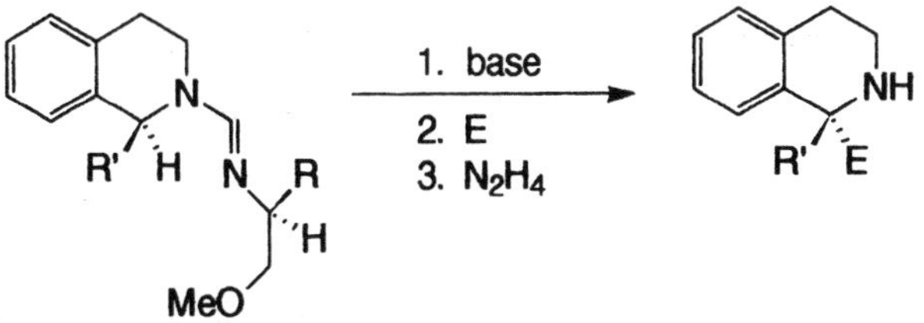

R = i-Pr, t-Bu
R' = Me, Et, $PhCH_2$
E = EtI, allyl Cl, CO_2

SCHEME 32

ACKNOWLEDGEMENTS

The authors would like to thank their colleagues at Reilly Industries, Inc. for their helpful comments. They are most especially grateful to Ms. Misty Bogle for her assistance with the preparation of this manuscript.

REFERENCES

91BSJ1325	M. Nitta, H. Soeda, S. Koyama, Y. Iino; *Bull. Chem. Soc. Jpn.*, **1991**, *64*, 1325.
91CC625	D. Gnecco, C. Marazano, B. C. Das; *J. Chem. Soc., Chem. Commun.*, **1991**, 625.
91CC634	P. R. Ashton, C. L. Brown, E. J. T. Chrystal, T. T. Goodnow, A. E. Kaifer, K. P. Parry, D. Philp, A. M. Z. Slawin, N. Spencer, J. F. Stoddart, D. J. Williams; *J. Chem. Soc., Chem. Commun.*, **1991**, 634.
91CC570	S. V. Kessar, P. Singh, K. N. Singh, M. Dutt; *J. Chem. Soc., Chem. Commun.*, **1991**, 570.
91CC1283	S. J. Geib, S. C. Hirst, C. Vicent, A. D. Hamilton; *J. Chem. Soc., Chem. Commun.*, **1991**, 1283.
91CC1677	P. R. Ashton, D. Philp, N. Spencer, J. F. Stoddart; *J. Chem. Soc., Chem. Commun.*, **1991**, 1677.
91CC1680	P. R. Ashton, D. Philp, M. V. Reddington, A. M. Z. Slawin, N. Spencer, J. F. Stoddart, D. J. Williams; *J. Chem. Soc., Chem. Commun.*, **1991**, 1680.
91CL1691	H. Togo, M. Aoki, M. Yokoyama; *Chem. Lett.*, **1991**, 1691.
91CL2063	H. Togo, K. Hayashi, M. Yokoyama; *Chem. Lett.*, **1991**, 2063.
91H879	V. Ratovelomanana, L. Vidal, J. Royer, H.-P. Husson; *Heterocycles*, **1991**, *32*, 879.
91H1869	D. L. Comins, M. M. Badawi; *Heterocycles*, **1991**, *32*, 1869.
91H2111	M.-J. Shiao, L.-H. Shih, W.-L. Chia, T.-Y. Chau; *Heterocycles*, **1991**, *32*, 2111.
91HCA1843	W. Zarges, J. Hall, J.-M. Lehn, C. Bolm; *Helv. Chim. Acta*, **1991**, *74*, 1843.
91JA2127	D. P. Curran, H. Liu; *J. Am. Chem. Soc.*, **1991**, *113*, 2127.
91JA6690	E. D. Edstrom; *J. Am. Chem. Soc.*, **1991**, *113*, 6690.
91JHC161	K. Shibata, I. Katsuyama, M. Matsui, H. Muramatsu; *J. Heterocycl. Chem.*, **1991**, *28*, 161.
91JHC321	A. R. Katritzky, Z. Wang, G. P. Savage; *J. Heterocycl. Chem.*, **1991**, *28*, 321.
91JHC1127	H. Sliwa, C. Randria-Raharimanana, G. Cordonnier; *J. Heterocycl. Chem.*, **1991**, *28*, 1127.
91JHC2059	H. Kojima, Y. Kinoshita, N. Matsumura, H. Inoue; *J. Heterocycl. Chem.*, **1991**, *28*, 2059.
91JOC2284	M. Gallant, M. T. P. Viet, J. D. Wuest; *J. Org. Chem.*, **1991**, *56*, 2284.
91JOC2751	T. T. Shawe, A. I. Meyers; *J. Org. Chem.*, **1991**, *56*, 2751.
91JOC2805	J. A. Zoltewicz, T. M. Oestreich; *J. Org. Chem.*, **1991**, *56*, 2805.
91JOC2866	F. Fontana, F. Minisci, M. C. N. Barbosa, E. Vismara; *J. Org. Chem.*, **1991**, *56*, 2866.
91JOC4858	M. Fourmiqué, H. Eggert, K. Bechgaard; *J. Org. Chem.*, **1991**, *56*, 4858.

91JOC5335 A. Citterio, R. Sebastiano, M. C. Carvayal; *J. Org. Chem.*, **1991**, *56*, 5335.

91JOC5812 Y. Fukuda, S. Matsubara, K. Utimoto; *J. Org. Chem.*, **1991**, *56*, 5812.

91JOC6034 R. D. Larsen, R. A. Reamer, E. G. Corley, P. Davis, E. J. J. Grabowski, P. J. Reider, I. Shinkai; *J. Org. Chem.*, **1991**, *56*, 6034.

91JOC6298 D. Hebel, S. Rozen; *J. Org. Chem.*, **1991**, *56*, 6298.

91JOC6697 D. H. R. Barton, R. H. Hesse, A. C. O'Sullivan, M. M. Pechet; *J. Org. Chem.*, **1991**, *56*, 6697.

91JOC6729 M. E. Jung, Y. M. Choi; *J. Org. Chem.*, **1991**; *56*, 6729.

91JOC6751 J. Barluenga, F. J. González, R. P. Carlón; *J. Org. Chem.*, **1991**, *56*, 6751.

91JOC6968 A. K. Bose, M. S. Manhas, M. Ghosh, M. Shah, V. S. Raju, S. S. Bari, S. N. Newaz, B. K. Banik, A. G. Chaudhary, K. J. Barakat; *J. Org. Chem.*, **1991**, *56*, 6968.

91JOC7197 D. L. Comins, H. Hong, J. M. Salvador; *J. Org. Chem.*, **1991**, *56*, 7197.

91MI20 J. A. Lepoivre; *Janssen Chim. Acta*, **1991**, *9*, 20.

91S975 Y. Fukuda, K. Utimoto; *Synthesis*, **1991**, 975.

91SL677 G. W. Gokel, J. C. Medina; C. Li, *SYNLETT*, **1991**, 677.

91SL695 S. Torii, L. H. Xu, H. Okumoto; *SYNLETT*, **1991**, 695.

91SL816 N. J. Sisti, F. W. Fowler, D. S. Grierson; *SYNLETT*, **1991**, 816.

91T5111 T. Sakamoto, Y. Kondo, D. Uchiyama, H. Yamanaka; *Tetrahedron*, **1991**, *47*, 5111.

91T6127 D. H. R. Barton, C. Chen, G. M. Wall; *Tetrahedron*, **1991**, *47*, 6127.

91T6851 R. P. Thummel; *Tetrahedron*, **1991**, *47*, 6851.

91TL237 S. Torii, H. Okumoto, L. H. Xu; *Tetrahedron Lett.*, **1991**, *32*, 237.

91TL2995 D. L. Comins, M. M. Badawi; *Tetrahedron Lett.*, **1991**, *32*, 2995.

91TL3199 R. S. Kusurkar, D. K. Bhosale; *Tetrahedron Lett.*, **1991**, *32*, 3199.

91TL3309 D. H. R. Barton, C.-Y. Chern, J. C. Jaszberenyi; *Tetrahedron Lett.*, **1991**, *32*, 3309.

91TL3321 D. H. R. Barton, J. C. Jaszberenyi, E. A. Theodorakis; *Tetrahedron Lett.*, **1991**, *32*, 3321.

91TL5235 R. J. Olsen; *Tetrahedron Lett.*, **1991**, *32*, 5235.

91TL5501 A. I. Meyers, M. A. Gonzales, V. Struzka, A. Akahane, J. Guiles, J. S. Warmus; *Tetrahedron Lett.*, **1991**, *32*, 5501.

91TL5505 A. I. Meyers, J. Guiles, J. S. Warmus, M. A. Gonzalez; *Tetrahedron Lett.*, **1991**, *32*, 5505.

91TL5509 A. I. Meyers, J. S. Warmus, M. A. Gonzales, J. Guiles, A. Akahane; *Tetrahedron Lett.*, **1991**, *32*, 5509.

91TL6895 T. Kumagai, S. Saito, T. Ehara; *Tetrahedron Lett.*, **1991**, *32*, 6895.

CHAPTER 6.2

Six-Membered Ring Systems: Diazines & Benzo Derivatives

T.J. KRESS & D.L. VARIE
Eli Lilly & Co, Indianapolis, IN, USA

6.2.1 INTRODUCTION

This chapter contains highlights of the chemistry of pyridazine, pyrimidine, pyrazine, and their benzo derivatives from the 1991 literature and articles that appeared in late 1990. Fused-ring derivatives are not covered. Each ring system will be considered separately with regard to new methods of ring construction, introduction and transformation of functional groups, and compounds containing interesting biological activity.

Advances in the chemistry of pyridazine have been reviewed up through mid 1988 [90AHC385]. Two reports, utilizing molecular orbital calculations to predict prototropic side chain tautomerism of a large number of amino- and hydroxypyrimidines [91AHC329] and a variety of nucleic acids and their N-methyl analogs [91JA1561] have appeared. In almost all cases, the same tautomer is predicted as most stable regardless of the calculation method used.

6.2.2 STRUCTURE

The solid-state structure of pyridazine at 100°K was determined by X-ray crystallography. Comparison of this structure with the structure of pyridazine in the liquid and vapor phases shows a slight shortening of the C-N and C(4)-C(5) bonds relative to the other ring bonds [91AX1933]. A ^{13}C NMR study of seventeen 3,6-disubstituted pyridazines was reported. The data include chemical shifts and ^{13}C-^{1}H coupling constants [91CJC972].

6.2.3 PYRIDAZINE, PHTHALAZINE

6.2.3.1 Synthesis

A number of pyridazine derivatives were prepared *via* 1, 2, 4, 5-tetrazine cycloadditions with dienophiles. The 4-stannylpyridazines **3** were prepared in 70-85% yield from tetrazines **la-c** and tributylstannyl acetylene (**2a**). Yields were modest with substituted acetylenes **2b, c** as dienophiles [91H1387]. Air oxidation

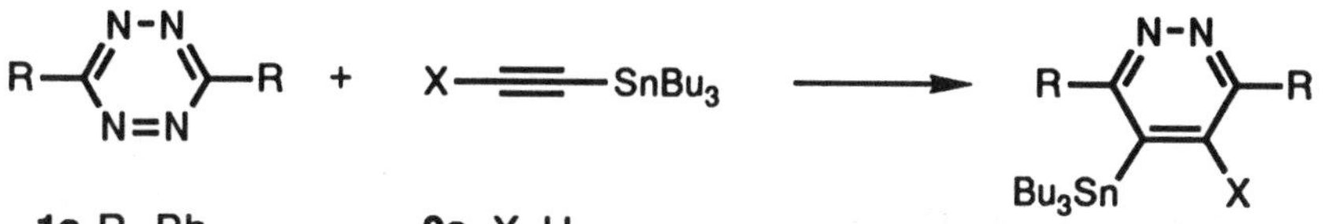

1a R=Ph
b R=Me
c R=CO_2Me

2a X=H
b X=Ph
c X=Bu

3

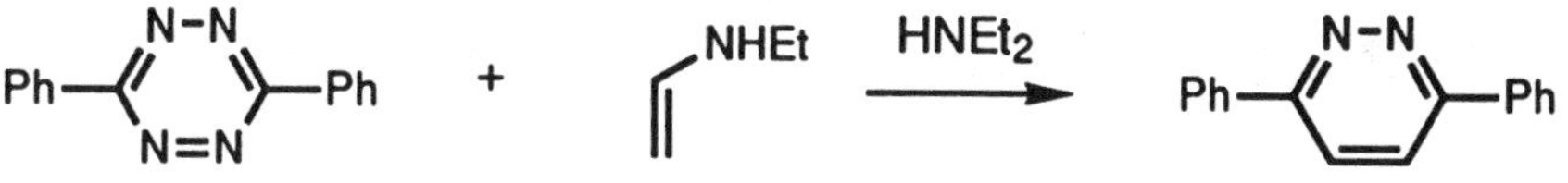

1a 4 5

of diethylamine provided enamine **4** which reacts with 3,6-diphenyltetrazine **1a** to give 3,6-diphenylpyridazine (**5**) in 82% yield [90KGS1545]. Reaction of tetrazine **1d** with cyclooctatetraene gave a 95% yield of dihydropyridazine **6** which at 111° C rearranged to **7** or pyridazine **8** in the presence of oxygen [91TL5949]. An extensive

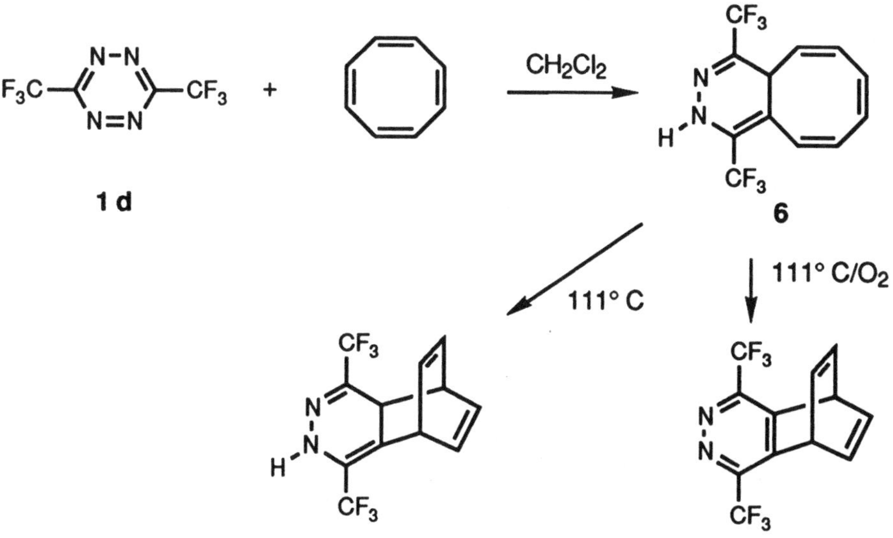

1d 6 7 8

kinetic study measuring the rates of cycloaddition of tetrazines **1c** and **1d** with a variety of simple olefins, enol ethers, enamines and other dienophiles was reported [90TL6851, 90TL6855]. Reaction of simple enamines **10** (R= alkyl, cycloalkyl) with 1,4-diarylpyridazino [4,5-d]pyridazines **9** is regioselective for the unsubstituted ring and provided phthalazines **11** in good yields [91T3959]. Phthalazine (**14**) was

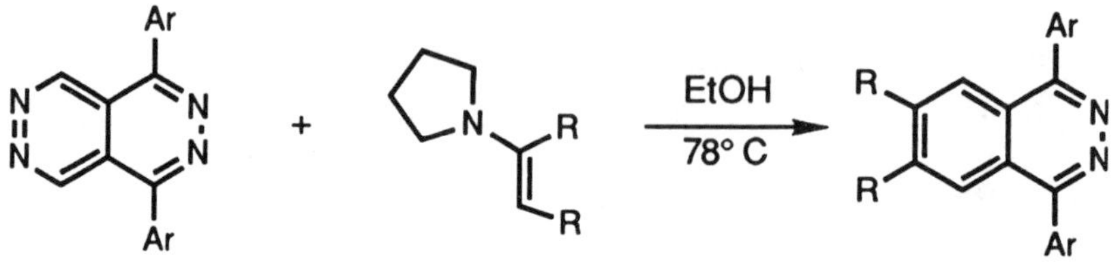

9 10 11

synthesized in 90% yield by the reaction of hydrazine with bromoalkylidene malonate **12** followed by DDQ oxidation of the intermediate dihydro compound [91MI183].

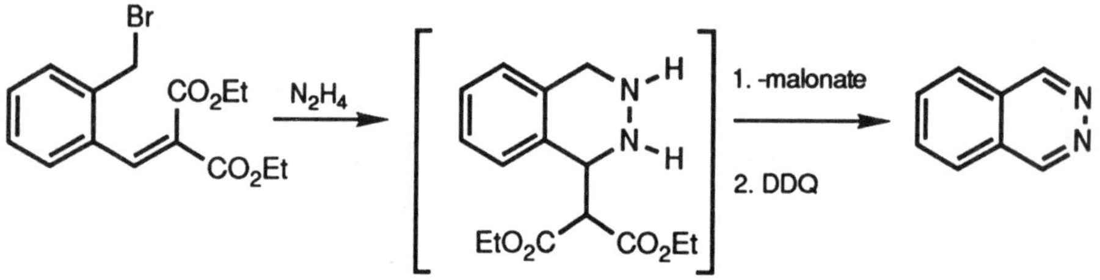

12 13 14

Pyridazinones **16** were synthesized from ketohydrazone **15** using a phosphonate ester [91SC1021] and Peterson olefination chemistry [91SC1935]. In most cases the latter conditions gave better yields of **16**.

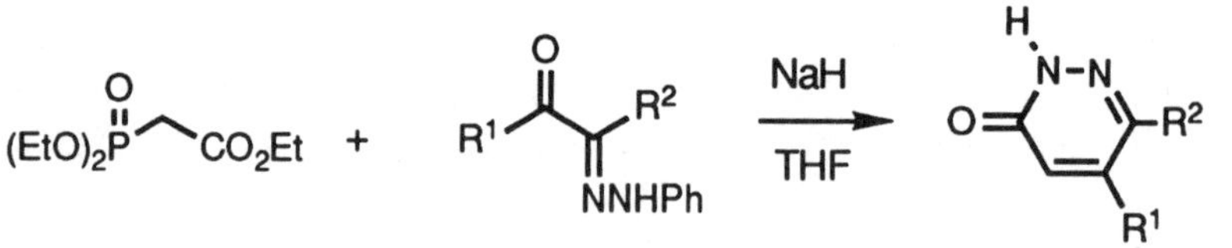

15 16

6.2.3.2 Reactions

The nucleophilic addition of a variety of alkyl radicals (Et, t-Bu, i-Pr, cyclohexyl) to 3-chloropyridazines **17** occurred regioselectively (>90:10 **18:19**) at the 4-position in all cases [91JHC583]. Treatment of 3,6-dichloropyridazine (**20**) with azide,

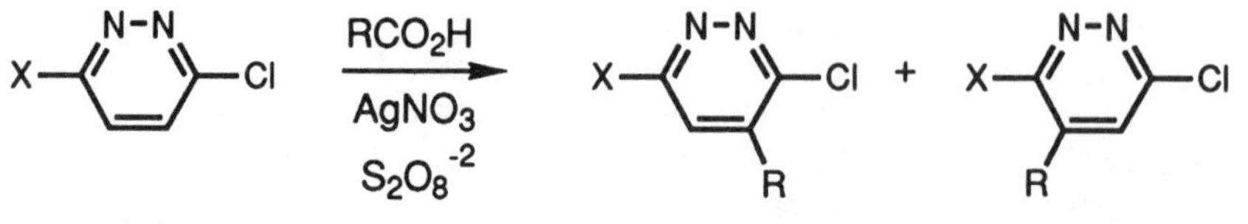

17 X=Me,Ph 18 19

20 21 22

followed by reaction with triphenylphosphine and hydrolysis, provided a new route to 3,6-diaminopyridazine (**22**) [91LA1225]. Cyanopyridazinone **23** reacts with phenylmagnesium chloride to give primarily (70% yield) 4-phenylpyridazinone **25** presumably *via* expulsion of cyanide from intermediate **26**. The expected reaction product, ketone **25**, was also obtained in 20% yield [91T8573]. The cycloaddition

23 24 25 26

reaction of pyridazinones **27** (R=alkyl, Ph, Cl, OMe) with 2-diazopropane was found to be highly regioselective. In all substrates examined, the pyrazolo isomer **28** was the major (>95%) or exclusive product [91JHC417]. (The initially formed dihydro compounds were isolated by precipitation if the reaction was performed in diethyl ether.) Substitution of the 4-position of pyridazinone **30** with a leaving group (Cl, SOPh, SPh) gave a single regioisomer (**31**) when reacted with 2-diazopropane. Placing the leaving group at the 5-position

27 **28** (major) **29** (minor)

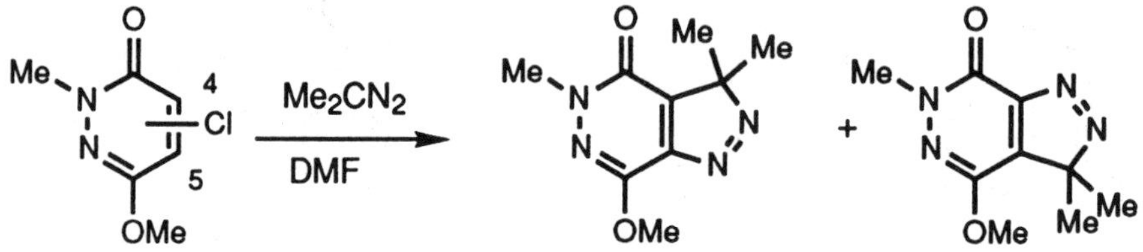

Me_2CN_2, DMF

30a 4-Cl
30b 5-Cl

31 (from **30a**) **32** (from **30b**)

gave only regioisomer **32** [91JHC369]. Pyridazine N-oxides **33** reacted with pyridynes (*e.g.*, **34**) to give modest (15-30%) yields of regioisomeric pyridooxepins **35** and **36** [90H1937]. The cycloaddition of phthalazines **37** with enamines **38** gives trisubstituted naphthalenes **39**. The rate and yield of the reaction was observed to increase with the electron withdrawing ability of the substituent on the phthalazine ring [90CPB3268].

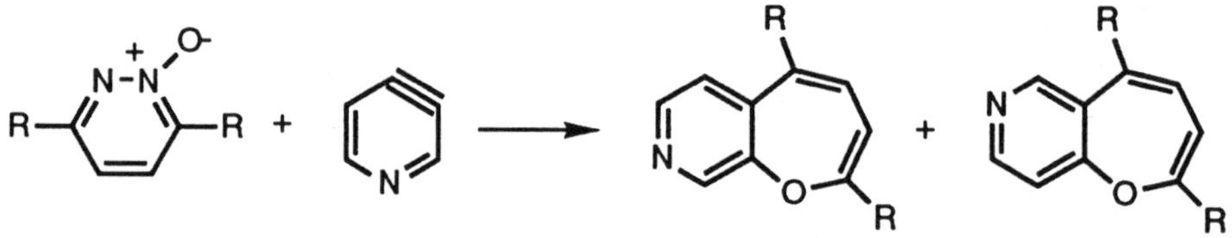

33 (R=H, Ph, Me) **34** **35** **36**

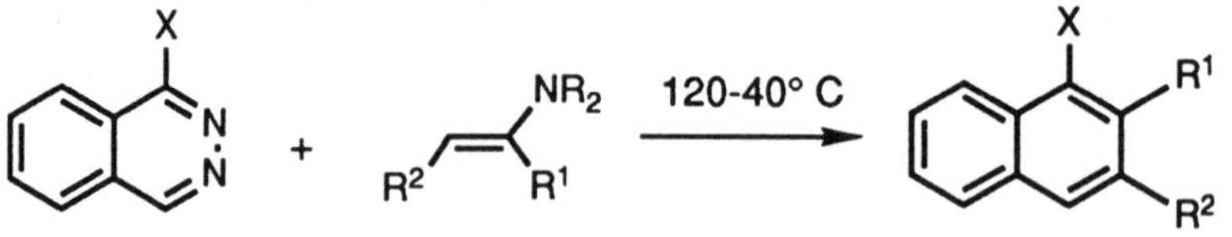

120-40° C

37 **38** **39**
X= CN, H, SO_2Me, Cl

6.2.3.3 Uses

Allenic pyridazines **40a,b** were key intermediates in the synthesis of the indole marine natural products cis- and trans-Trikentrin A (**41a, b**) [91JA4230]. A series of twenty pyridazinone ketones such as **42** were prepared and found to have modest *in vitro* anti-tumor activity against several human tumor cell-lines [91FES873]. Pyridazinone **43** was shown to have twice the *in vivo* anxiolytic activity of diazepam with reduced diazepam-type side effects [91CPB2556].

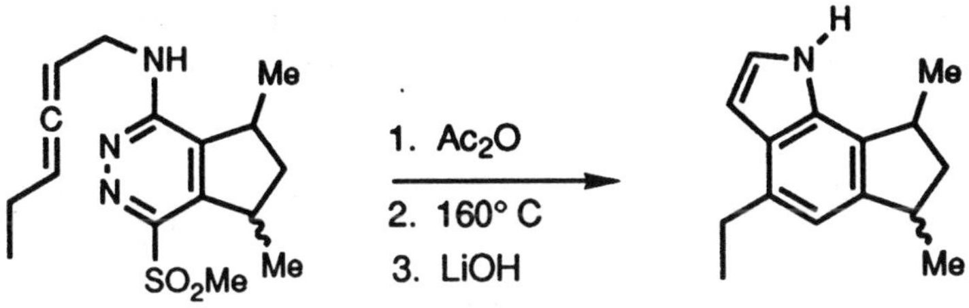

1. Ac_2O
2. 160° C
3. LiOH

40a,b **41a,b**

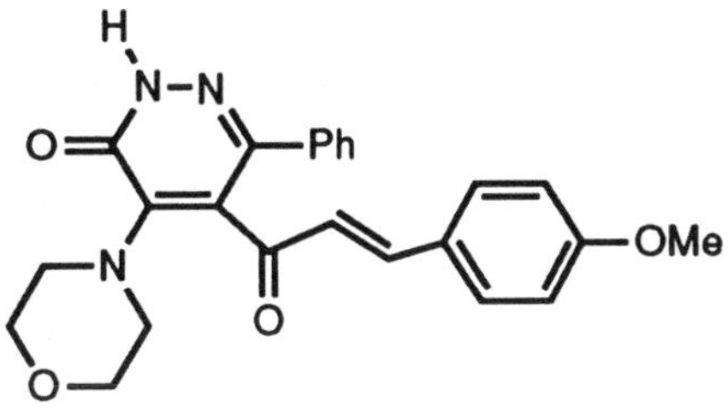

42 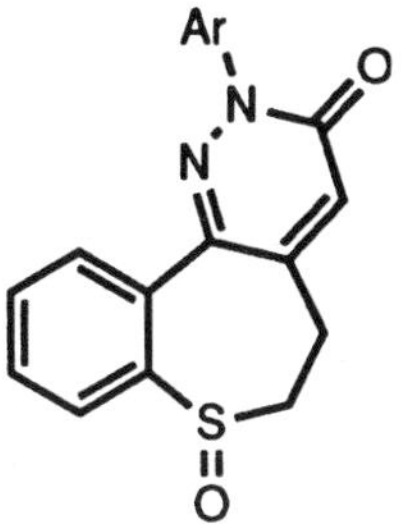 **43**

6.2.4 PYRIMIDINE, QUINAZOLINE

6.2.4.1 Synthesis

Two syntheses have exploited the pyrimidine ring in the preparation of acyclic nitrogen compounds. A stereoselective synthesis of 1,3-diamines has been achieved by reduction of 1-butyl-1,2-dihydropyrimidine **45**. Intermediate **45**, prepared by reaction of an aldehyde (R^2= alkyl, aryl) with azabutadiene **44** (R^1= alkyl, aryl), was reduced with an excess of $NaBH_4$ to produce the N,N'-1,3-diamine **46** as a single diastereomer in 75-80% yield [91MI821]. The synthesis of (R)-β-tyrosine, a

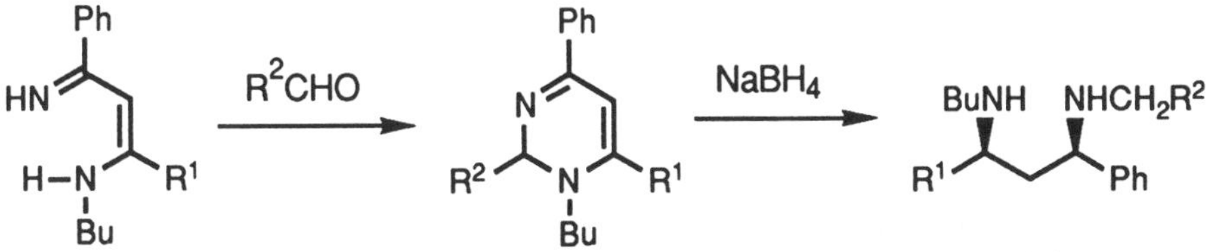

R^2CHO $NaBH_4$

44 **45** **46**

key intermediate in the synthesis of the marine natural product Jasplakinolide, was accomplished utilizing the dihydropyrimidine **47**. Enantiomerically pure **47** was prepared in three steps starting from (S)-asparagine and pivalic acid. Stereospecific palladium catalyzed coupling of **47** with iodoanisole gave **48** as a crystalline product. Reduction of **48** with $NaBH_4$ followed by hydrolysis produced the desired β–amino acid **49** [91JOC1355, 5196]. A straightforward and efficient route to

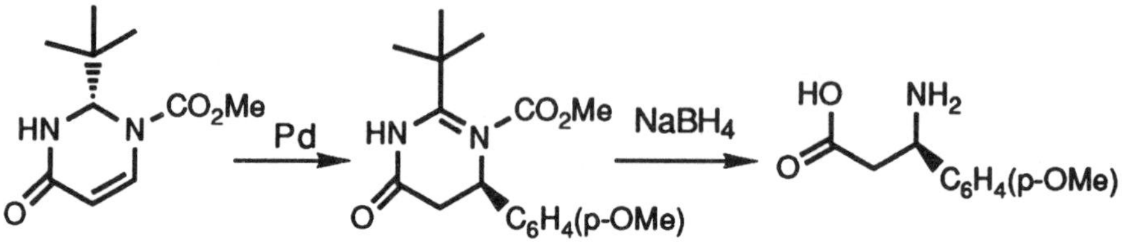

47 **48** **49**

2-substituted 5-formylpyrimidines (**52**) is now available by reaction of 2-iminovinamidinium perchlorate (**50**) with an amidine. For example, amidines (**51**, R = Me, Ph, NH_2, OMe, SMe) afforded **52** in 84-99% yield [91JHC1281]. Based

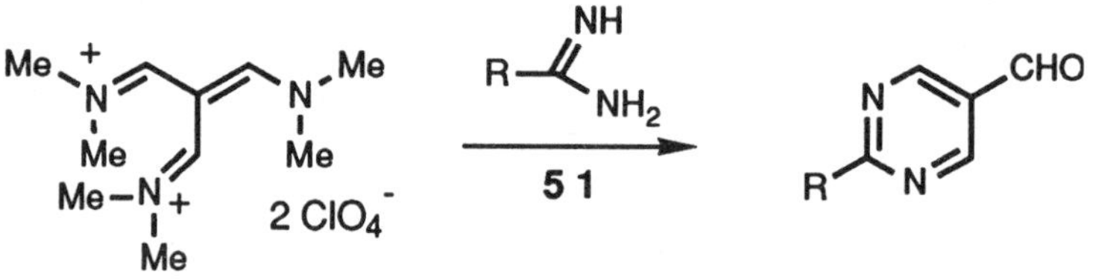

50 **52**

on the known conversion of oxazolines into cytosines, the carbocyclic analog of cytidine (**55**) was constructed in a one pot procedure from the unprotected tetrol **53**.

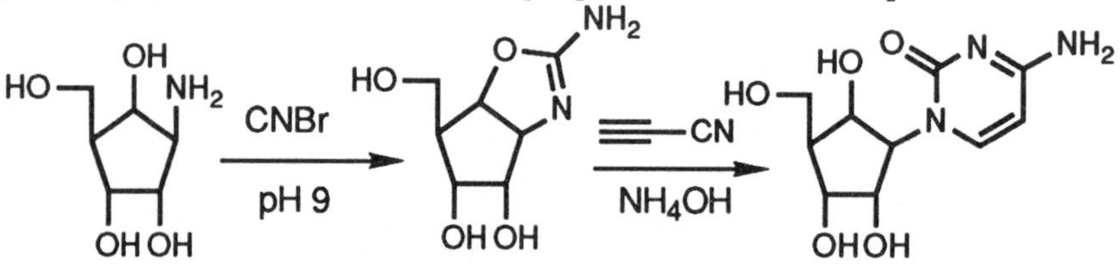

53 **54** **55**

Reaction of **53** with cyanogen bromide gave the oxazoline **54** which on treatment with cyanoacetylene afforded **55** in 40% overall yield [91TL6281]. The tetrasubstituted pyrimidine **58** (X = OEt, pyrrolidino, piperidino) was obtained in

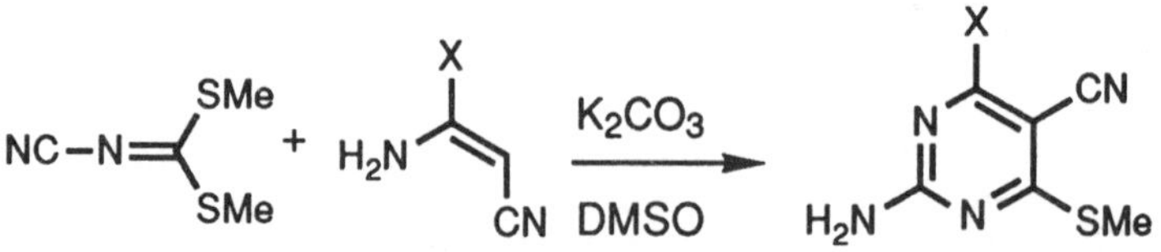

56 **57** **58**

good yield by base induced condensation of propenenitrile **57** with dithiocyanamide **56** [91S529]. An unusual cascade of ring transformations occurred on

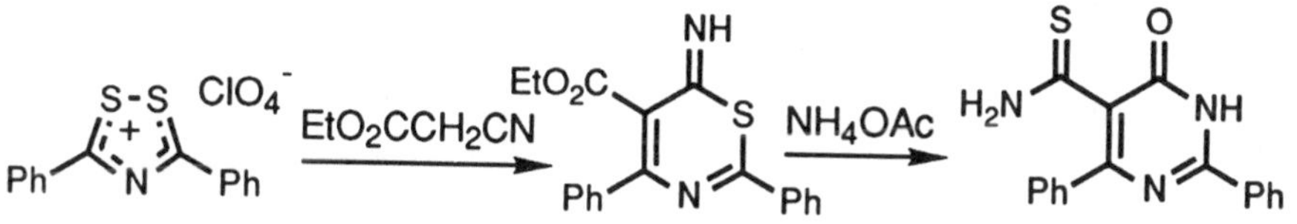

59 60 61

stepwise treatment of the dithiazolium salt **59** with ethyl cyanoacetate followed by ammonium acetate and ultimately resulted in the thioamide **61**. The authors have suggested the thiazoline **60** as an intermediate [91LA345]. α–Chloroketones usually afford imidazoles on reaction with amidines. However, the trifluoromethyl ketone **62**

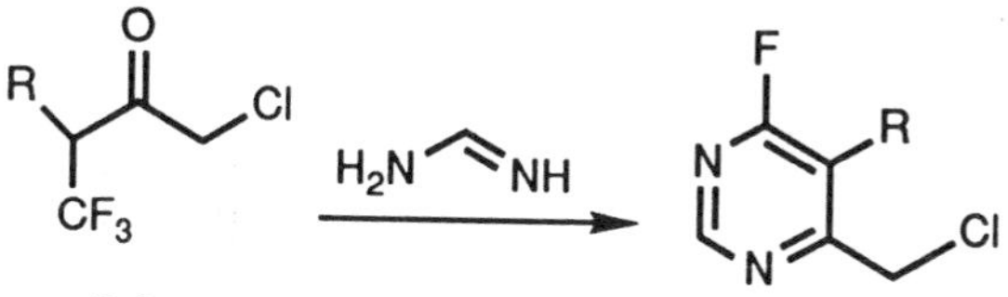

62 63

(R = Ph, cyclopentyl) undergoes a novel cyclization in which two fluorine atoms are displaced giving the 4,5,6-trisubstituted pyrimidine **63** [91TL2467]. The methanaminium salt **64** has been found to be an efficient reagent for *in situ* generation of ketene **66** from carboxylic acids. Ketene **66a** underwent cycloaddition with 1,3-diaza-1,3-butadiene **67** to afford pyrimidone **68**. Similar reaction of **66b** did not give a pyrimidone but instead gave the azetidinone **69** [91S1026].

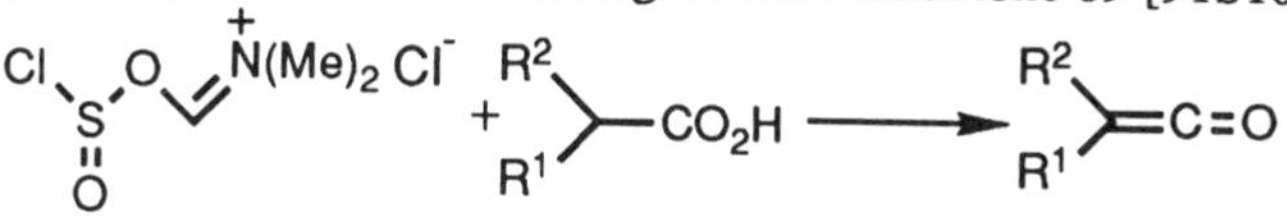

64 65 **66a**, R1=H, R2=Ph
66b, R1=R2=Ph

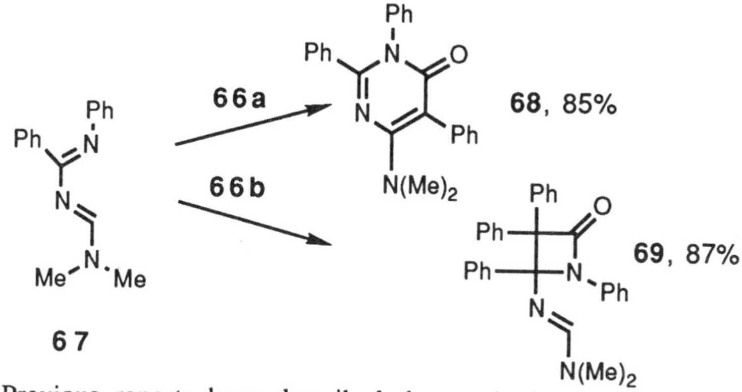

67 68, 85% 69, 87%

Previous reports have described the synthesis of quinazolines from 2-fluorobenzonitriles and guanidine. This approach has been extended to include 7-substituted 2,4-diaminoquinazolines [91JHC1357]. A general and efficient pyrimidine annulation process has been developed from o-aminonitriles and esters. The procedure was successfully applied to the fusion of the pyrimidine ring to a variety of

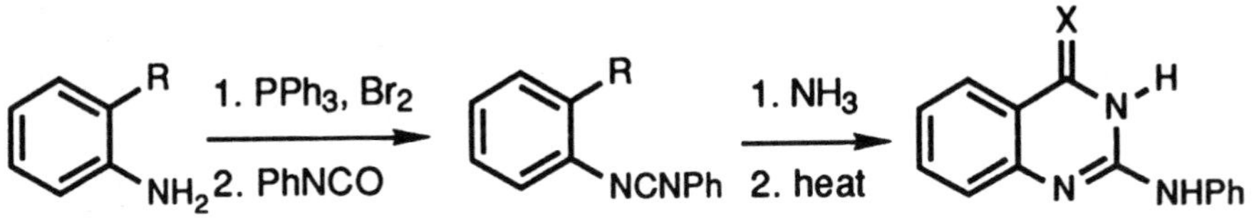

70 **71** **72 (X=O, NH)**

heterocycles and is illustrated here for the synthesis of quinazoline **72** [91JHC1357]. Carbodiimide **71**, prepared in two steps from **70** (R=CN, CO_2Et), upon treatment with ammonia affords a guanidino intermediate which undergoes intramolecular cyclization generating **72** in good yield. Irradiation of 5-phenyl-1,2,4-oxadiazole **73**,

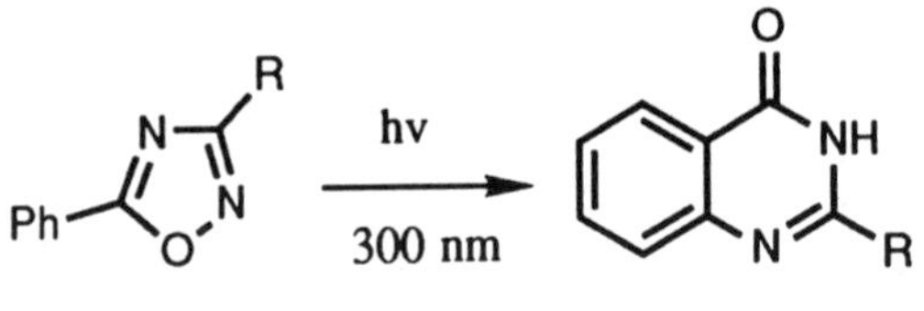

73 **74**

(R=Me, Ph) resulted in cleavage of the ring O-N bond and ring closure to quinazolinone **74** (R=Me, Ph) in 40% yield [91JCS(P2)187].

6.2.4.2 Reactions

Palladium catalyzed coupling reactions have proven to be the method of choice for the synthesis of 5-substituted pyrimidines. Silyl protected 5-iodocytosine **75** served as a convenient starting material for coupling with tin reagents affording **76**

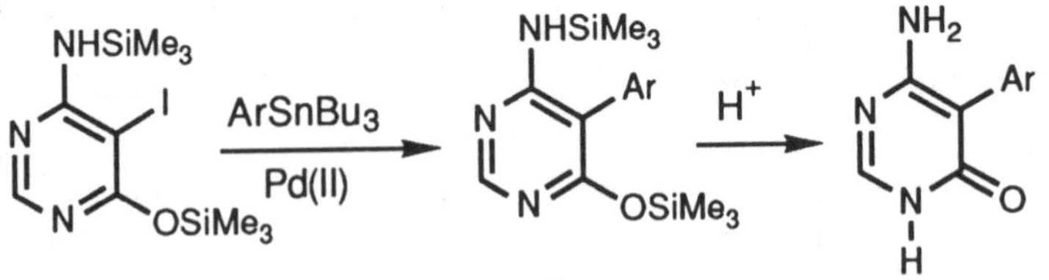

75 **76** **77**

which after deprotection furnished a large number of cytosines (**77**, Ar=2,3-furyl, 2,3,4-pyridyl, 2,3-thienyl) [91JHC1613, 1629]. Phenyl substituted boronic acids **79**

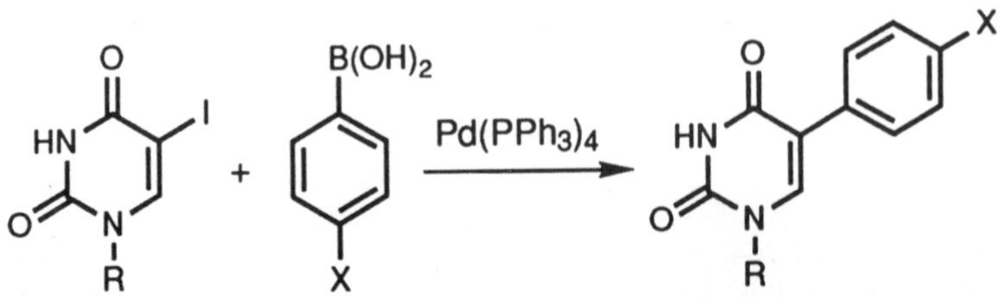

78 **79** **80**

bearing electron releasing groups couple with **78** (R=3',5'-cyclic disiloxanyl-2'-deoxyuridine) giving uracils **80** (X=H, OMe, NMe_2) in 37-55% yield [91MI763]. 2-Halopyrimidines **81** (X=Cl, Br) dimerize in the presence of nickel tetrakistriphenylphosphine to the corresponding 2,2'-bipyrimidines **82** in 60% yield [91SC901].

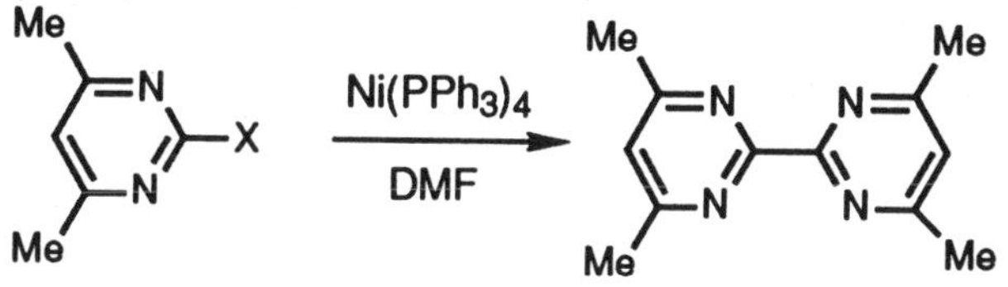

Me
N
X
Ni(PPh3)4
DMF
Me

81 **82**

The efficiency of *ortho*-directed lithiation of halo- (**83a**, R^2=R^4=Cl) and methoxypyrimidines (**83b**, R^2=R^4=OMe) with either LDA or LTMP was demonstrated by trapping intermediate **84** with numerous electrophiles.

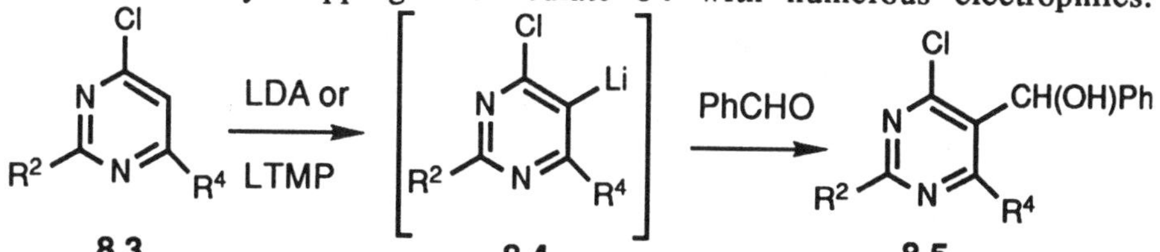

Cl
LDA or
LTMP
Li
PhCHO
CH(OH)Ph
R2
R4

83 **84** **85**

Representative of this series is formation of 5-benzyl alcohol **85** in 85-90% yield on reaction of benzaldehyde with **84** [91JOC4793, 91JHC283]. Dropwise addition of 2-furyllithium **87** to **86** did not result in **89** as expected but instead gave

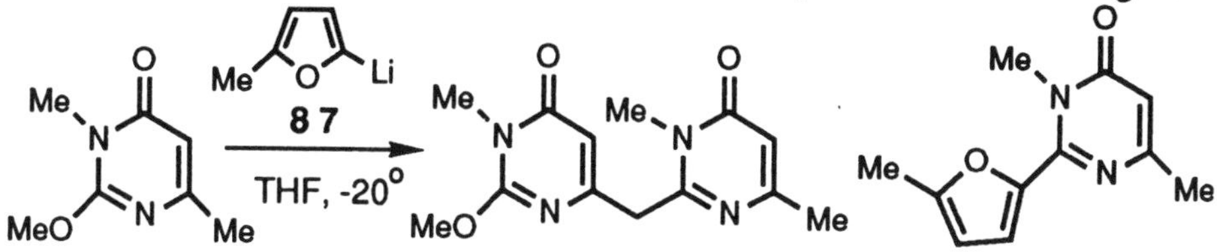

Me
O
Li
87
THF, -20°
MeO

86 **88** **89**

88 as the sole product. Presumably, **88** is formed by transmetalation between **87** and the C-6 methyl group of **86** and subsequent displacement of methoxide from another molecule of **86** [91H1537]. The regiochemical outcome of the bromination

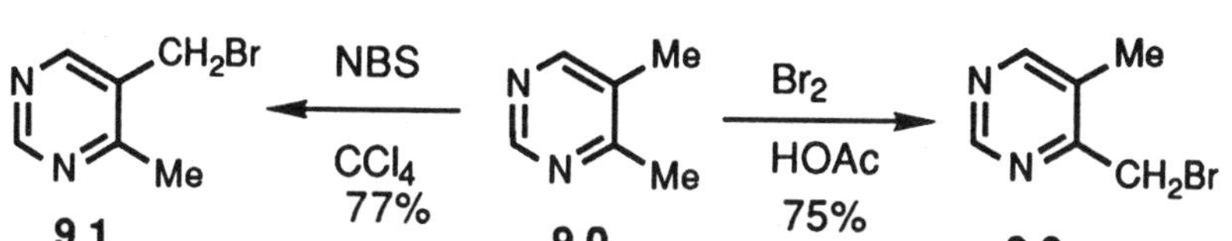

CH2Br
NBS
CCl4
77%
Me
Br2
HOAc
75%

91 **90** **92**

of 4,5-dimethylpyrimidine (**90**) could be controlled by varying the reaction conditions. Bromination of **90** under ionic conditions favored **92** while under radical conditions substitution predominated at the C-5 methyl group affording **91** [91JOC5610]. Chiral barbiturates such as **94** have been obtained in 50% yield (92%ee) by lipase-catalyzed hydrolysis of **93**. Alkylation of **94** followed by hydrolysis provided chiral (R)-(-)methobarbital **95** in 90% yield (92%ee)[91TL6763].

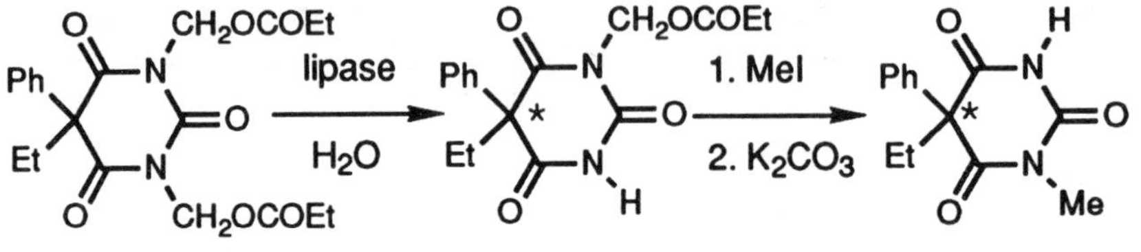

Alkylation of **96** with methyl iodide occurs regioselectively at N-1. Both ^{13}C NMR and X-ray studies have shown that the polymethine structure **97b** is preferred [91JPR561]. In general, alkylation of silylated uracils takes place at N-1. However

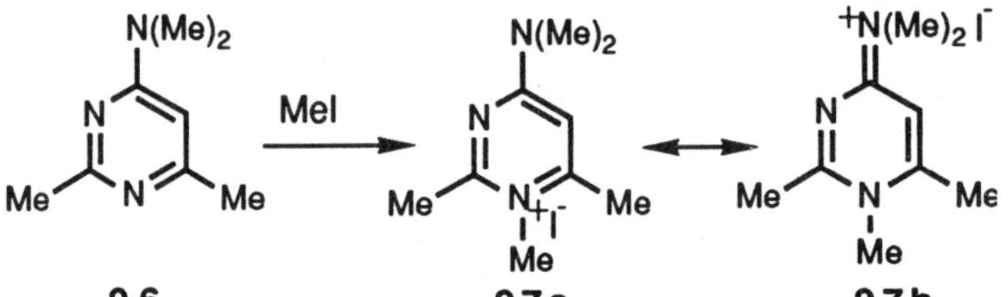

6-aminouracil **98** alkylates specifically at N-3 giving a regioselective route to N-3 substituted 6-aminouracil **99** [91TL6539]. Dimethyluracil **100** undergoes

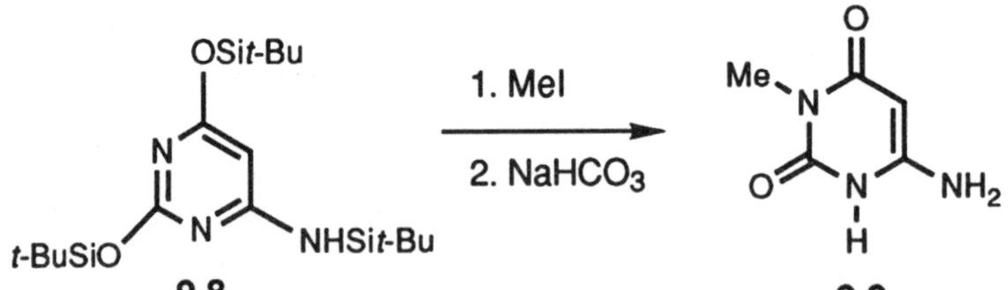

regioselective electrophilic sulfenylation and selenation at the 5-position in the presence of silver reagents. Both **101** (X=S, Se) were obtained on reaction of **100** with phenylsulfenyl chloride or phenylselenyl chloride in 89 and 97% yields, respectively. [91TL2401].

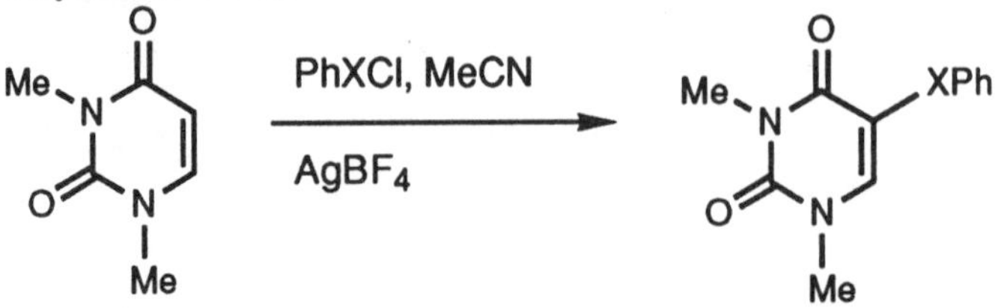

6.2.4.2 Uses

Analogs of 2-pyrimidone **102** [91ACS731] and quinazoline continue to be evaluated as inhibitors of cancer cell growth [91JMC978, 2209, 91JHC1981]. A variety of 2,4-dioxoquinazolineacetic acids (e.g. **103**) were synthesized and evaluated as reductase inhibitors [91JMC1492]. A new class of boronated pyrimidines (**104**) are of interest as agents for boron neutron capture therapy in treatment of cancer patients [91JOC2391].

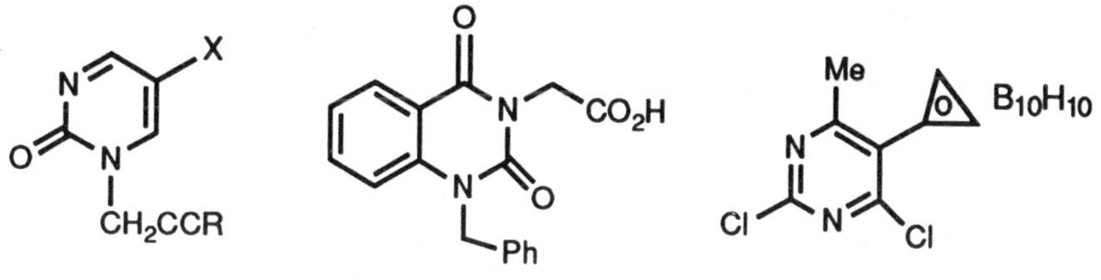

102 **103** **104**

From a series of 5-acetylpyrimidines, **105** showed significant *in vitro* activity against Salmonella [91MI26]. 1,4-Dihydropyrimidines **106** continue to be of interest in the treatment of hypertension [91JMC809]. A number of quinazolinones (**107**), based on an asperlicin substructure have been shown to be cholecystokinin-B receptor antagonists [91JMC1505].

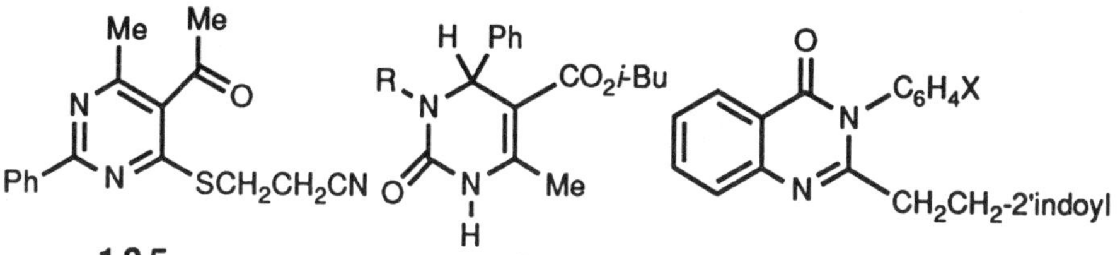

105 **106** **107**

6.2.4 PYRAZINE

6.2.4.1 Synthesis

A new and apparently general approach to alkylpyrazines is now available. Typical of this methodology is the condensation of the oximido carbonyl compound **108** with allylamine followed by isomerism and acylation affording **109**. Ring closure was performed by pyrolysis of a toluene solution of **109** at 300°C affording

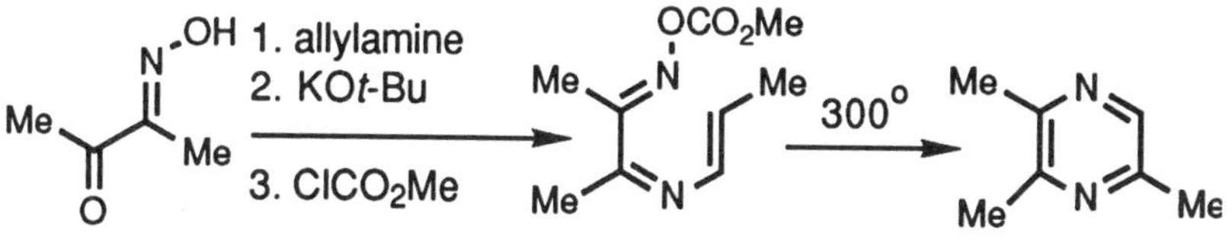

1. allylamine
2. KO*t*-Bu
3. $ClCO_2Me$
300°

108 **109** **110**

110 in an overall yield of 60% [91JOC2605]. The chiral pyrazine **113**, a key intermediate in the total synthesis of (+) Septorine (**114**), was prepared in four steps from L-isoleucine and **111**. The remainder of the sequence required eleven additional steps [91H923]. A number of trisubstituted alkyl and aryl pyrazinones (**118**) were

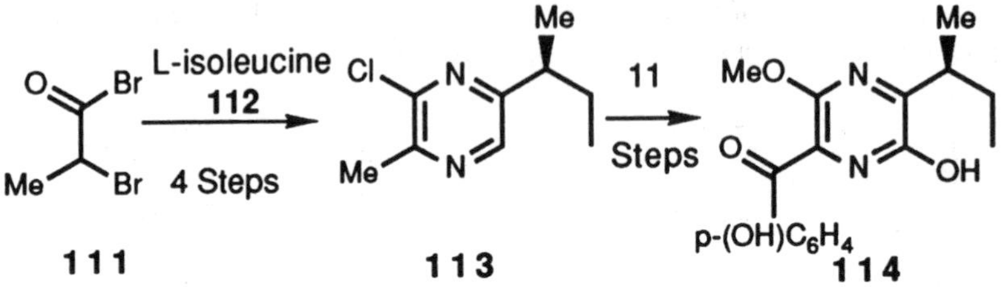

O Br Me Br — L-isoleucine **112**, 4 Steps → Cl, Me, N, N, Me — 11 Steps → MeO, N, N, OH, Me, O, p-(OH)C_6H_4

111 **113** **114**

prepared from aminoisoxazolones. For example, treatment of the amine **115** with the α-oxo acid chloride **116** gave **117** which after hydrogenolysis of the N-O bond

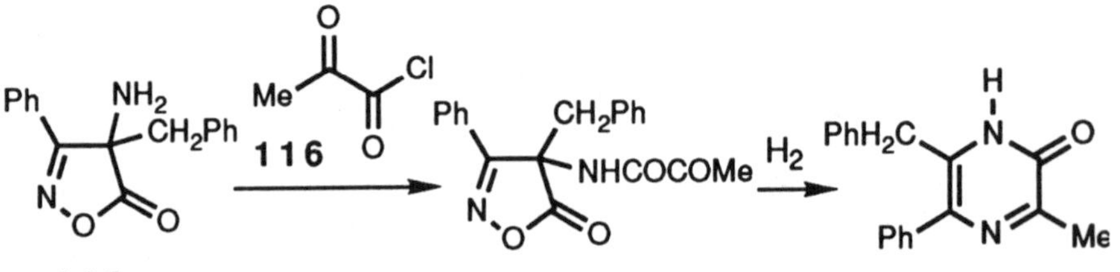

Ph, NH_2, CH_2Ph, N, O, O — Me, O, Cl, O (**116**) → Ph, CH_2Ph, NHCOCOMe, N, O, O — H_2 → PhH_2C, H, N, O, Ph, N, Me

115 **117** **118**

undergoes decarboxylation and recyclization producing **118** [91S861]. Deoxyfructosazine **120** was formed in 45% yield by dimerization of D-glucosamine **119** in the presence of L-cysteine as a reducing agent [91CPB792].

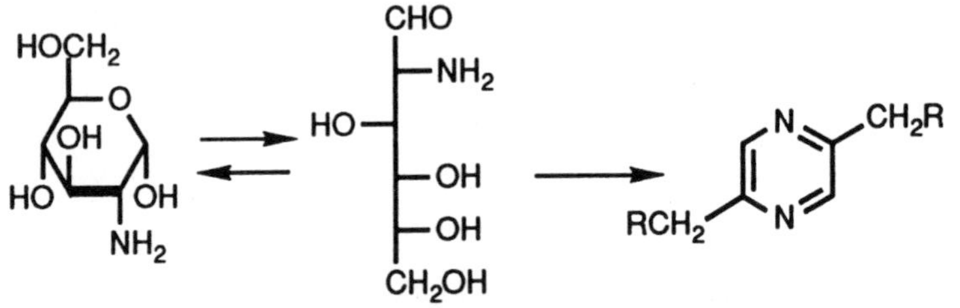

$HOCH_2$, O, OH, HO, OH, NH_2 ⇌ CHO, NH_2, HO, OH, OH, CH_2OH → N, CH_2R, RCH_2, N

119a **119b** **120** R=CH(OH)CH(OH)CH_2OH

6.2.4.2 Reactions

Symmetrically substituted pyrazines (**121**, R=Cl, OMe, SMe) undergo *ortho*-metalation with LTMP and afford good yields of alcohols **122** on quenching with an aldehyde (R^3=alkyl, aryl) [91JOM301]. The phenyl ketone **123** was prepared

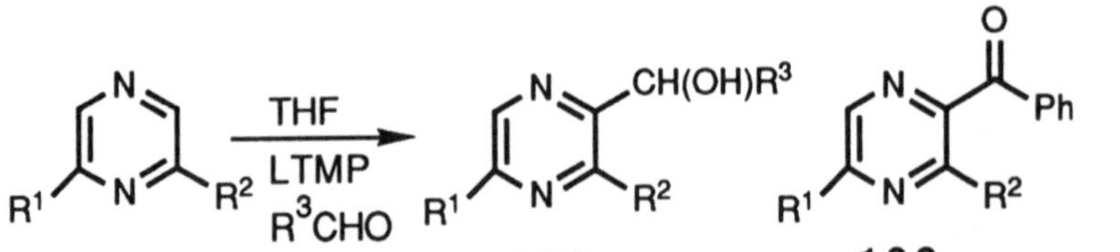

N, R^1, N, R^2 — THF, LTMP, R^3CHO → N, CH(OH)R^3, R^1, N, R^2 ; N, O, Ph, R^1, N, R^2

121 **122** **123**

directly by treatment of **121** (R^1=H, R^2=Cl, OMe, SMe) with LTMP followed by N-methoxy-N'-methylbenzamide [91JHC765]. 2,5-Dialkylpyrazine 1-oxide **124** (R=*sec*-butyl) gave an 82% yield of 2-formylpyrazine **125** on reaction with methyl formate in the presence of LTMP. The product **125** was readily deoxygenated to **126** with phosphorus tribromide [91H735]. One of the problems with homolytic

1. LTMP
2. HCO_2Me
PBr_3

124 **125** **126**

acylations has been the inability to control multiple substitution. Acyl radicals generated by silver catalyzed decarboxylation of α-ketoacids have been determined to be more effective at monoacylation when multiple sites were available [91JOC2866]. Yields in the radical acylation of pryazine **127** were markedly improved by

RCHO
t-BuOOH
$FeSO_4$

127 **128**

performing the reaction in an ultrasound bath. Ketones **128** (R=Me, Ph) were obtained in 74-85% yield in 30 minutes [91S581]. Cyanation of pyrazine N-oxides **129** were determined to be very sensitive to the electronic nature of the substituent

TMSiCN
Et_3N
MeCN

129 **130**

at the 3-position. Electron donating groups (X=NH_2, OMe, Ph, Me) gave good yields of **130** with high regioselectivity for the 2-position. Electron withdrawing groups suppressed the conversion [91JCS(P1)2877].

Alkyl phenyl sulfoxides normally undergo the Pummerer rearrangement when treated with carboxylic anhydrides to give α-acyloxyalkyl phenyl sulfides. Application of this procedure using trifluoroacetic anhydride and the secondary alkyl pyrazinyl sulfoxide **131** did not yield the desired α-acyloxy product **132** but

$(CF_3CO)_2O$

131 **132**

produced a mixture of olefin **134** (37%) and hydroxypyrazine **135** (68%) [91H937]. A proposed mechanism for the formation of either **134** or **135** was based on the instability of acyloxypyrazine **133** towards base.

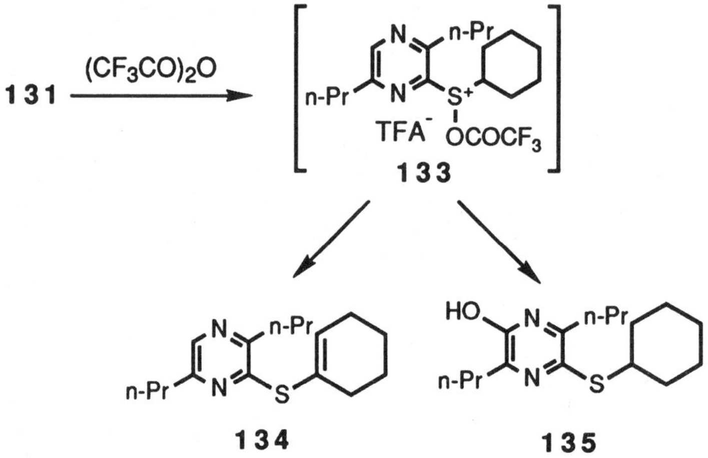

131 —$(CF_3CO)_2O$→ [133: TFA⁻, $OCOCF_3$] → 134 + 135

6.2.4.3 Uses

The pyrazine analog **136** of nicotine has been synthesized in racemic form and its binding properties in the rat brain were found to be similar to nicotine at two of three receptor binding sites [91JHC1147]. A common structural feature of the spicy

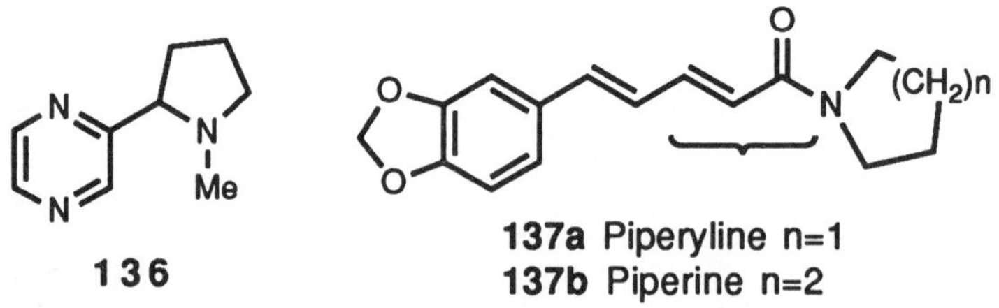

136

137a Piperyline n=1
137b Piperine n=2

component of piperaceous plants (**137**) is the (E)-α,β–unsaturated amide function. This class of olefins has been synthesized taking advantage of the highly stereoselective β–elimination of the pyrazinylsulfinyl group. In the general process, pyrazine sulfoxide **138**, constructed in two steps from an α–chloroamide and 2-pyrazinethiol, was readily alkylated giving **139** which undergoes β–elimination affording exclusive formation of the (E)-olefin **140** [91H965].

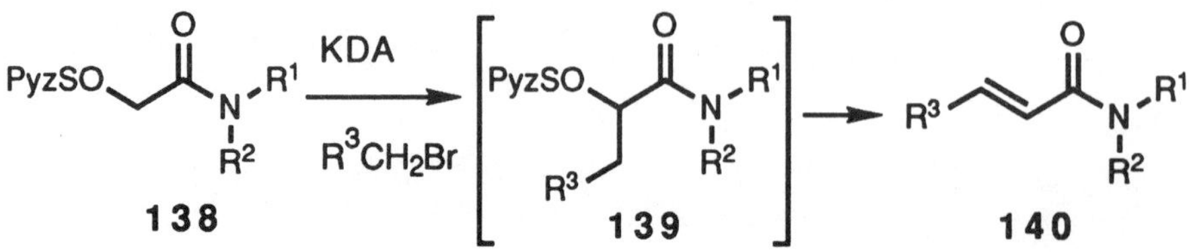

138 —KDA, R^3CH_2Br→ [139] → 140

REFERENCES

90AHC385	M. Tisler and B. Stanovnik, *Adv. in Heterocycl. Chem.*, 1990, **49**, 385.
90CPB3268	E. Oishi, N. Taido, K. Iwamoto, A. Miyashita and T. Higashino, *Chem. Pharm. Bull.*, 1990, **38**, 3268.
90H1437	J. Kurita, N. Kakusawa, S. Yasuike and T. Tsuchiya, *Heterocycles*, 1990, **31**, 1437.
90KGS1545	S. V. Shorshnev, S. E. Esipov, V. V. Kuzmenko, A. V. Gulevskaya, A. F. Pozharskii, A. I. Chernyshev, G. G. Aleksandrov and V. N. Doronkin, *Khim. Geterotsikl. Soedin.*, 1990, 1545.
90TL6851	F. Thalhammer, U. Wallfahrer and J. Sauer, *Tetrahedron Lett.*, 1990, **30**, 6851.
90TL6855	A. Meier and J. Sauer, *Tetrahedron Lett.*, 1990, **30**, 6855.
91ACS731	T. Benneche, G. Keilen, J. Solberg, R. Oftebro and K. Undheim, *Acta Chem. Scand.*, 1991, **45**, 731.
91AX1933	A. J. Blake and D. W. H. Rankin, *Acta Cryst.*, 1991, **C47**, 1933.
91CPB792	K. Sumoto, M. Irie, N. Mibu, S. Miyano, Y. Nakashima, K. Watanabe and T. Yamaguchi, *Chem. Pharm. Bull.*, 1991, **39**, 792.
91CPB2556	T. Nakao, M. Obata, M. Kawakami, K. Morita, H. Tanaka, Y. Morimoto, S. Takehara, T. Yakushiji and T. Tahara, *Chem. Pharm. Bull.*, 1991, **39**, 2556.
91CJC972	G. Heinisch and W. Holzer, *Can. J. Chem.*, 1991, **69**, 972.
91FES873	G. Ciciani, V. D. Diaz and M. P. Giovannoni, *Farmaco Ed. Sci.*, 1991, **46**, 873.
91H329	A. R. Katritzky, M. Karelson and P. A. Harris, *Heterocycles*, 1991, **32**, 329.
91H735	Y. Aoyagi, A. Maeda, M. Inoue, M. Shiraishi, Y. Sakakibara, Y. Fukui and A. Ohta, *Heterocycles*, 1991, **32**, 735.
91H923	A. Ohta, A. Kojima, T. Saito, K. Kobayashi, H. Saito, K. Wakabayashi, S. Honma, C. Sakuma and Y. Aoyagi, *Heterocycles*, 1991, **32**, 923.
91H937	M. Shimazaki, M. Hikita, T. Hosoda and A. Ohta, *Heterocycles*, 1991, **32**, 937.
91H965	A. Ohta, Y. Tonomura, J. Sawaki, N. Sato, H. Akiike, M. Ikuta and M. Shimazaki, *Heterocycles*, 1991, **32**, 965.
91H1387	T. Sakamoto, N. Funami, Y. Kondo and H. Yamanaka, *Heterocycles*, 1991, **32**, 1387.
91H1537	M. Botta, *Heterocycles*, 1991, **32**, 1537.
91JA1561	A. R. Katritzky and M. Karelson, *J. Am. Chem. Soc.*, 1991, **113**, 1561.
91JA4230	D. L. Boger and M. Zhang, *J. Am. Chem. Soc.*, 1991, **113**, 4230.
91JCS(P1)2877	N. Sato, Y. Shimomura, Y. Ohwaki and R. Takeuchi, *J. Chem. Soc. Perkin Trans. 1*, 1991, 2877.

91JCS(P2)187	S. Buscemi and N. Vivona, *J. Chem. Soc. Perkin Trans. 2*, 1991, 187.
91JHC283	N. Ple, A. Turck, E. Fiquet and G. Queguiner, *J. Heterocycl. Chem.*, 1991, **28**, 283.
91JHC369	B. Jelen, A. Stimac, B. Stanovnik and M. Tisler, *J. Heterocycl. Chem.*, 1991, **28**, 369.
91JHC417	A. Stimac, B. Stanovnik, M. Tisler, *J. Heterocycl. Chem.*, 1991, **28**, 417.
91JHC583	J. G. Samaritoni and G. Babbitt, *J. Heterocycl. Chem.*, 1991, **28**, 583.
91JHC765	J. S. Ward and L. Merritt, *J. Heterocycl. Chem.*, 1991, **28**, 765.
91JHC1147	A. R. Howell, W. R. Martin, J. W. Sloan and W. T. Smith, Jr., *J. Heterocycl. Chem.*, 1991, **28**, 1147.
91JHC1281	J. T. Gupton, J. E. Gall, S. W. Riesinger, S. Q. Smith, K. M. Bevirt, J. A. Sikorski, M. L. Dahl and Z. Arnold, *J. Heterocycl. Chem.*, 1991, **28**, 1281.
91JHC1357	J. B. Hynes, A. Tomazic, C. A. Parrish and O. S. Fetzer, *J. Heterocycl. Chem.*, 1991, **28**, 1357.
91JHC1613	D. Peters, A. Hornfeldt and S. Gronowitz, *J.Heterocycl. Chem.*, 1991, **28**, 1613.
91JHC1629	D. Peters, A. Hornfeldt and S. Gronowitz, *J.Heterocycl. Chem.*, 1991, **28**, 1629.
91JHC1857	E. C. Taylor and M. Patel, *J. Heterocycl. Chem.*, 1991, **28**, 1857.
91JHC1981	J. B. Hynes, A. Tomazic, A. Kumar, V. Kumar and J. H. Freisheim, *J.Heterocycl. Chem.*, 1991, **28**, 1981.
91JMC809	K. S. Atwal, B. N. Swanson, S. E. Unger, D. M. Floyd, S. Moreland, A. Hedberg and B. C. O'Reilly. *J. Med. Chem.*, 1991, **34**, 809.
91JMC978	T. J. Thornton, T. R. Jones, A. L. Jackman, A. Flinn, B. M. O'Connor, P. Warner and A. H. Calvert, *J. Med. Chem.*, 1991, **34**, 978.
91JMC1492	M. S. Malamas and J. Millen, *J. Med. Chem.*, 1991, **34**, 1492.
91JMC1505	M. J. Yu, K. J. Trasher, J. R. McCowan, N. R. Mason and L. G. Mendelsohn, *J. Med. Chem.*, 1991, **34**, 1505.
91JMC2209	P. R. Marsham, A. L. Jackman, A. J. Hayter, M. R. Daw, J. L. Snowden, B. M. O'Connor, J. A. M. Bishop, A. H. Calvert and L. R. Hughs, *J. Med. Chem.*, 1991, **34**, 2209.
91JOC1355	J. P. Konopelski, K. S. Chu and G. R. Negrete, *J. Med. Chem.*, 1991, **34**, 1355.
91JOC2391	R. C. Reynolds, T. W. Trask and W. D. Sedwick, *J. Org. Chem.*, 1991, **56**, 2391.
91JOC2605	G. Buchi and J. Galindo, *J. Org. Chem.*, 1991, **56**, 2605.
91JOC2866	F. Fontana, F. Minisci, M. C. N. Barbosa and E. Vismara, *J. Org. Chem.*, 1991, **56**, 2866.
91JOC4793	R. Radinov, C. Chanev and M. Haimova, *J. Org. Chem.*, 1991, **56**, 4793.

91JOC5196 K. S. Chu, G. Negrete and J. P. Konopelski, *J. Org. Chem.*, 1991, **56**, 5196.

91JOC5610 L. Strekowski, R. L. Wydra, L. Janda and D. B. Harden, *J. Org. Chem.*, 1991, **56**, 5610.

91JOM301 A. Turck, D. Trohay, L. Mojovic, N. Ple and G. Queguiner, *J. Organomet. Chem.*, 1991, **412**, 301.

91JPR561 F. Zeuner, H. J. Niclas, G. Reck and H. Schenk, *J. Prakt. Chem.*, 1991, **333**, 561.

91LA345 D. Briel, *Liebigs Ann. Chem.*, 1991, 345.

91LA1225 A. Deeb, H. Sterk and T. Kappe, *Liebigs Ann. Chem.*, 1991, 1225.

91MI26 S. El-Bahaie, A. El-Deeb and M. C. Assy, *Pharmazie*, 1991, **46**, 26.

91MI183 C. K. Sha, C. P. Tsou, *J. Chin. Chem. Soc.*, 1991, **38**, 183.

91MI763 B. L. Flynn, V. Macolino and G. T. Crisp, *Nucleosides & Nucleotides*, 1991, **10**, 763.

91MI1821 J. Barluenga, M. Tomas, V. Kouznetsov, J. Jardon and E. Rubio, *Synlett.*, 1991, 821.

91S529 M. T. Cocco, C. Congiu, V. Onnis and A. Maccioni, *Synthesis*, 1991, 529.

91S581 W. Ried and T. Russ, *Synthesis*, 1991, 581.

91S861 E. M. Beccalli and A. Marchesini, *Synthesis*, 1991, 861.

91S1026 S. P. Singh, A. R. Mahajan, D. Prajapati and J. S. Sandhu, *Synthesis*, 1991, 1026.

91SC901 J. Nasielski, A. Standaert and R. N. Hinkens, *Synth. Commun.*, 1991, **21**, 901.

91SC1021 H. V. Patel, K. A. Vyas, S. P. Pandey, F. Tavares and P. S. Fernandes, *Synth. Commun.*, 1991, **21**, 1021.

91SC1935 H. V. Patel, K. A. Vyas, S. P. Pandey, F. Tavares and P. S. Fernandes, *Synth. Commun.*, 1991, **21**, 1935.

91T3959 N. Haider, *Tetrahedron*, 1991, **47**, 3959.

91T8573 N. Haider, G. Heinisch and J. Moshuber, *Tetrahedron*, 1991, **47**, 8573.

91TL2401 C. H. Lee and Y. H. Kim, *Tetrahedron Lett.*, 1991, **32**, 2401.

91TL2467 G. de Nanteuil, *Tetrahedron Lett.*, 1991, **32**, 2467.

91TL5949 L. Bauman and G. Seitz, *Tetrahedron Lett.*, 1991, **32**, 5949.

91TL6281 D. H. Desai, A. B. Cheikh and J. Zemlicka, *Tetrahedron Lett.*, 1991, **32**, 6281.

91TL6539 C. E. Muller, *Tetrahedron Lett.*, 1991, **32**, 6539.

91TL6763 M. Murata and K. Achiwa, *Tetrahedron Lett.*, 1991, **32**, 6763.

CHAPTER 6.3
Six-Membered Ring Systems: Triazines, Tetrazines and Fused Ring Polyaza Systems

DEREK T. HURST
Kingston Polytechnic, Kingston upon Thames, UK

6.3.1 SYNTHESIS

6.3.1.1 Triazines and Benzoderivatives

The use of azomalonates (1) in the synthesis of heterocycles (2) by a one-step reaction, first reported in 1987 (87HCA2118), has now been extended to the synthesis of valuable, novel, 1,2,4-triazin-5-one derivatives (3). Some reactions, including the elaboration of these products to triazino[1,6-*a*]indoles is also described (90HCA1700) (Scheme 1).

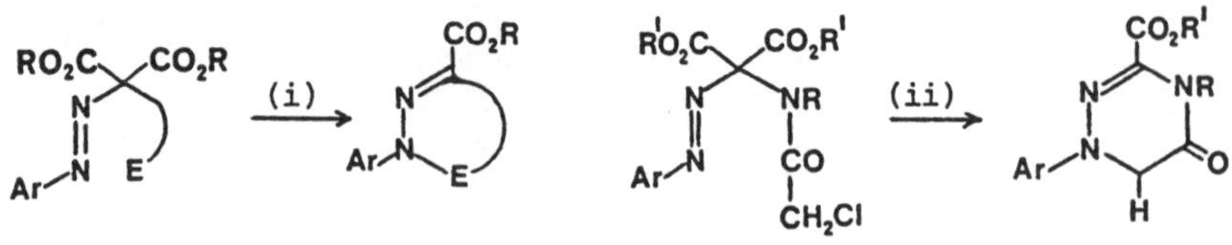

(1) (2) (3)

(Ar=Ph;R=Ph,R^1=Et, and many other examples

Reagents: (i) NaOH,ROH; (ii) $NaOR^1$,R^1OH heat

Scheme 1

The reaction of 1,2,3-triazole 1-oxides with dialkyl acetylenedicarboxylates provides a new route to 1,2,3-triazines via a multistep sequence of reactions involving cycloaddition, sigmatropic rearrangement, and ring expansion (Scheme 2) [90JCS(P1)3321].

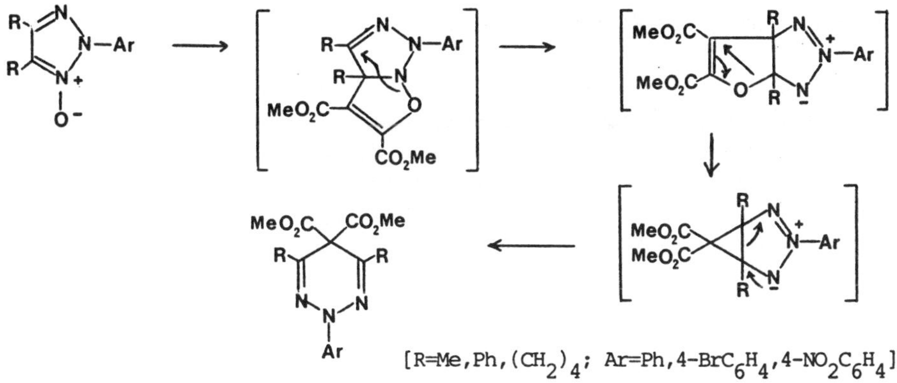

[R=Me,Ph,$(CH_2)_4$; Ar=Ph,4-BrC_6H_4,4-$NO_2C_6H_4$]

Scheme 2

6.3.1.2 Tetrazines and Benzoderivatives

The thermolysis of 2-hydrazonyltetrazoles (4) gives *N*-hydrazonylimides (5) which can either cyclize to give triazoles (6) or to give tetrazines (7), according to the conditions of the reaction. In pure, dry, hexanol the triazoles are favoured, but in the presence of small amounts of water or acid, the tetrazines are formed [91JCR(S)272; see also 87BSB675] (Scheme 3).

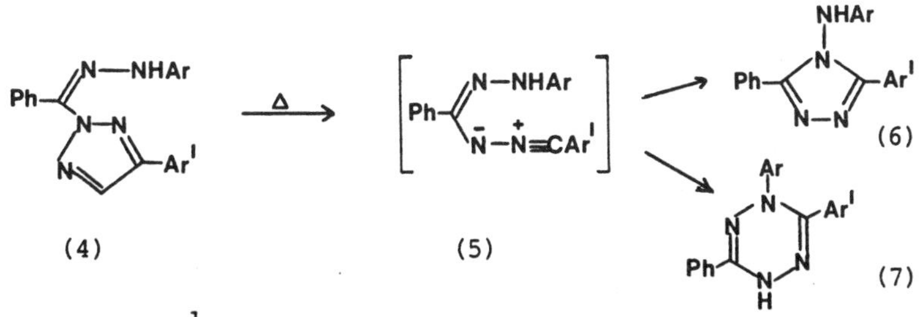

(4) (5) (6) (7)

(Ar=4-$NO_2C_6H_4$; Ar^1=Ph, and 5 other examples)

Scheme 3

6.3.1.3 Purines and Related Compounds

The 1,3-dipolar cycloaddition of 2-diazopropane to 2-methyl-6-phenylpyridazin-3[2*H*]-one (8) gives the cycloadduct (9) which was not originally characterised (69TL1063). It has now been found that compound (9) can be isolated below 0°C and characterized. However it isomerizes to (10) in the presence of acid and readily oxidizes to (11) in air. Other transformations which can occur include the formation of (12) and (13) (Scheme 4) (90T6915).

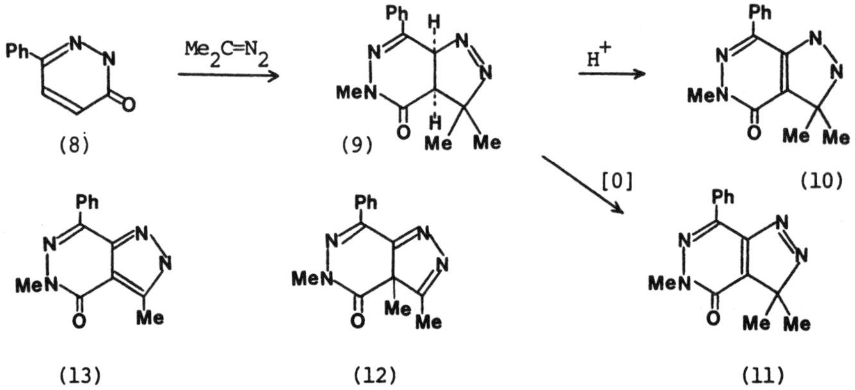

Scheme 4

This reaction has been investigated further and a number of novel 3*H*-pyrazolo[3,4-*d*]pyridazin-4(5*H*)-one and -7(6*H*)-one derivatives have been obtained (91JHC417,425).

Pyrazoles (14) have been synthesised and have been reacted with aldehydes to give pyrazolo[3,4-*d*]pyridazines (15). The use of ethyl orthoformate has given some pyrazolopyridazones (16) (91AP585).

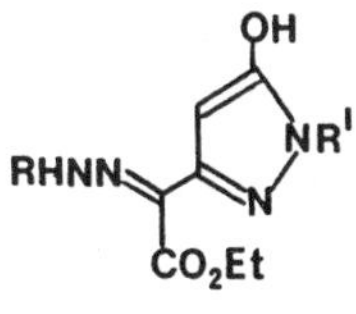

(14)

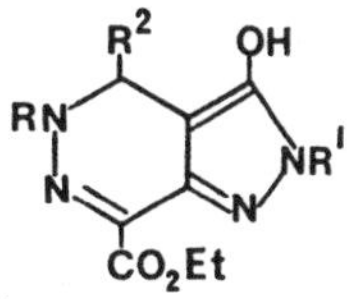

(15)

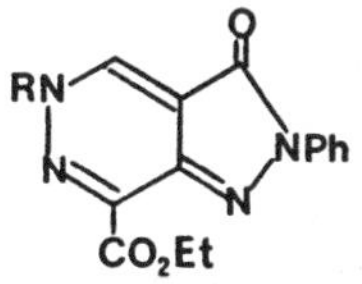

(16) (R=4-ClC$_6$H$_4$,4-MeC$_6$H$_4$; R^1,R^2=H,Ph)

1-Amino-2-carbonylmethylthiopyrimidinium salts (17) react with phenacyl halides to give pyrimidothiadiazonium salts (18) which undergo ring contractions when treated with base to give high yields of mercaptopyrazolopyrimidines (19) (90JPR470).

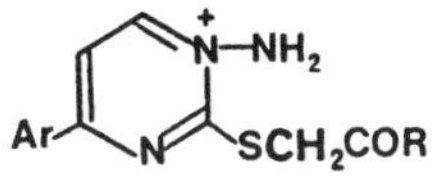

(17)

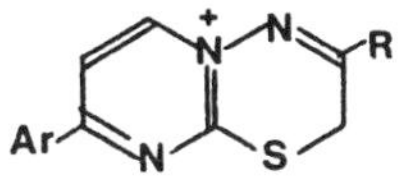

(18)

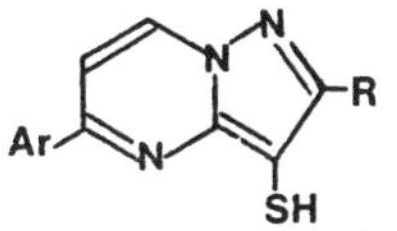

(19) (Ar,R=various)

Six novel ring systems based on the 1,2,4-triazolo[1,5-*a*]pyrimidine system have been obtained by reacting 5-amino-1,2,4-triazoles (20) with thiacyclohexane *β*-oxoesters (21). These are the thiopyrano[4,3-*d*], [3,4-*e*], [3,4-*d*], [4,3-*e*], [3,2-*d*], and [2,3-*e*] systems (Scheme 5). Some other compounds of this type were also obtained (91JHC561).

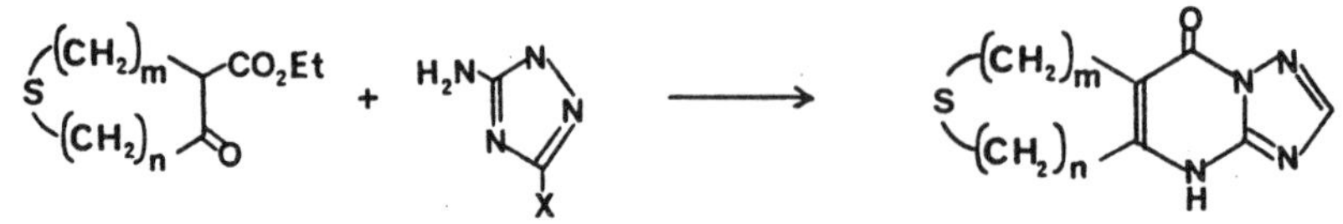

(21) (m,n=1,2) (20) (X=MeS,morpholino)

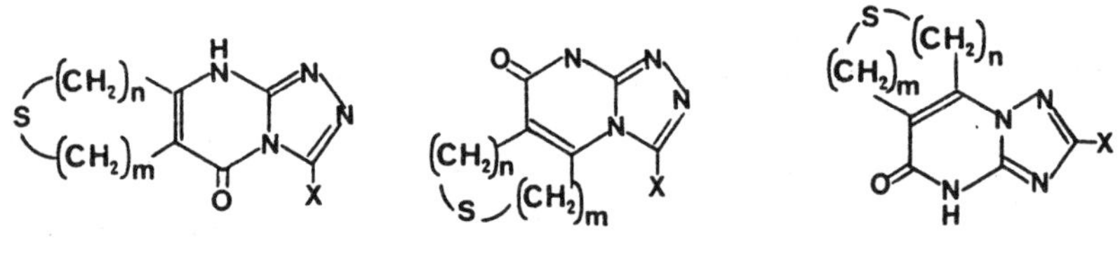

Scheme 5

A very simple synthesis of triazolo (and tetrazolo) pyrimidines, and other related ring systems, involves merely heating ethyl arylazobenzoyl acetates with the aminoazole in ethanol (Scheme 6) (91CCC1560). This represents part of a general study of the synthesis of such compounds [see also 90CCC2795; 90M281; 91H895; 91ZN(B)835].

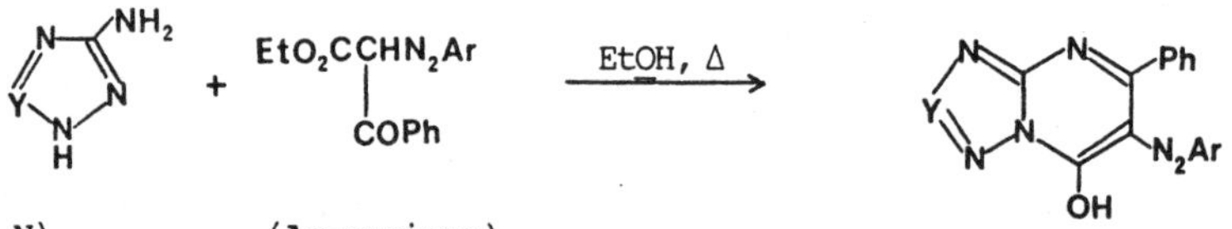

EtOH, Δ

(Y=CH,N) (Ar=various)

Scheme 6

The condensation of 5-amino-1-methyl-4-pyrazolin-3-one with *N*-acyl and *N*-ethoxycarbonylimidates provides a new route for the synthesis of pyrimidopyrazolinones (e.g. 22). The use of 4-amino-1,2,4-triazole gives triazolotriazines, whilst a related reaction using hydrazides and imidates leads to 1,2,4-triazolo[1,5-*c*]pyrimidines and 1,2,4-triazolo[1,5-*c*]quinazolines (89MI3,9).

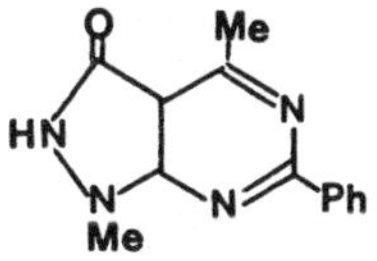

(22)

6.3.1.4 Pteridines, other Diazinodiazines, and Related Compounds

A novel approach to the pyrido[4,3-*c*]pyridazine system starts from ethyl 3-methyl-4-pyridazine carboxylate using the route shown in Scheme 7. This method offers good potential for further development (91JHC1043).

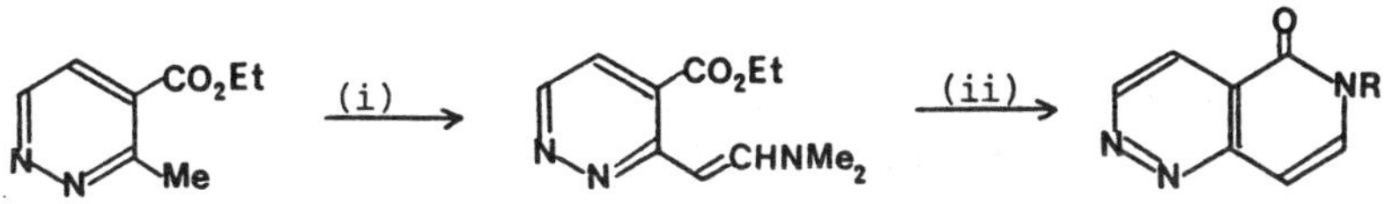

(R=H, various Ar)

Reagents: (i) $Me_3COCH(NMe)_2$, DMF, 110°, 9h; (ii) $ArNH_2$, AcOH, r.t., overnight.
(or $NH_4^+AcO^-$, EtOH, reflux, 2d)

Scheme 7

An efficient and easy synthesis of some novel 2-arylpyrido-[2,3-*d*]pyrimidin-5(8*H*)-ones (23) is provided by the thermolysis of the pyrimidinyl-aminomethylene derivatives of Meldrum's acid (24), prepared from the pyrimidinamine and the methoxy derivative of Meldrum's acid (90MI549). This reaction may show further potential for such annelations.

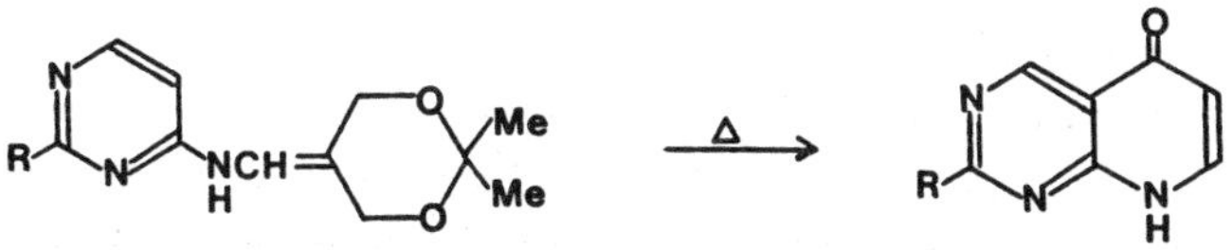

(24) (R=4-, 3-pyridyl, 4-$MeOC_6H_4$)

(23)

The reaction of ethyl-2-formylpropionate with guanidine to give 2-amino-5-methylpyrimidin-4-one is well known, having been first reported in 1905 (05MI130; see also 47JCS41). However a recent study of this reaction has shown that, in addition to the pyrimidine, small amounts of the two fused ring compounds (25,26) are formed as well as 2-amino-6-ethyl-5-methylpyrimidin-4-one (91JHC231).

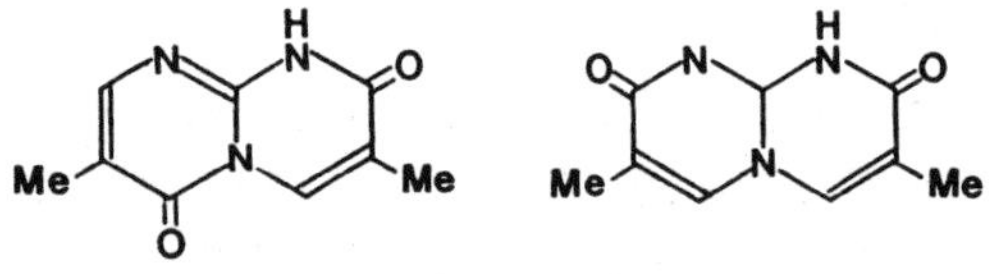

(25) (26)

The novel pyridopyridazine rings having fused thieno rings (27,28) have been obtained from the thienopyridines (29,30) by treatment with hydrazine in methanol. Reaction with phosphorus oxychloride only converts the oxo function in the pyridazine ring of the products to the chloro group (91JHC205).

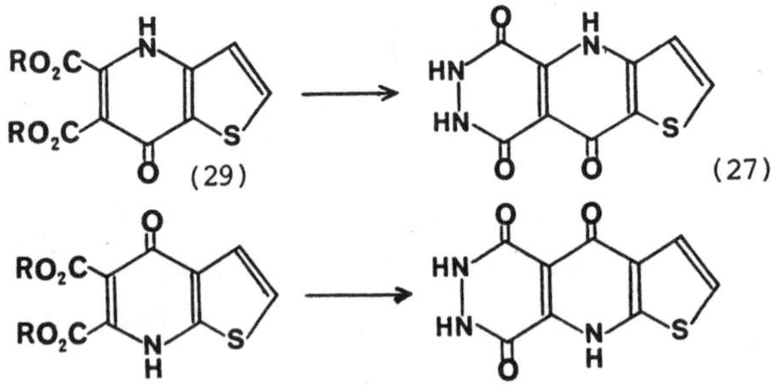

(29) (27)

(30) (28)

Some other novel thieno fused heterocycles have also been obtained. The reaction of 2-aminothiophene-3-carbonitriles with phenyl isothiocyanate does not give the expected thieno[2,3-*d*]pyrimidines, but instead gives dithieno-[2',3':4,3][2',3':8,9]pyrimido[3,4-*a*]pyrimidine-7-thiones (31) (91JPR229).

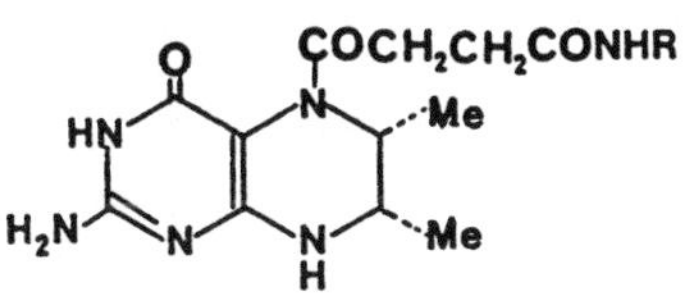

(32) (Various R)

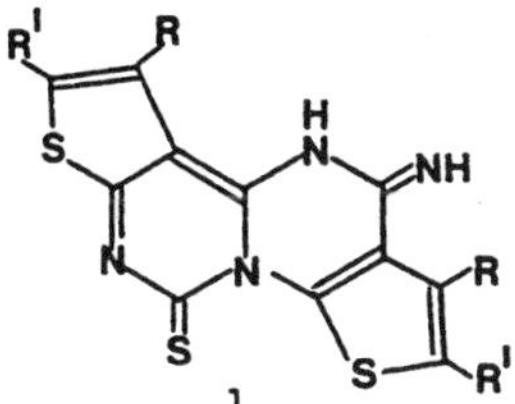

(31) [R,R^1=H,Me,$(CH_2)_4$]

A series of new *N*-5-acyl-5,6,7,8-tetrahydropterins (32) has been obtained, with the acyl function being a variety of amidosuccinyl substituents, starting from the N^2-isobutyryl compounds, with subsequent deprotection (89MI199).

5-Hydrazinopyridazin-3(2*H*)-ones (33) react with dimethyl acetylene dicarboxylate to give 4,6-dihydropyridazino[4,5-*c*]pyridazin-5-(1*H*)ones (34) by cyclization with dehydrogenation. However the reactions of 4-bromo-5-hydrazino and 5-bromo-4-hydrazinopyridazin-3(2*H*)-ones (35,37) with DMAD result in cyclization and rearrangement to give (36) and (34) respectively as well as the expected cyclization products. Heating compound (38) with acetic acid in tetralin gives compounds (39) and (40) [91JCS(P1)991].

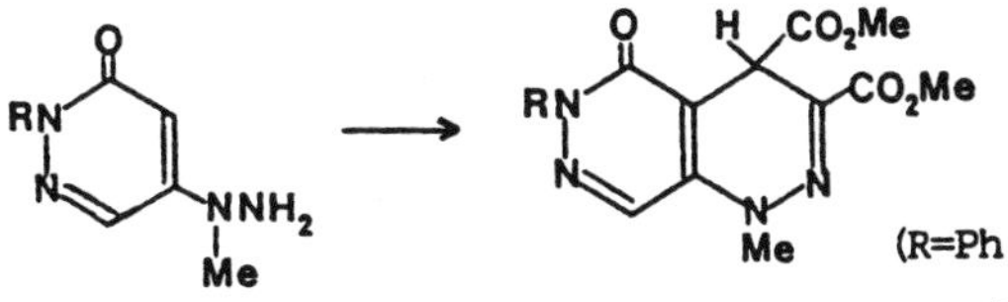

(33) (34) (R=Ph,H,Me,$PhCH_2$,4-$ClC_6H_4CH_2$)

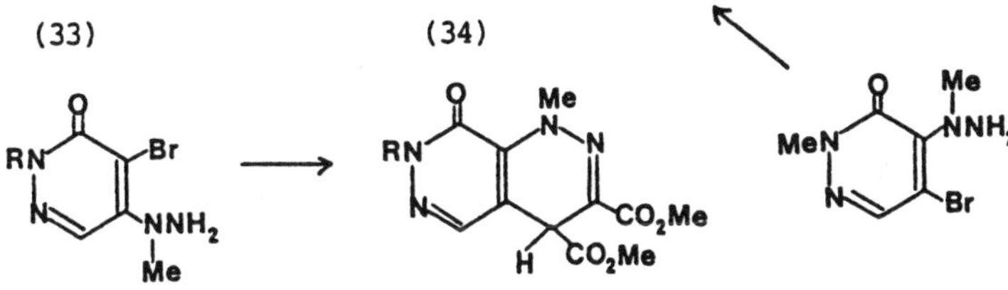

(35) (36) (37)

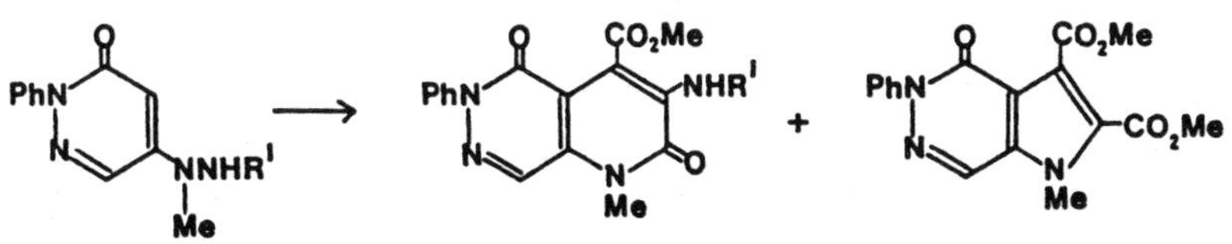

(38) (39) (R^1=Et,Pr) (40)

The reaction of 1,2-diazepino[3,4-*b*]quinoxalines (e.g. 41) with *N*-bromosuccinimide in water gives a convenient synthesis of 1,4-dihydro-4-oxopyridazino[3,4-*b*]quinoxalines (42) (Scheme 8) (91JHC199).

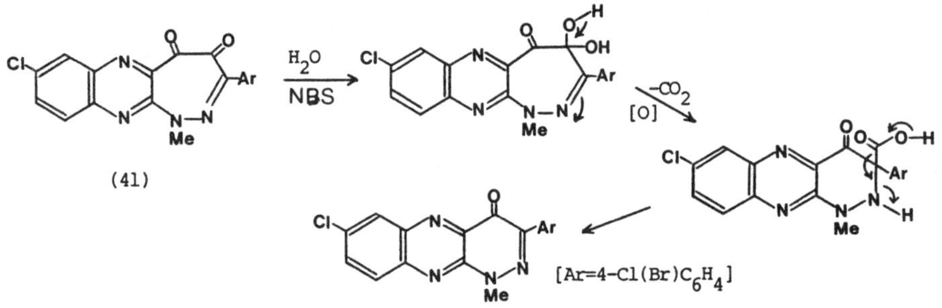

(41)

(42)

[Ar=4-Cl(Br)C_6H_4]

Scheme 8

6.3.1.5 Triazines and Fused Rings

The cyclization of triazolylamidines (43) with trichloromethyl nitrile/sodium methoxide gives the triazolotriazine (5-azapurine) (44a). These products react with amines to give 5-azaadenine derivatives (44b) (91IZV241).

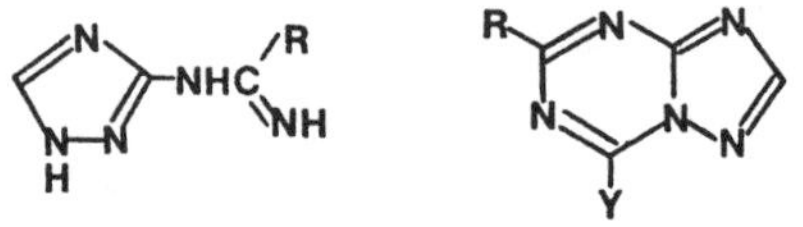

(a) Y=CCl_3
(b) Y=piperidino,morpholino, 3-methylpropylamino.

(43) (44)

An interesting reaction to give the 1*H*-pyrrolo[2,3-*e*]-1,2,4-triazine ring system (45) results from interaction between 3,5,6-trichloro-1,2,4-triazine and 1-pyrrolidin-1-ylcycloheptene, the mechanism for which is given in Scheme 9 [91JCS(P1)1762].

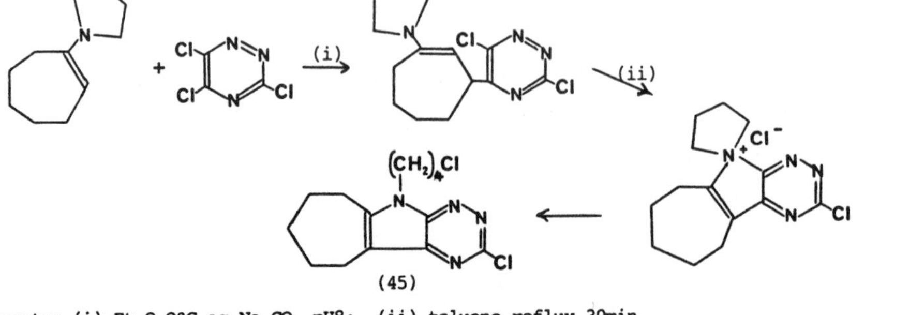

(45)

Reagents: (i) Et_2O,0°C,aq.Na_2CO_3,pH8; (ii) toluene,reflux,30min.

Scheme 9

Another new ring system which has been synthesised is the triazinobenzothiazine (46) formed by cyclization of the benzothiazinone with isocyanates (90H1393).

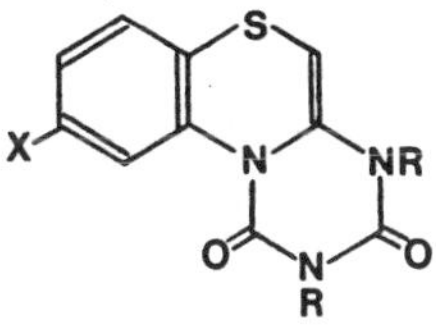

(46) (X=Cl,H; R=Me,Et,Bu,Ph,CH_2CO_2Et)

6.3.1.6 Miscellaneous Ring Systems

The preliminary observation [88JCS(C)1608] that the oxidation of 1-amino-5-phenyl-1,5-dihydro-[1,2,3]triazolo[4,5-*d*]triazole (47) with lead tetraacetate gives 2-phenyl-2*H*-[1,2,3]triazolo[4,5-*e*][1,2,3,4]tetrazine (48) has now been detailed, whilst the oxidation of the 2-amino rather than the 1-amino-triazolotriazole has been shown to give 1,3-dicyano-2-phenyltriazenimide (49) under similar conditions [91JCS(P1)2045].

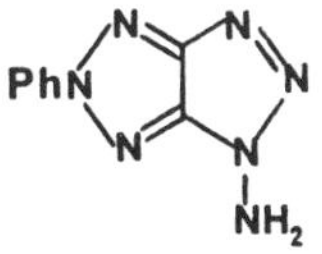

(47) 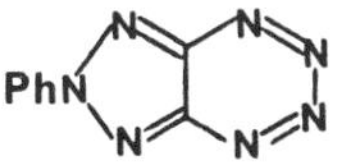 (48) 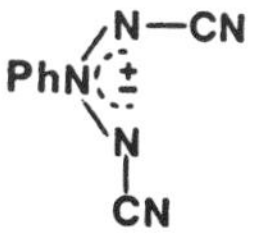 (49)

A number of new multiring polyaza systems have been reported recently. Some interesting examples of these are given below.

The pyrazolo[1,5-*a*]pyrido[3,4-*e*]pyrimidine system (50) has been obtained by cyclization reactions of 6-acetyl-7-(2-dimethylaminovinyl)pyrazolo[1,5-*a*]-pyrimidine (51) with hydroxylamine, followed by deoxygenation of the pyridine *N*-oxide intermediate (90H1635).

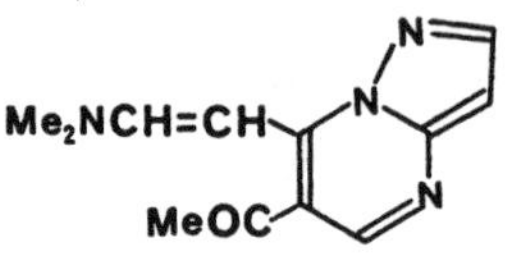

(51) 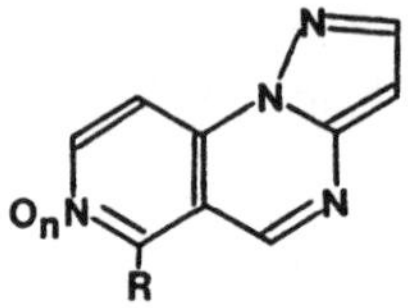 (50) (n=1,0;R=H,Me,and 4 others)

The pyrimido[2,1-*a*]phthalazine systems which also have fused thiophene rings (52,53) have been synthesised by elaboration of the thienopyrimidine systems. The benzothienopyrimidotriazolophthalazine system (54) has also been obtained (91JHC545).

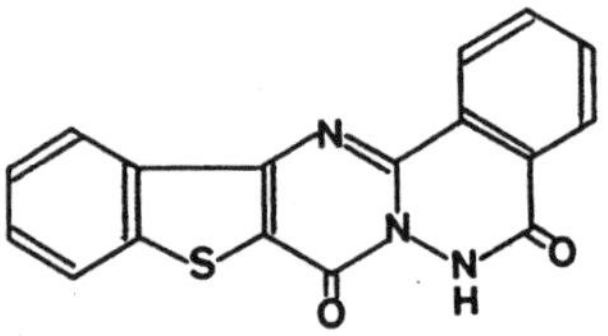

(52) 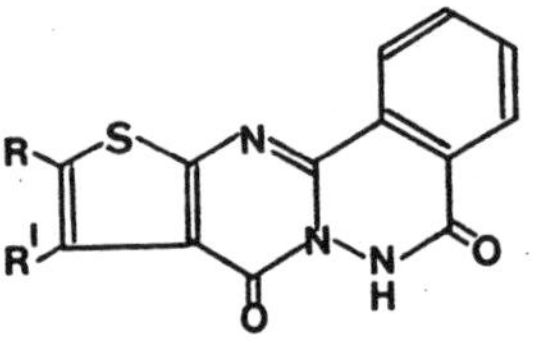 (53) [R,R^1=$(CH_2)_4$; R=Ph,R^1=H] 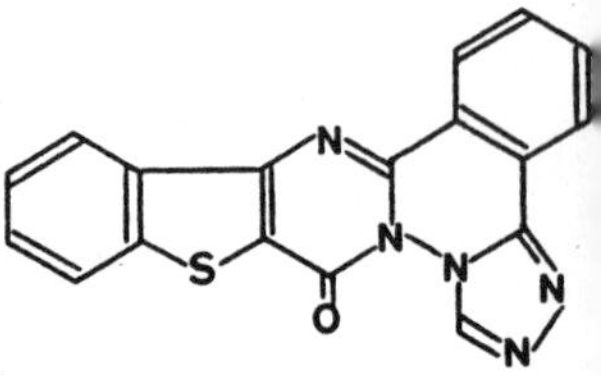 (54)

Pyrazolo[3,4-*e*][1,3,4]thiadiazolo[3,2-*a*]pyrimidine (55) and pyrazolo-[4',3':5,6]pyrimido[2,1-*b*]benzothiazole derivatives (56) have also been synthesised (91G113).

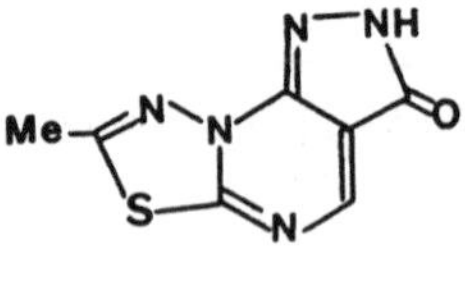

(55) 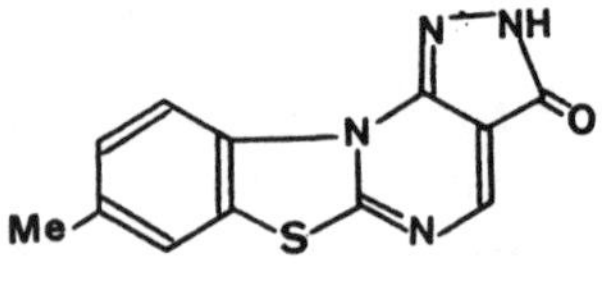 (56)

The simple reaction of refluxing 2-methyl-1*H*-benzimidazole with ethoxy carbonyl isocyanate in bromobenzene affords the tetraazafluoranthrene system

(57). Reaction of the pyrimidobenzimidazole (58) with malonic esters leads to the triazafluoranthrene system (59) (91JHC995).

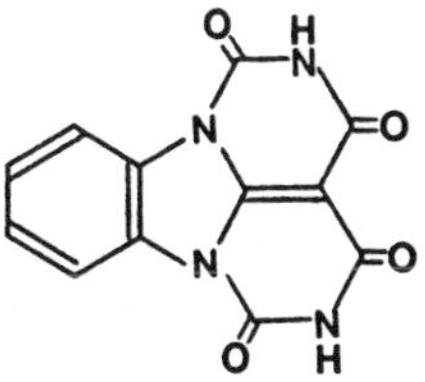

(57) 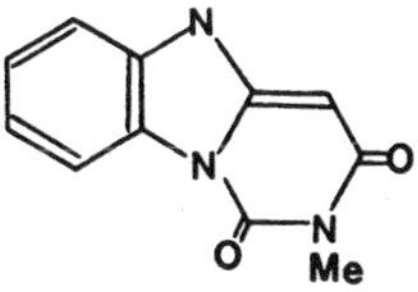 (58) 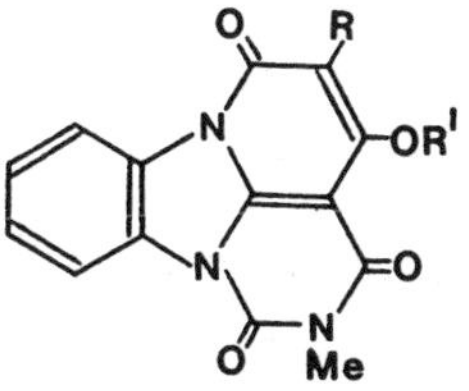 (59) (R,R^1=various)

Diazotized 3,6-diamino-4-phenylpyrazolo[3,4-*b*]pyridine-5-carbonitrile reacts with active methylene reagents to give pyridopyrazolotriazine derivatives (60). The pyridodipyrazolotriazine (61) has also been obtained (91G209).

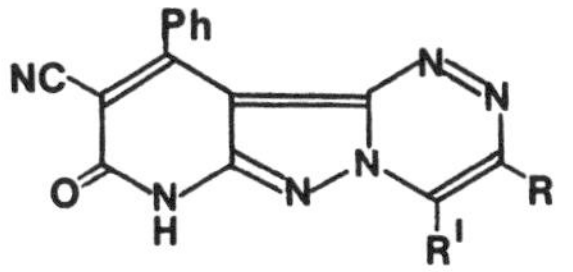

(60) (R=CN,CO_2Et,COMe; R^1=NH_2OH,Me) 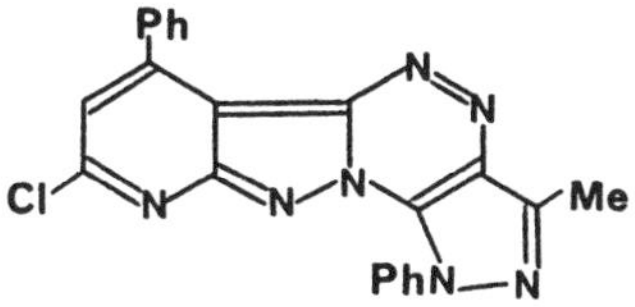 (61)

3-Amino-1,2,4-triazolo[4,3-*a*]quinoline (62) has proved to be a useful starting material for the development of novel multiring *N*-heterocycles including 9(*H*)-oxoquinazolino[2',3':1,5][1,2,4]triazolo[4,3-*a*]quinoline (63a) and 9(*H*)-oxopyrido[2",3":4',5']pyrimido[1',2':1,5][1,2,4]triazolo[4,3-*a*]-quinoline (63b) [90JHC981; 91IJC(B)710].

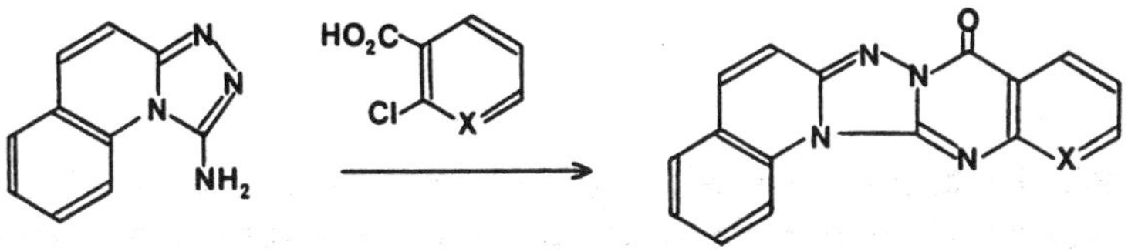

(62) (63) (a,X=CH; b,X=N)

The stable azomethine imines (64) can be obtained and these undergo thermal rearrangement to yield the novel 9,10-dihydro-8*H*-tetrazolo[1',5':1,6]pyridazino[4,5-*c*][1,2]diazepines (65) (90MI707).

(64) (65)

R	R^1	R^2	R^3
Me	Me	Me	Me
Me	Me	CD_3	CD_3
Me	Et	Me	Me

Shaw and coworkers have continued to develop methods for the synthesis of polycyclic polyaza heterocycles and have now prepared the following novel systems:

13*H*-1,3,7,8,12a,13c-hexaazabenzo[*d,e*]naphthacene (66);
1,3,7,8,11b,12,14,14d-octaazadibenzo[*de,hi*]naphthacene (67);
8,13-dihydro-1,3,7,8,13c-pentaazabenzo[*de*]naphthacene (68);
7*H*-1,3,7,8,11b,12,14-heptaazadibenzo[*de,hi*]naphthacene (69); and
8*H*-1,3,7,8,13,14c-hexaazabenzo[4,5]cyclohepta[1,2-*a*]phenalene (70)
(90JHC1591; 91JHC987)

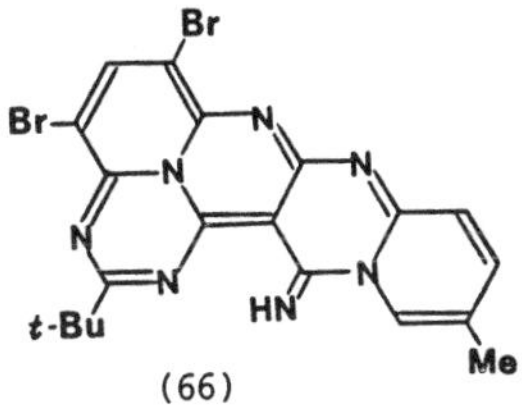

(66) 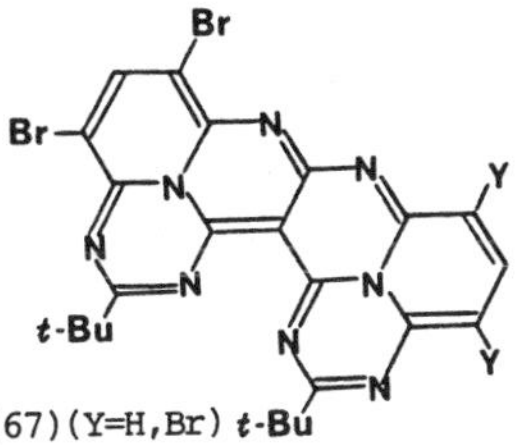 (67) (Y=H,Br)

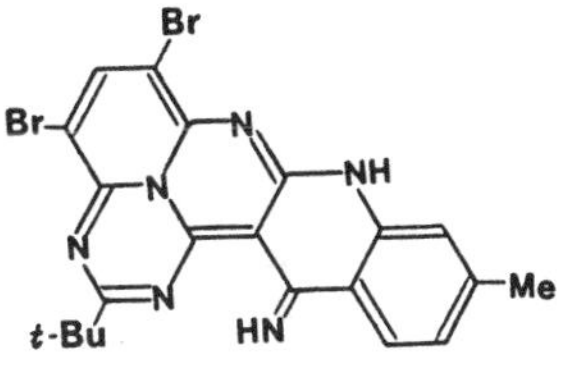

(68) 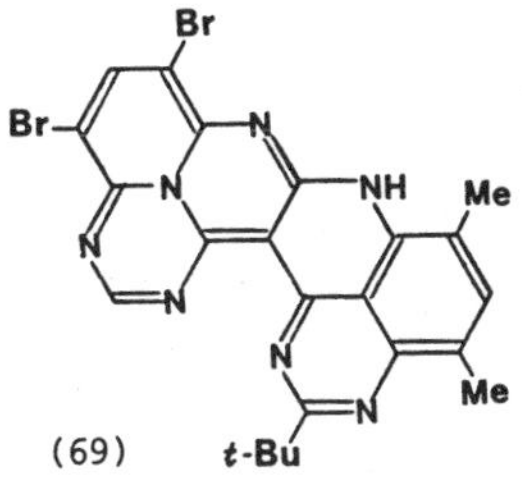 (69) 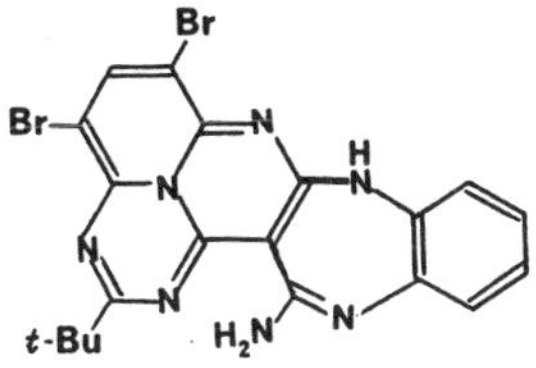 (70)

6.3.2 REACTIONS

6.3.2.1 Triazines, Tetrazines and Derivatives

There have been differing views on the structure of the products of *N*-oxidation of 1,2,3-triazines and both *N*-2 and *N*-1 derivatives have been proposed. X-ray crystallography and spectral data have now confirmed the products of *m*-chloroperoxybenzoic acid oxidation of 4,5,6-triazyl-1,2,3-triazines to be the 2-*N*-oxides as the sole products (91CPB2117).

The reactions of 1,2,3-triazines with electron-rich dienophiles have been studied. The products which are obtained vary with the triazine substitution but pyridines and pyridazines are formed.

The reaction of the triazine imine (71) with electron-deficient dienophiles has been reported to produce pyrazolo[2,3-*b*][1,2,4]triazines (83CPB3759). However with an electron-rich dienophile the 1,2,3-triazolo-[1,4-*a*]pyrimidines (72) are formed (Scheme 10) (90CPB2108).

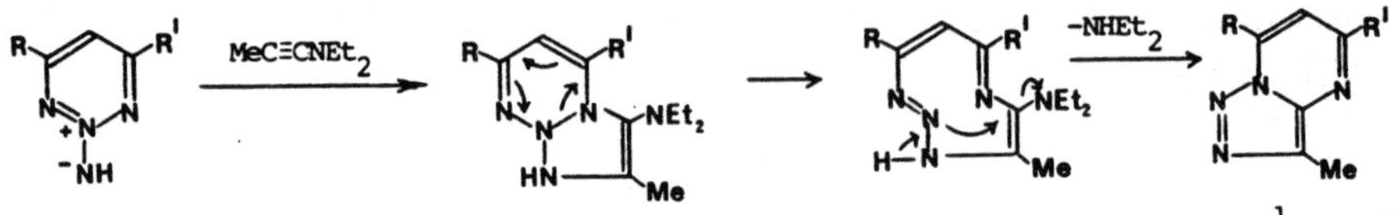

(71)

(72) (R,R[1]=Me,Ph)

Scheme 10

Although usually stable to nucleophilic displacement in *N*-heterocyclic systems, some examples of nitrile group displacement are known. In 1,2,4-triazines, regardless of its position, the nitrile group is displaced, for example by carbanions and by Grignard reagents, to give the substituted triazine (91CPB486).

The nucleophilic radical carbamoylation of the triazinium dicyanomethylides (73) occurs at the 5-position, with subsequent elimination of dicyanomethylene, to give the 5-carbamoyl-1,2,3-triazines (74), but the parent triazines do not undergo this reaction (91H855).

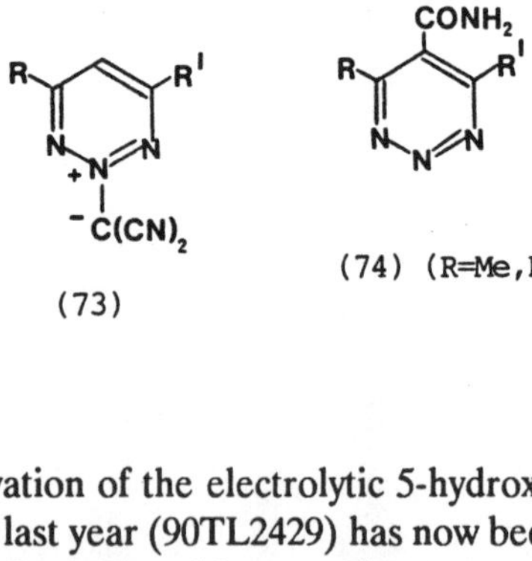

(73)

(74) (R=Me,Et,Ph)

The observation of the electrolytic 5-hydroxylation of 5-halogeno-1,2,3-triazines reported last year (90TL2429) has now been investigated further. The reaction seems to be initiated by one electron transfer and to be specific for electrolytically generated superoxide since the action of hydroxide ion, or potassium superoxide gives a complex mixture of products (91T4317).

A new type of dimerisation in azaaromatic chemistry has been observed. The reaction of 1-ethyl-3-alkylthio-5-phenyl-1,2,4-triazinium tetrafluoroborates (75) with triethylamine in ethanolic solution results in the formation of 4a,4b,9,10-tetrahydro-1,3,6,8,8a,10a-hexaazaphenanthrenes (76). The suggested mechanism for the reaction involves the intermediary of the biradical (77) (90TL7665).

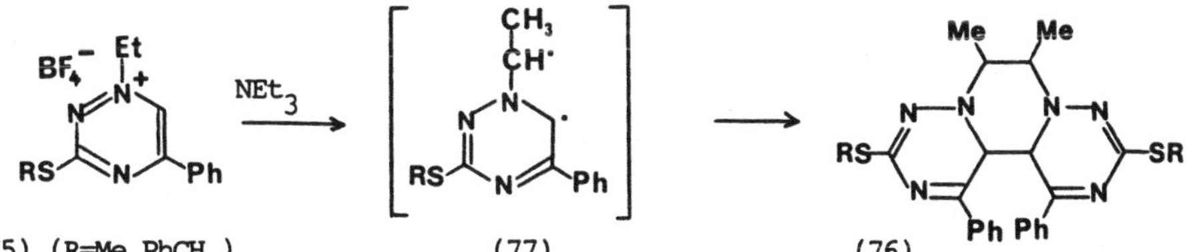

(75) (R=Me, $PhCH_2$) (77) (76)

The diazobistetrazines (78a) undergo loss of nitrogen on heating in DMF to give the bis-*sym*-tetrazines (78b) (90KGS1691).

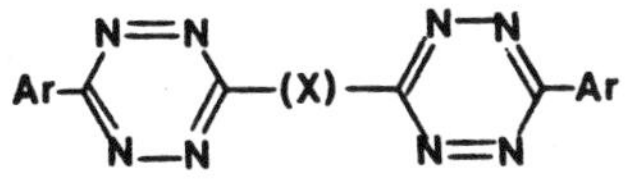

(78) (a, X=N_2; b, X=bond) (Ar=Ph, 4-ClC_6H_4) 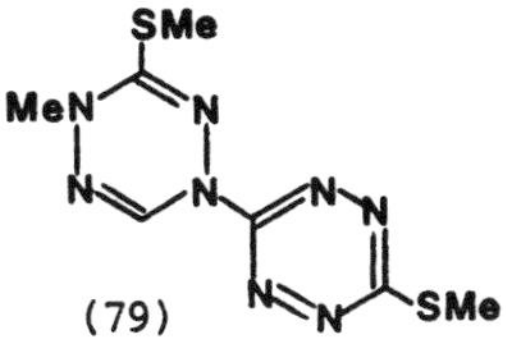 (79)

3,6-Bismethylthio-1,2,4,5-tetrazine undergoes *N*-alkylation with methyllithium or with Grignard reagents, but the "reverse addition" of methyllithium leads to the formation of the tetrazinyltetrazine (79). The reaction of the bismethylthiotetrazine with *N*-nucleophiles results in the displacement of the methylthio group (91JHC1163).

6.3.2.2 Purines and Related Systems

A series of eleven 6-(substituted)-2-aminopurines has been alkylated with 2-acetoxymethyl-4-iodobutyl acetate. The ratio of *N*-9 to *N*-7 alkylation varied from 1.8:1 to 25:1 as the group was changed. A correlation was found between the log of the ratio with a combination of resonance and lipophilicity factors of the 6-substituent. This is the first such study and the results should be similar in the case of other purine alkylations (90T6903).

The annelations of 8-aminoxanthines with dimethyl acetylenedicarboxylate give different products according to the conditions, and products involving 1,2, and also 3 molecules of DMAD have been isolated (Scheme 11) (91CPB270).

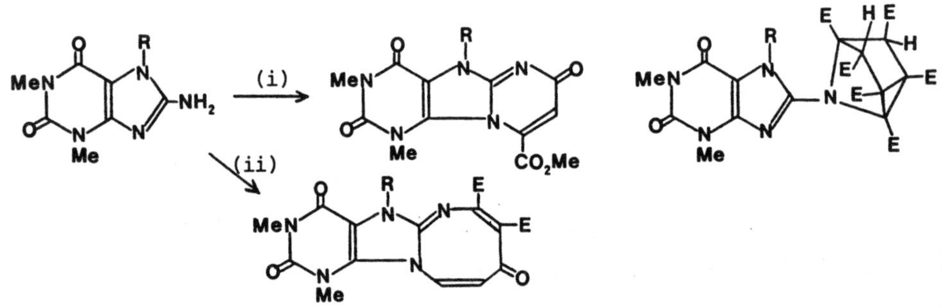

Reagents: (i) DMAD,MeOH,reflux,19-30h; (ii) DMAD,DMF,70°C,17h.

Scheme 11

The triazolopyrimidine (80) undergoes ring transformation on reaction with enamines to give the triazolopyridines (81). However (80), and some substituted compounds, undergoes reaction with the ynamine (82) to give the triazolodiazocines (83) (91CPB282).

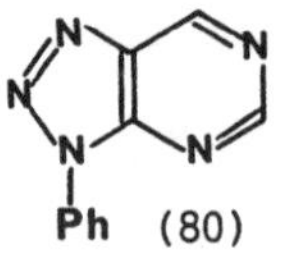

(80) 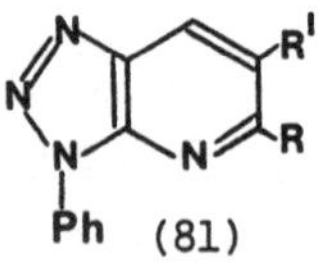 (81) [R=Ph,E;R^1=H,Me;R,R^1=$(CH_2)_3$,$(CH_2)_4$]

$R^1C{\equiv}CNEt_2$ (82)

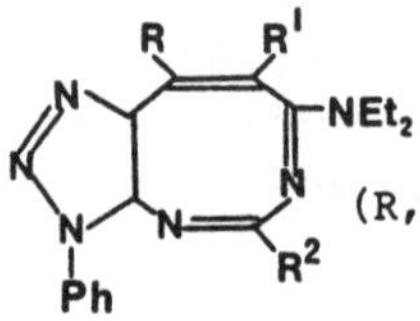

(83) (R,R^2=H;R^1=Me, and 8 other examples)

An example of the elimination of nitrogen from a triazolopyrimidine to result in the formation of a pyrimidine has been demonstrated. The pyrimidine

(84) has been obtained by treating compound (85) with butyllithium and 2-bromopropionamide (90TL6103).

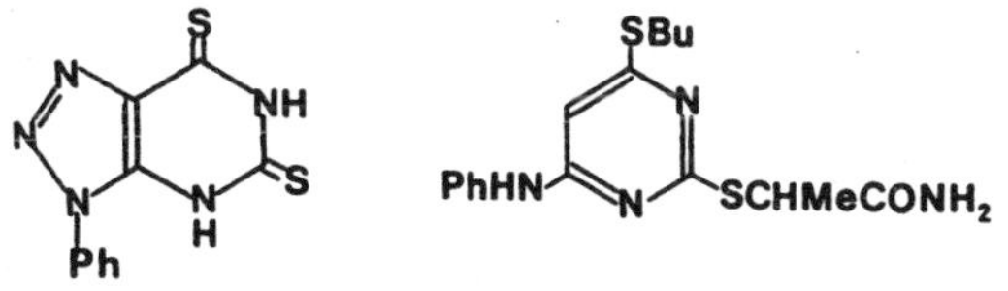

(85)

(84)

6.3.2.3 Pteridines and Related Heterocycles

The 1,4-diarylpyridazino[4,5-*d*]pyridazine derivatives (86) undergo [4+2] cycloadditions with a variety of electron-rich dienophiles to give good yields of phthalazines (87) (91T3959).

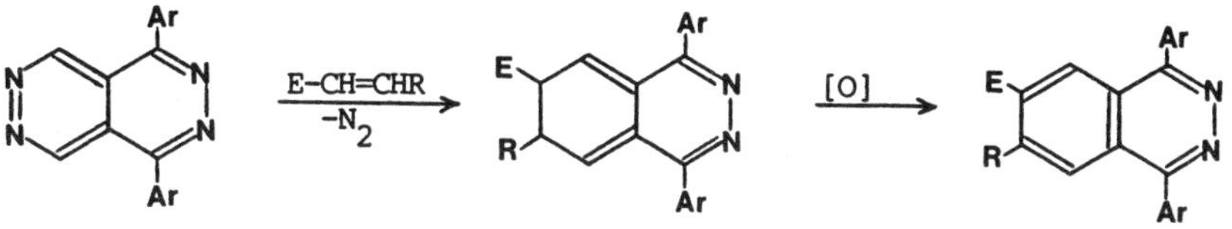

(86)

(87) (Ar=Ph, 4-MeOC_6H_4; E,R=various)

The reaction of fervenulin 4-oxide (88) with alkylamines at room temperature gives imidrazones (89), and *C*-nucleophiles behave similarly to give related products. However 1,3-dimethyllumazine 5-oxide (90) reacts with aklylamines to give the 6-substituted product (91) (Scheme 12)(91MC46).

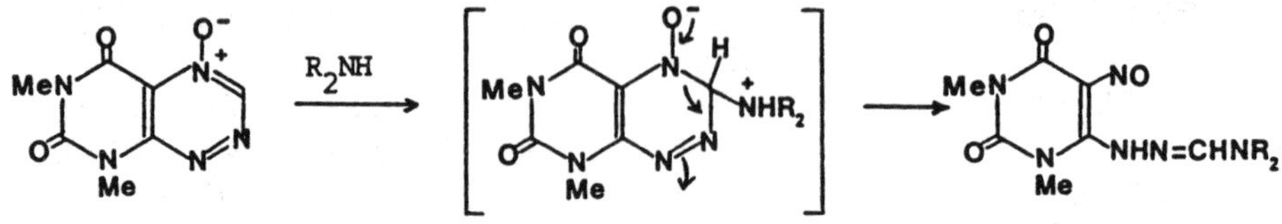

(88) (89)

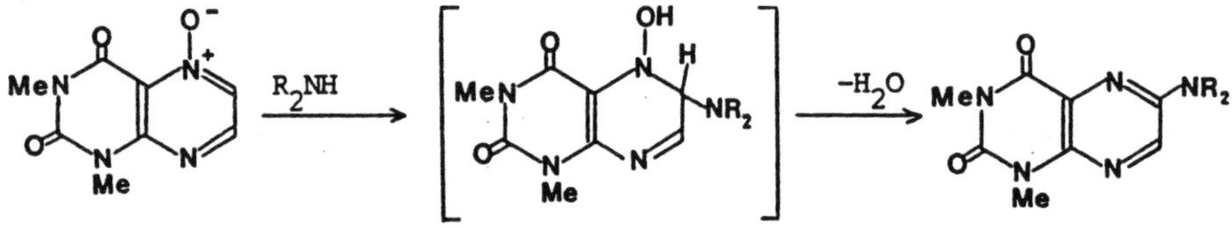

(90) (91)

Scheme 12

REFERENCES

05MI130 T. B. Johnson and S. H. Clapp, *Am. Chem J.*, 1905, **32**, 130.

47JCS41 R. Hull, B. J. Lovell, H. T. Openshaw, and A. R. Todd, *J. Chem Soc.*, 1947, 41.

69TL1063 M. Franck-Neumann and G. Leclerc, *Tetrahedron Lett.*, 1969, 1063.

83CPB3759 A. Ohsawa, H. Arai, K. Yamaguchi, H. Igeta, and Y. Iitaka, *Chem. Pharm. Bull.*, 1983, **31**, 3759.

87BSB675 I. Plenkiewicz and T. Zdrojewski, *Bull. Chem. Soc. Belg.*, 1987, **96**, 675.

87HCA2118 R. Heckendorn, *Helv. Chim. Acta*, 1987, **70**, 2118.

88JCS(C)1608 T. Kaihoh, T. Itoh, K. Yamaguchi, and A. Ohsawa, *J. Chem. Soc. Chem Commun.*, 1988, 1608.

89MI3 M. L. Benkhoud, H. Mraihl, and B. Baccar, *J. Soc. Chim. Tunis,* 1989, **2**, 3; *Chem Abstr.*, 1991, **114**, 81750.

89MI9 M. T. Kaddachi, B. Hajjem, and B. Baccar, *J. Soc. Chim. Tunis,* 1989, **2**, 9; *Chem Abstr.*, 1991, **114**, 81751.

89MI199 R. J. Lockart and W. Pfleiderer, *Pteridines,* 1989, **1**, 199.

90CCC2795	A. Deeb, *Collect. Czech. Chem. Commun.*, 1990, **55**, 2795.
90CPB2108	T. Itoh, M. Okada, K. Nagata, K. Yamaguchi, and A. Ohsawa, *Chem. Pharm. Bull.*, 1990, **38**, 2108.
90H1393	J. H. Musser, S. C. Lewis, and R. H. W. Bender, *Heterocycles*, 1990, **31**, 1393.
90H1635	F. Bruni, S. Chimichi, B. Cosimelli, A. Costanzo, G. Guerrini, and S. Selleri, *Heterocycles*, 1990, **31**, 1635.
90HCA1700	R. Heckendorn, *Helv. Chim. Acta*, 1990, **73**, 1700.
90JCS(P1)3321	R. N. Butler, D. Cunningham, E. G. Marren, and P. McArdle, *J. Chem. Soc. Perkin Trans. 1*, 1990 3321.
90JHC981	T. Ramalingam, M. S. R. Murty, Y. V. D. Nageswar, and P. B. Sattur, *J. Heterocycl. Chem.*, 1990, **27**, 981.
90JHC1591	J. T. Shaw, M. F. Egler, V. S. Peciulis, R. Pustover, and W. C. Ruth, *J. Heterocycl. Chem.*, 1990, **27**, 1591.
90JPR470	J. Liebscher and A. Hassoun, *J. Prakt. Chem.*, 1990, **332**, 470.
90KGS1691	E. G. Kovalev, V. A. Annfriev, and G. L. Rusinov, *Khim. Geterotsikl. Soedin.*, 1990, 1691; *Chem. Abstr.*, 1991, **114**, 207209.
90M281	A. Deeb, A. Essawy, A. M. El-Gendy, and A. Shaban, *Monatsh*, 1990, **121**, 281.
90MI549	B. Singh, S. C. Laskowski, and G. Y. Lesher, *Synlett*, 1990, 549.
90MI707	M. Zlicar, B. Huc, B. Stanovnik, and M. Tisler, *Synlett*, 1990, 707.
90T6903	G. R. Green, T. J. Grinter, P. M. Kincey, and R. L. Jarvest, *Tetrahedron*, 1990, **46**, 6903.
90T6915	A. Stimac, B. Stanovnik, and M. Tisler, *Tetrahedron*, 1990, **46**, 6915.
90TL2429	T. Itoh, K. Nagata, M. Okada, and A. Ohsawa, *Tetrahedron Lett.*, 1990, **31**, 2429.
90TL6103	M. J. Dooley, R. J. Quinn, W. C. Patalinghug, and A. H. White, *Tetrahedron Lett.*, 1990, **31**, 6103.
90TL7665	O. N. Chupakin, B. V. Rudakov, S, G. Alexeev, V. N. Charushin, and V. A. Chertkov, *Tetrahedron Lett.*, 1990, **31**, 7665.

91AP585	M. A. F. Sharaf, F. A. Abd El-Aal, G. E. H. Elgemeie, and A. A. El-Damaty, *Arch. Pharm. (Weinheim, Ger.)*, 1991, **324**, 585.
91CCC1560	A. Deeb, *Collect. Czech. Chem. Commun.*, 1991, **56**, 1560.
91CPB270	T. Heda, Y. Kawabata, N. Murakami, S. Nagai, J. Sakakibara, and M. Goto, *Chem. Pharm. Bull.*, 1991, **39**, 270.
91CPB282	A. Miyashita, N. Taido, S. Sato, K. Yamamoto, H. Ishida, and T. Higashino, *Chem. Pharm. Bull.*, 1991, **39**, 282.
91CPB486	S. Ohta, S. Konno, and Y. Yamanaka, *Chem. Pharm. Bull.*, 1991, **39**, 486.
91CPB2117	A. Ohsawa, T. Itoh, K. Yamaguchi, and C. Kawabata, *Chem. Pharm. Bull.*, 1991, **39**, 2117.
91G113	M. B. El-Ashmawy, I. A. Shehata, H. I. El-Subbagh, and A. A. El-Emann, *Gazz. Chim. Ital.*, 1991, **121**, 113.
91G209	M. El-Mobayed, A. N. Essawy, A. El-Bahnasawi, and A. M. Amer, *Gazz. Chim. Ital.*, 1991, **121**, 209.
91H855	K. Nagata, T. Itoh, M. Okada, H. Takahishi, and A. Ohsawa, *Heterocycles*, 1991, **32**, 855.
91H895	A. Deeb, B. Bayoumy, A. E. Essawy, and R. Fikry, *Heterocycles*, 1991, **32**, 895.
91IJC(B)710	T. Ramalingam and M. S. R. Murty, *Indian J. Chem. Sect. B*, 1991, **30**, 710.
91IZV241	V. A. Dorokhov, A. R. Amamchyan, V. S. Bogdanov, and B. I. Ugrak, *Izv. Akad. Nauk SSSR, Ser. Khim.*, 1991, 241; *Chem. Abstr.*, 1991, **115**, 8734.
91JCR(S)272	R. N. Butler, D. Cunningham, K. J. Fitzgerald, P. McArdle, and K. F. Quinn, *J. Chem Res. (S)*, 1991, 272.
91JCS(P1)991	T. Yamasaki, E. Kawaminami, T. Yamada, T. Okawara, and M. Furukawa, *J. Chem Soc. Perkin Trans. 1*, 1991, 991.
91JCS(P1)1762	A. S. Wells, P. W. Sheldrake, I. Lantos, and D. S. Eggleston, *J. Chem Soc. Perkin Trans. 1*, 1991, 1762.
91JCS(P1)2045	T. Kaihoh, T. Itoh, K. Yamaguchi, and A. Ohsawa, *J. Chem Soc. Perkin Trans. 1*, 1991, 2045.
91JHC199	Y. Kurusawa, H. S. Kim, T. Kawano, R. Katoh, and A. Takada, *J. Heterocycl. Chem.*, 1991, **28**, 199.

91JHC205	J. K. Luo and R. N. Castle, *J. Heterocycl. Chem.*, 1991, **28**, 205.
91JHC231	P. Stoss, E. Kaes, G. Eibel, and V. Thewalt, *J. Heterocycl. Chem.*, 1991, **28**, 231.
91JHC417,425	A. Stimac, B. Stanovnik, and M. Tisler, *J. Heterocycl. Chem.*, 1991, **28**, 425.
91JHC545	A. Santagati, M. Santagati, and F. Russo, *J. Heterocycl. Chem.*, 1991, **28**, 545.
91JHC561	J. Reiter and K. Esses-Reiter, *J. Heterocycl. Chem.*, 1991, **28**, 561.
91JHC987	J. T. Shaw, W. L. Corbett, G. D. Cuny, F. M. Egler, and V. S. Peciulis, *J. Heterocycl. Chem.*, 1991, **28**, 987.
91JHC995	E. A. M. Badawey and T. Kappe, *J. Heterocycl. Chem.*, 1991, **28**, 995.
91JHC1043	J.-P. Vors, *J. Heterocycl. Chem.*, 1991, **28**, 1043.
91JHC1163	M. C. Wilkes, *J. Heterocycl. Chem.*, 1991, **28**, 1163.
91JPR229	K. Gewald, T. Jeschke, and M. Gruner, *J. Prakt. Chem.*, 1991, **333**, 229.
91MC46	A. V. Gulevskaya, A. F. Pozharskii, S. V. Shorshnev, and V. V. Kuz'menko, *Mendeleev Commun.*, 1991, 46.
91T3959	N. Haider, *Tetrahedron*, 1991, **47**, 3959.
91T4317	T. Itoh, K. Nagata, M. Okada, H. Takahishi, and A. Ohsawa, *Tetrahedron*, 1991, **47**, 4317.
91ZN(B)835	A. Deeb, A. N. Essawy, F. Yasine, and R. Fikry, *Z. Naturforsch., B. Chem. Sci.*, 1991, **46**, 835.

CHAPTER 6.4

Six-Membered Ring Systems: With O and/or S Atoms

JOHN D. HEPWORTH
Lancashire Polytechnic, Preston, UK

6.4.1 HETEROCYCLES CONTAINING ONE OXYGEN ATOM

6.4.1.1 Pyrans

Once again, the Diels-Alder reaction features predominantly in the construction of the pyran ring system, with emphasis being placed on stereoselectivity. The reaction between methyl glyoxylate and 1-methoxybutadiene gives the *cis*-adduct with 94% ee when catalysed by a chiral Ti complex (91TL935). A similar cycloaddition is efficiently conducted under pressure in a microwave (91TL1723).

Enaminoketones form synthetically useful 4-amino-3,4-dihydropyrans with electron rich alkenes. An electron withdrawing function is necessary at the 2-position of the 1-oxabutadiene (91CB881).

NR^2R^3 R^1 O + R^4 → NR^2R^3 R^1 O R^4 + NR^2R^3 R^1 O R^4

2-Arylidene-1,3-indandiones also function as an oxabutadiene system reacting with ynamine esters or nitriles to form indeno[1,2-b]pyrans (1) (91TL4051).

O O R + N EWG —THF, 20°C→ N O EWG O R

(1)

Cyclopenta[b]pyrans (2), red in colour and isoelectronic with azulenes, are formed when (β-dimethylaminovinyl)carbene chromium complexes are heated with either phenylacetylene or pent-1-yne presumably by an intramolecular [4+2]-cycloaddition (91AG(E)1658).

$(CO)_5Cr$ OEt NMe$_2$ OEt Me Me → R OEt R O OEt Me Me (2)

A [2+2]-photocycloaddition features in the reaction of spirocyclic 1,3-dioxin-4-ones with cyclopentenes which can lead to a single diastereoisomer. Iridoids, cyclopenta[c]dihydropyrans, (3) are readily obtained from the adducts (Scheme 1) (91JHC241).

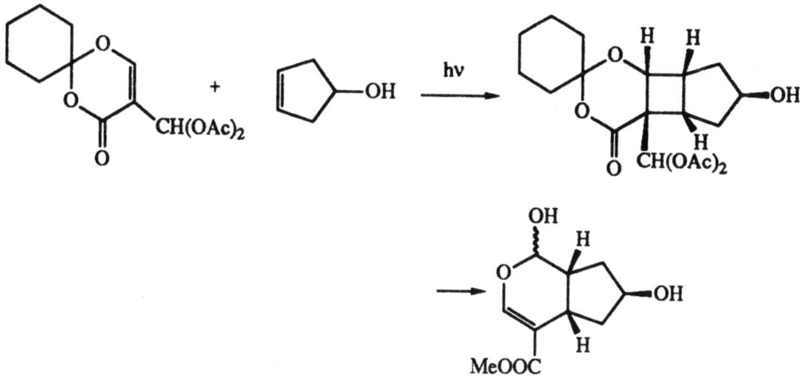

O O O CH(OAc)$_2$ + OH hν → O O O H H OH H CH(OAc)$_2$

→ OH H O OH H MeOOC (3)

Scheme 1

Deprotonation at the allylic site in 5-bromo-3,4-dihydro-2H-pyran leads to 1-oxa-2,3-cyclohexadiene which can be trapped by alkenes or dienes. The observed regioselectivity can be rationalised on the basis of the electron poor 3,4-double bond acting as a dienophile in a [4+2]-cycloaddition, whilst the electron rich 2,3-double bond takes part in [2+2]-reactions (Scheme 2) (91T4603).

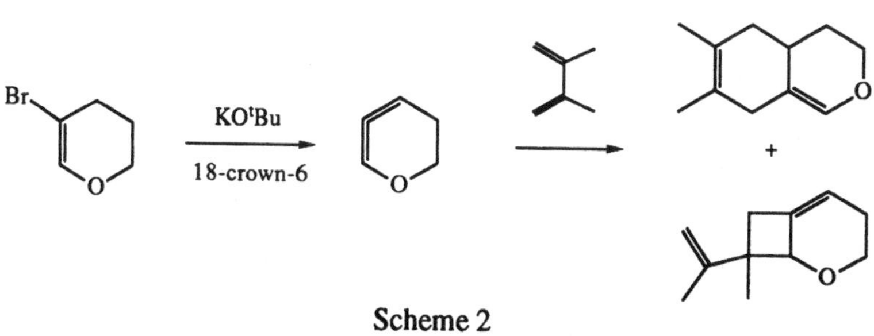

Br O KO^{t}Bu 18-crown-6 → O → O + O

Scheme 2

Stereochemical control plays an important role in a number of syntheses of tetrahydropyrans which proceed through an intramolecular cyclisation. Catalysis by the Lewis acid system $TiCl_4$-PPh_3 effects the cyclisation of ω-organometallic ether acetals (4) to 2,3-disubstituted tetrahydropyrans with high diastereoselectivity (91SL823) and the stereochemistry of the radical cyclisation of the alkenyl ethers (5) can be controlled to give the *trans*-2,3-disubstituted pyran system through the introduction of a bulky terminal (Z)-silyl substituent, which energetically differentiates between the possible transition states (91JA2336).

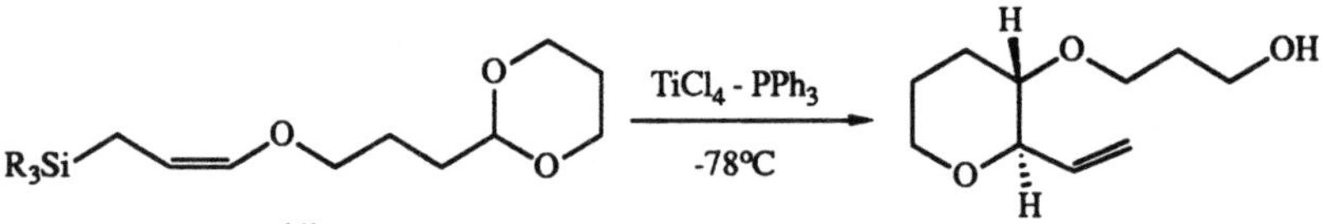

R_3Si (4) $TiCl_4$ - PPh_3 -78°C H O OH H

$SiPh_2{}^tBu$ SPh COO^tBu (5) $HSnBu_3$ AIBN benzene , Δ $SiPh_2{}^tBu$ COO^tBu

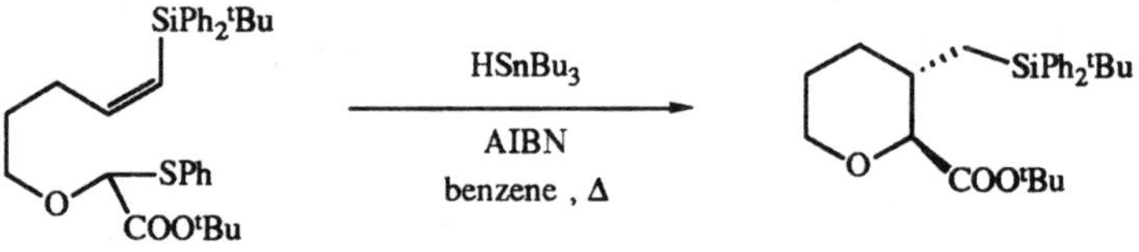

Control of the intramolecular displacement of the N-phenylcarbamate moiety in (6) is achieved by acetylation of the secondary alcohol function, preventing formation of a tetrahydrofuran ring and forcing the cyclisation to involve the tertiary hydroxy group. The consequence is a neat route to all four optically pure enantiomers of 2,2,6-trimethyl-6-vinyl-2H̲-tetrahydropyran-3-ols (Scheme 3) (91S681).

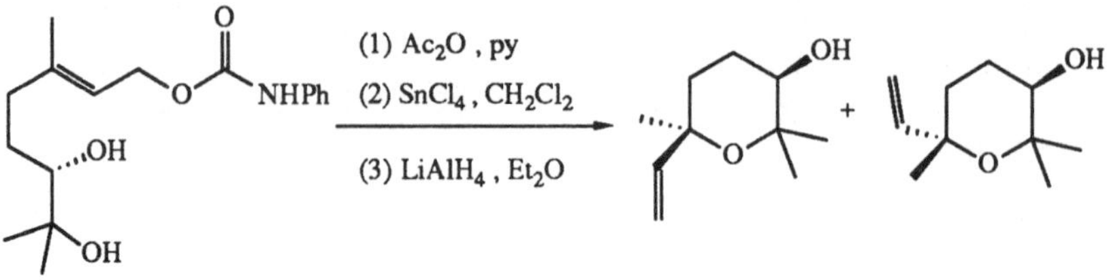

O NHPh OH OH (6) (1) Ac_2O , py (2) $SnCl_4$, CH_2Cl_2 (3) $LiAlH_4$, Et_2O OH + OH

Scheme 3

Intramolecular nucleophilic capture of the cations produced when the 1,5-diols derived from iron tricarbonyl-butadiene complexes leads to a separable mixture of substituted tetrahydropyrans which can easily be decomplexed (Scheme 4) (91SL195). Related methodology is utilised in a leukotriene synthesis via a substituted dihydropyran-4-one (91SL895). Manganese pentacarbonyl features in a synthesis of the pheromones of both the common wasp and the olive fruit fly from a common precursor (91JOC3207).

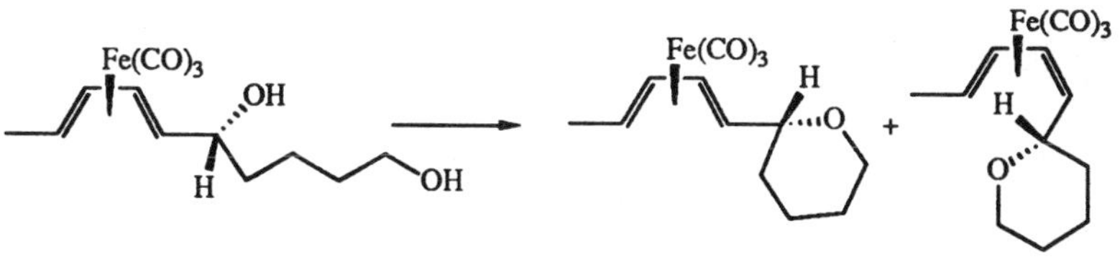

Scheme 4

The Pd-catalysed spiro-cyclisation of 1-(4-hydroxyalkyl)-1,3-cycloalkadienes can be controlled to give either *cis*- or *trans*-1,4-addition across the unsaturated system (Scheme 5) (91JOC2274).

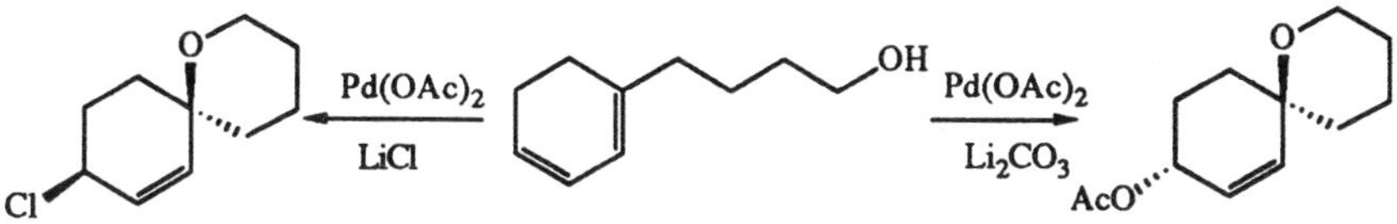

Scheme 5

Developments in the chemistry of the potent anti-parasitic macrolides, the avermectins and milbemycins, since 1985 are covered in a two part wide ranging review (91CSR211 and 271). Progress in these areas since 1990 include a highly convergent total synthesis of avermectin B1a which is applicable to the milbemycin series (91JCS(P1)667). A facile total synthesis of milbemycin α1 involves cyclisation of the seco-acid without isomerisation (91JA1830) and a similar reaction has been discussed vis-a-vis dihydroavermectin Blb (91SL618). A total synthesis of the aglycone of this avermectin has been described (91SL611 and 614).

6.4.1.2 Chromenes

2-Iodophenols undergo a Pd-catalysed heteroannulation with tertiary allylic alcohols providing a direct route to 2,2-disubstituted 2H-chromenes (91TL7739),whilst precocenes 1 and 2 result from the reaction of the corresponding phenol with 3-methylbut-2-enal in pyridine (91TL7251). 4-Halogenochromenes are produced when chroman-4-ones are boiled with phosphorus trihalide (91TL827). 2-Hydroxyacetophenones react with the carbanion derived from phenyl vinyl sulphoxide and the resulting β-

hydroxysulphoxides cyclise under basic conditions to the 4-hydroxymethyl-2H-chromene (91S569).

Acid catalysed ring closure of the bis(2-oxoalkyl)ethers of 2,6-dialkyl-1,5-naphthalenediols affords 1,6-dioxapyrenes (7) which are good electron donors, giving charge transfer salts with tetracyanoquinodimethane as well as cation radical salts upon electrochemical oxidation. In the absence of the 2- and 6-alkyl substituents, furonaphthofuran formation competes very successfully with pyran ring construction (91JOC7055).

Four macrocyclic chromenes have been isolated from the Seychelles sponge, *Smenospongia sp.*, and their structures elucidated. These unusual compounds, which may be derived from farnesyl hydroquinone, are typified by (8) in which the vinyl methyl group lies under the aromatic system and the resulting ^{1}H nmr signal is shifted upfield to 1.19 ppm (91JOC6271).

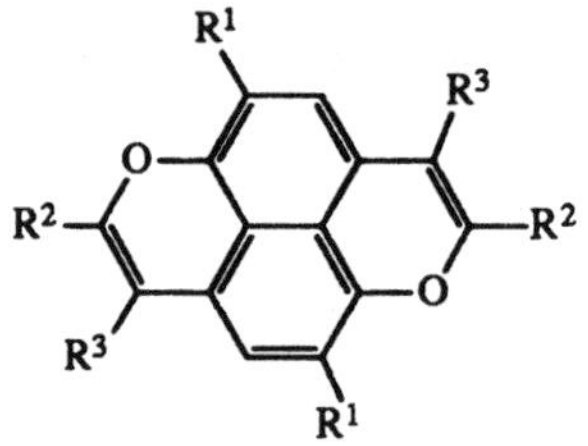

(7) 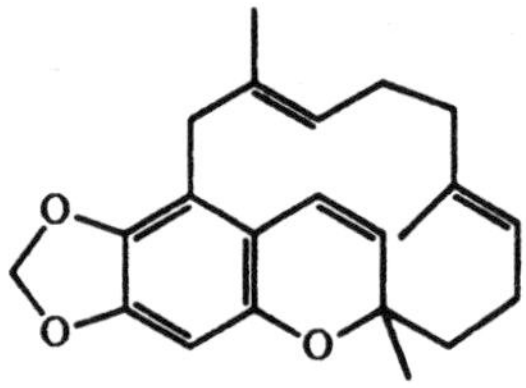 (8)

6.4.1.3 Chromans

The asymmetric epoxidation of 2,2-dimethylchromenes can be achieved using commercial bleach as the oxygen source and an enantiomer of Mn(III) salen as the catalyst. A simplified route to (-)cromakalim (9), a potassium channel activator, results (91TL5055). Dimethyl substitution at C-2 appears to be essential for optimal activity of cromakalim as a smooth-muscle relaxant, since both the monomethyl and the unsubstituted chromanol show reduced potency (91JCS(P1)2763).

The chroman-4-ol (10) obtained via the pyrrolidine-catalysed reaction of ketones with 2-hydroxyacetophenone undergoes an acid catalysed dehydration with concomitant cyclisation to the oxazepine (11) (91AG(E)1709).

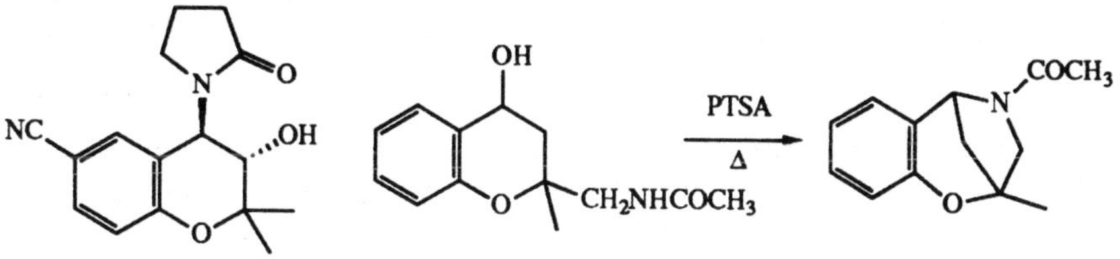

(9) (10) (11)

Chroman-3-ols result directly from the reaction of *o*-allylphenols with mCPBA, through the intermediacy of the 2',3'-epoxide (91SC1455).

The continued use of marijuana as a drug together with the known pharmacological activity of cannabinoid derivatives ensure that interest in Δ^9-THC is maintained. An improved synthesis of Δ^9-THC (13, R=Me) which appears to eliminate most of the by-products and resinous material associated with other routes, involves the reaction of olivetol with *cis*-p-menth-2-en-1,8-diol at room temperature in the presence of pTSA. The THC-precursor (12) can be isolated (91SL553).

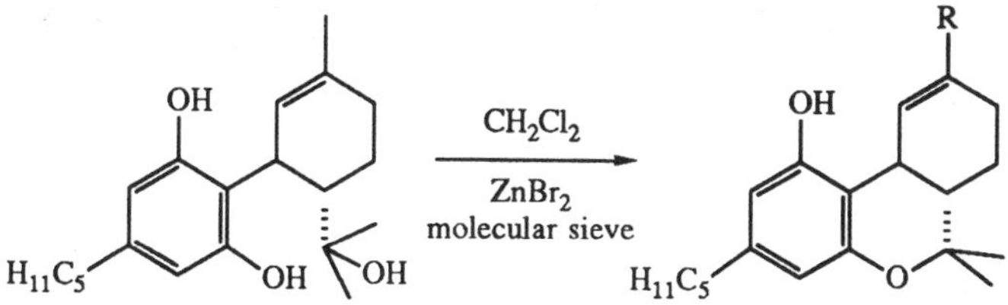

(12) (13)

A range of 11-substituted metabolites of Δ^9-THC has been prepared in both racemic and optically active forms from the cyclohexenol (14). The dithiane unit not only gives access to 11-oxo derivatives without recourse to troublesome oxidative conditions, but also inhibits isomerisation of the C-9 double bond (91JOC6865). Both of these problems are also addressed in a seven step synthesis of the major human metabolite (13, R = COOH) of Δ^9-THC. The key step is the regioselective condensation of the aryl lithium derived from the bis MOM ether of olivetol with (+)-apoverbenone, precluding formation of iso-cannabinoids (91JOC1481).

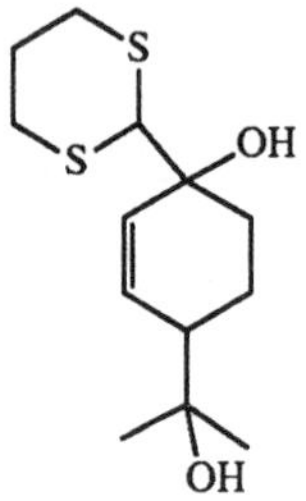

(14) 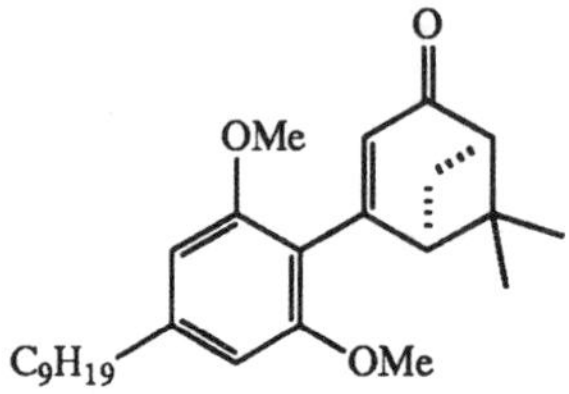 (15)

An improved route which avoids the need for MOM ethers also offers the advantage that both the natural (6aR, 10aR) and unnatural (6aS, 10aS) cannabinoid series are accessible from the common intermediate (15). Both enantiomers of nabilone, a synthetic cannabinoid with anti-emetic properties, have been obtained from the enone (91JOC2081).

Isomerisation of the Δ^9-double bond is not observed during oxidation of the 9-formyl group to a carboxylic acid function by sodium chlorite and using 2-methylbut-2-ene as the scavenger for hypochlorous acid (91S851).

Two rotationally restricted THC ethers (16) and (17) have been prepared from Δ^8-THC derivatives to help assess the importance of the orientation of the lone pairs of electrons on the phenolic oxygen atom on the pharmacological activity (91JOC1549).

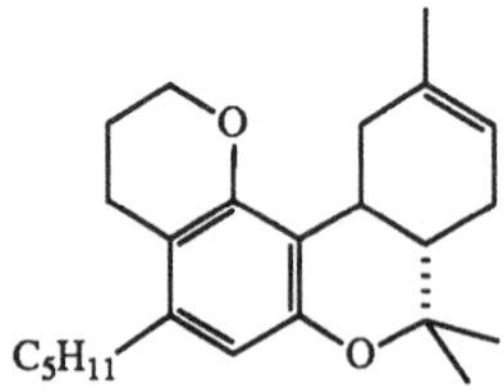

(16) 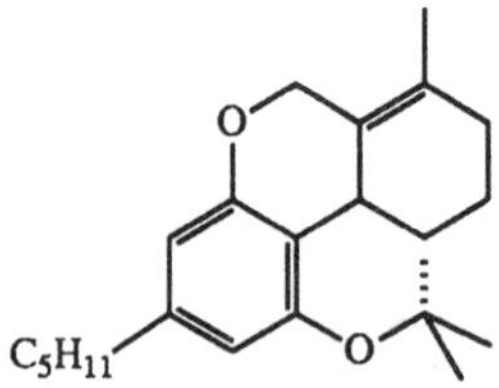 (17)

The formation of chroman derivatives (18) by a one-pot combination of the Dotz benzannulation of vinylchromium carbene complexes and a Mitsunobu cyclisation has been extended to decalin-derived carbene complexes from which the oxasteroid skeleton can be generated (91TL7759).

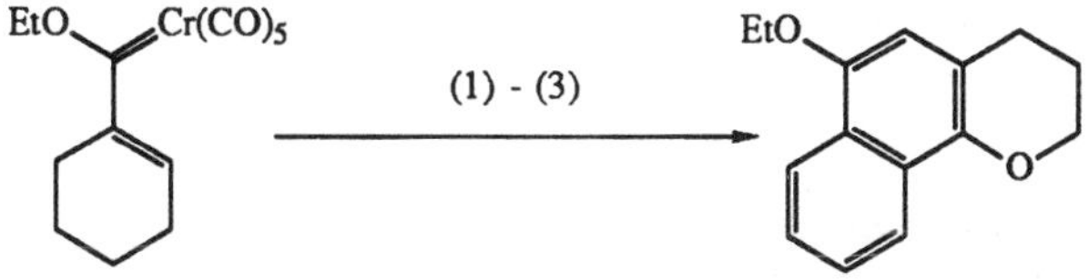
EtO
Cr(CO)5
(1) - (3)
EtO
O

(18)

Reagents : (1) 4-pentynol , THF , Δ ;(2) Ph_3P , DEAD , THF , RT ; (3) hν , PhH , RT.

Phenyl and naphthyl substituted dioxolanes (19) undergo an intramolecular Mukaiyama reaction on treatment with $TiCl_4$ to yield isochromans (20) and the angular benzologues, respectively (91CC287). Cyclopentane annulated derivatives of isochroman result from the Pd-catalysed tricyclisation of the 4- and 10-oxa derivatives of 2-bromotetradeca-1-ene-6,12-diynes (Scheme 6) (91SL777).

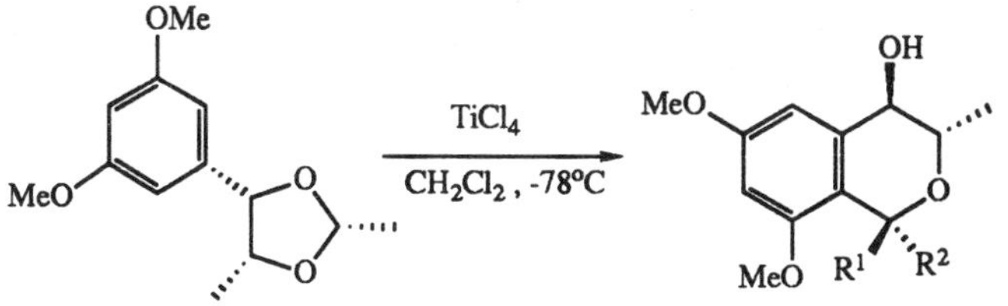
OMe
MeO
O
O
TiCl4
CH2Cl2 , -78°C
OH
MeO
O
MeO
R1
R2

(19) (20)

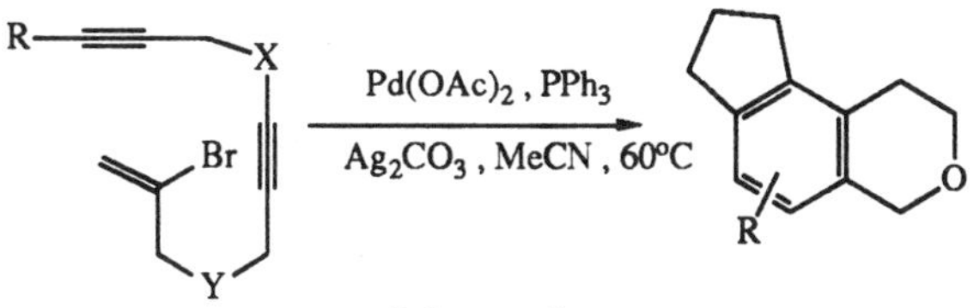
R
X
Br
Y
Pd(OAc)2 , PPh3
Ag2CO3 , MeCN , 60°C
O
R

Scheme 6

6.4.1.4 Pyranones

Unlike the parent compound, 3-bromopyran-2-one undergoes regiospecific Diels-Alder reactions with both electron rich and electron deficient dienophiles on moderate heating and at atmospheric pressure. Since subsequent reductive debromination is readily accomplished, the compound is a viable synthetic equivalent to 2-pyranone itself and its value is enhanced by a convenient new synthesis (Scheme 7) (91TL5295).

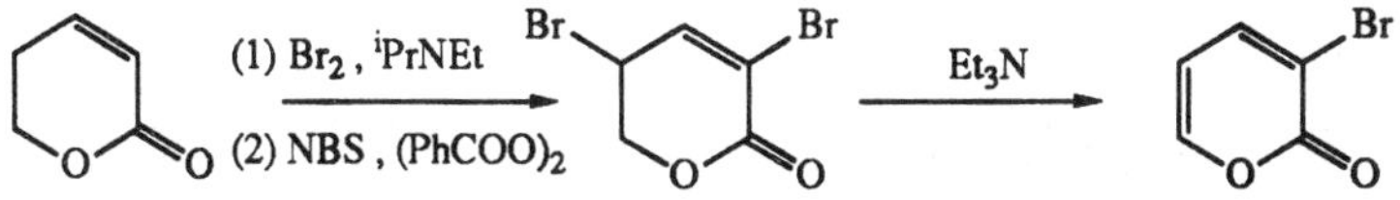
O
O
(1) Br2 , iPrNEt
(2) NBS , (PhCOO)2
Br
Br
O
O
Et3N
Br
O
O

Scheme 7

In the presence of t-butyldimethylsilyl triflate, both 6-methyl-2H-pyran-2-one and 2H-1-benzopyran-2-one (coumarin) act as electron deficient dienophiles and undergo cycloaddition reactions with 2-(trialkylsilyloxy)dienes. Introduction of a 3-ester function enhances their reactivity which is a result of their conversion to the corresponding pyrylium salt (91JOC5052).

Photolysis of 4-(ω-alkenyloxy)-6-methylpyran-2-ones (21) results in intramolecular [2+2]-cycloadditions to give tricyclic derivatives (22) in the presence of a photosensitiser. In its absence, cyclobutene carboxylic acids result (91JOC7150).

hν, Ph_2CO

(21) (22)

A synthesis of the pyran-2,5-dione unit, from which the grevillin fungal pigments can be derived, is based on the cyclisation of oxalates (24) prepared from unsymmetrically substituted benzylacyloins (23) (91JCS(P1)2363).

DBU, DMF, -15°C

(23) (24)

Quenching of the anion derived from ethyl 2-(dimethylaminomethylene)-acetoacetate with acid chlorides leads to 6-substituted 3-carbethoxypyran-4-ones on acidic work up (91JOC4963). The enol acetate of ethyl 4,4,4-trifluoro-3-oxobutanoate self condenses under Lewis acid catalysis to give 2,6-bis(trifluoromethyl)pyran-4-ones (91JHC819).

Cyclisation of the acetylenic ketones (25) leads to dihydropyran-4-ones which are a useful *in situ* source of 2H-pyrans (26) from which highly substituted aromatic compounds can be obtained by reaction with electron deficient alkynes (Scheme 8) (91HCA27).

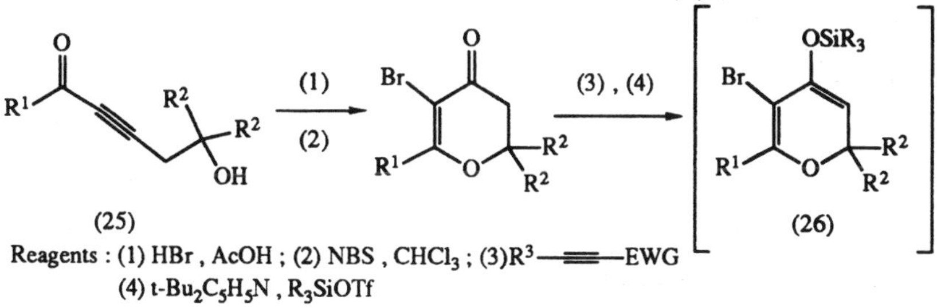

(25) (26)

Reagents : (1) HBr , AcOH ; (2) NBS , $CHCl_3$; (3)R^3—≡—EWG
(4) t-$Bu_2C_5H_3N$, R_3SiOTf

Scheme 8

A metal carbonyl complex is used as a control element in the hetero-Diels-Alder reaction between the dienal, sorbaldehyde, and Danishefsky's diene. The diastereochemistry of the resulting dihydropyranone is influenced by the catalyst, $BF_3.Et_2O$ leading predominantly to (27) and $TiCl_4$ to (28) (91JOC4563).

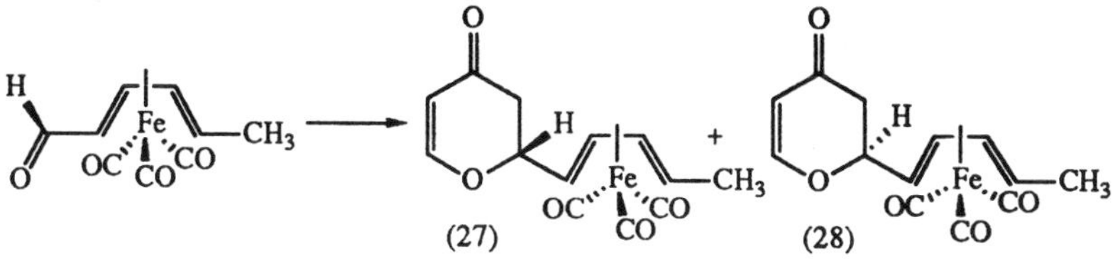

(27) (28)

6.4.1.5 Coumarins

When treated with SF_4 in an attempt to prepare the 4-fluorocoumarin, 6-chloro-4-hydroxycoumarin formed a bis(benzopyrano)oxathiazole in good yield. 4-Fluorocoumarins can be obtained by halogen exchange (91S937) and coumarins are regioselectively perfluoroalkylated at the 3-position by reaction with bis(perfluoroalkanoyl)peroxides (91SL113). TFAA effects the conversion of the potassium salts of *o*-methoxybenzylidene malonic acids into coumarin-3-carboxylic acids without the need for either isolation of the free malonic acid or demethylation (91JCS(P1)219).

3-(1,1-Dimethylallyl)coumarins can be obtained from dihydrocoumarins by opening the lactone ring with 3-methylbut-2-en-1-ol followed by an Ireland-Claisen rearrangement and ring closure of the resulting acid (91TL3209).

A combination of the directed ortho-metallation of benzamides and the Pd-catalysed arylboronic acid-aryl bromide cross coupling sequence provides a general regiospecific synthesis of dibenzo[b,d]pyran-6-ones (29) (91JOC3763). The same condensed system results from the arylation of phenols by azosulphides. An ortho-cyano function in the sulphide ensures that lactonisation follows an $S_{RN}1$ process (91T9297).

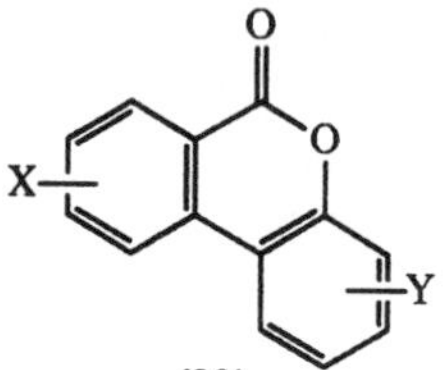

(29) 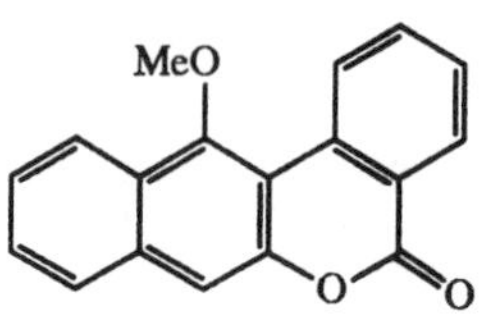 (30)

The benzo[d]naphtho[2,3-b]pyran-5-one system (30) is formed directly when homophthalic anhydride is treated successively with methanol and a methylating agent. The reaction is considered to proceed via the two half esters of homophthalic acid (91JCR(S)279).

A Pd-catalysed cross coupling reaction also features in a synthesis of isocoumarins which can be readily adapted to the formation of pyranopyridine derivatives (91T1877).

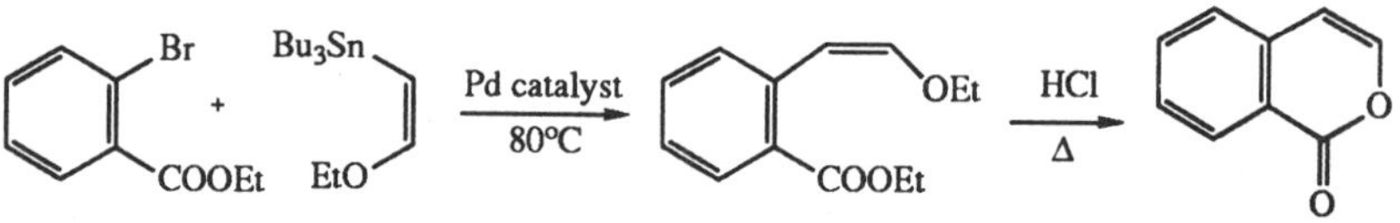

Br + Bu_3Sn COOEt EtO —Pd catalyst, 80°C→ OEt COOEt —HCl, Δ→

A highly substituted isocoumarin resulted from attempts to acylate 2,3,6-trimethoxytoluene with (E)-2-methylbut-2-enoic acid in the presence of TFAA. The additional oxygen atom is derived from air and is thought to be incorporated through rearrangement of an indanone hydroperoxide (91JCS(P1)89).

2-Benzopyran-3-ones are reactive species which are present in solutions of o-formylphenylacetic acids in acetic anhydride. The steroid nucleus has been constructed from them by a sequence of reactions initiated by a Diels-Alder reaction with a cyclopentane based dienophile (91JCS(P1)1683). This benzopyranone system is stabilised by a 6-alkoxy function presumably by a donor-acceptor interaction involving this substituent and the carbonyl group (91JCS(P1)169) and indeed the 2-benzopyran-3-one derived from 2-acetyl-4,5-dimethoxyphenylacetic acid by treatment with DCC and 2-hydroxypyridine in acetonitrile can be isolated. Its ^1H chemical shifts are indicative of a lack of aromatic character (91JCS(P1)639).

6.4.1.6 Chromones

Substitution at the 2-position of chromones can be achieved through the intermediacy of the 4-silyloxy-1-benzopyrylium salt. Subsequent reaction with silyl enol ethers or allyl organometallic reagents in the presence of 2,6-lutidine results in nucleophilic attack at C-2 and the formation of the 4-silyloxychromene (31) and/or the derived chromanone (32). The former compounds can undergo electrophilic attack at C-3, giving (33). Combination of these two modes of attack by the use of α,β-unsaturated ketones results in a cyclohexene annulation and formation of tetrahydroxanthone derivatives (34) (Scheme 9) (91JOC2058).

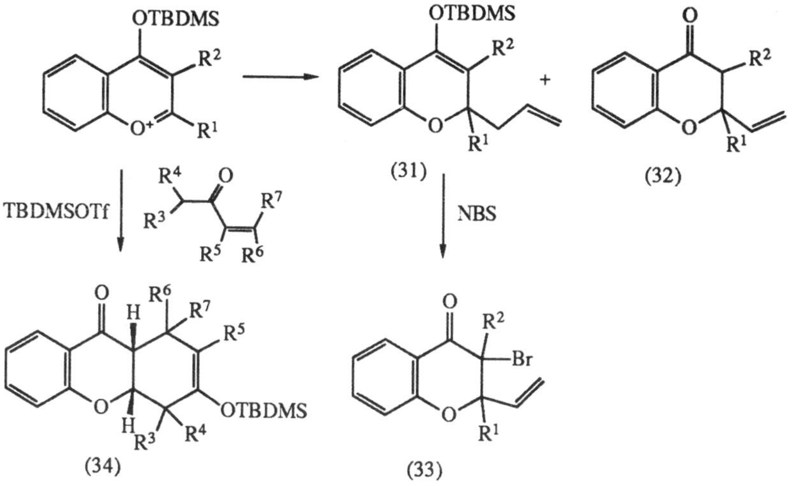

Scheme 9

The aldol condensation of 2-hydroxyacetophenone with 4-chlorobenzaldehyde in ethanol yields the 1,3-butadiene derivative (35) in addition to the expected chalcone. The two additional carbon atoms are derived from the solvent which is presumably oxidised to acetaldehyde during the reaction. Subsequent cyclisation of the butadiene provides a new route to 2-styrylchromones (91CL445).

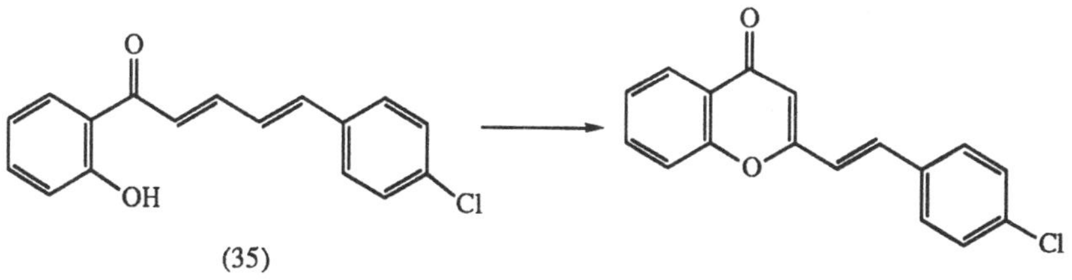

Under Mukaiyama aldol conditions, enolisable aldehydes and ketones react with the bis-silyl enol ether derived from 2-hydroxyacetophenone to give β-hydroxyketones, from which 2-substituted chroman-4-ones are readily available (91JOC1325). 2-Spiro-annulated chromanones result from the Lewis acid catalysed reaction of resorcinols with cycloalkylidene acetic acids (91T4775) and undergo an oxidative rearrangement to 2,3-fused chromones on reaction with thallium(III) nitrate (91TL5619).

Unlike simple chroman-4-ones, their 3-phenylsulphonyl derivatives undergo an efficient C-alkylation, providing routes to both 3-alkylchromanones by reductive desulphonilation and 3-alkylchromones by elimination of phenylsulphinic acid (91TL7727).

Routes to the little studied chroman-3-ones have been reviewed (91BSF189) and the hydroboration-PCC oxidation of 4-substituted 2H-chromenes is a recommended route to them (91S879).

A synthesis of the 5-thiorotenoid system (37) is based on formation of the alkene (36) by a Wadsworth-Emmons reaction followed by a Mukaiyama cyclisation (91CC972).

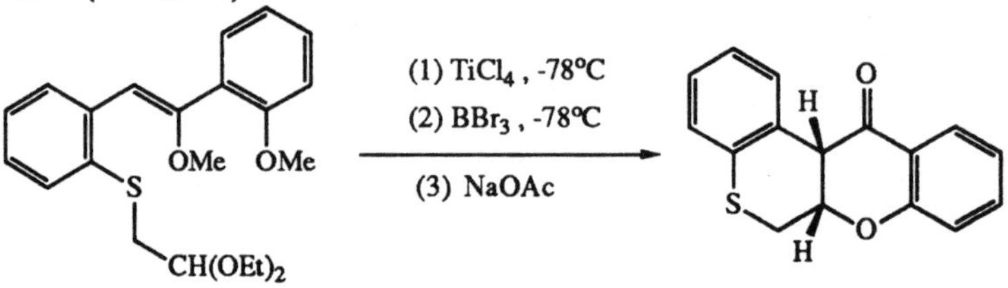

(1) $TiCl_4$, -78°C
(2) BBr_3, -78°C
(3) NaOAc

(36) (37)

6.4.1.7 Flavonoids

Flavones with hydroxy substituents in both the A and C rings are readily prepared by the reaction of the lithium polyanions of hydroxylated acetophenones with O-silyloxylated methyl benzoate derivatives. Acid catalysed cyclisation of the resulting propan-1,3-diones occurs with concomitant cleavage of the protecting groups (91JOC4884, 91T5071).

Polyhydroxy isoflavones are accessible by cyclisation of the deoxybenzoins formed from the Lewis acid catalysed reaction of phenols with phenylacetic acids. Dimethyl formamide acts as the source of the C-2 atom in the isoflavone (91JCS(P1)3005).

On treatment with dimethyldioxirane, flavones are converted into their relatively unstable epoxides, which rearrange at room temperature to 3-hydroxyflavones. With methanol, the 3-membered ring is opened yielding substituted flavanones (Scheme 10) (91JOC7292).

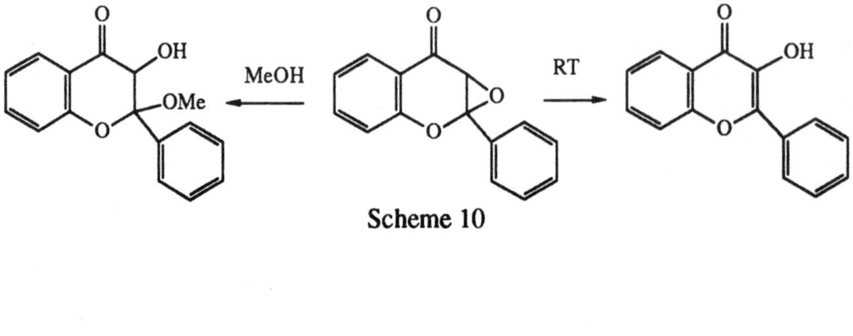

MeOH RT

Scheme 10

The wide range of colours of flowers and fruits is largely a consequence of the presence of anthocyanins. It has long been thought that the pH of the sap had a major influence on the colour of the anthocyanin. However, the reasons are more complex, involving for example intra- and inter-molecular stacking, co-pigmentation and chelation, all of which affect the equilibria between the flavylium cation, its carbinol base, a quinonoid anhydro-base and the ring-opened form, the chalcone (91AG(E)17, 91JCS(P2)891 and 1287).

6.4.1.8 Xanthones

In aqueous acetonitrile, xanthene photoisomerises to 6H-dibenzo[b,d]pyran. The intermediacy of a singlet biradical is proposed, recombination of which at the *ipso* benzylic position affords a cyclohexadienone and thence a quinone methide (91JOC5437).

The pink colour associated with the fibrous clay sepiolite, quincyte, dating from the Eocene period, is caused by a number of organic pigments. Their structures are all based on the *peri*-xanthenoxanthenequinone nucleus. The major component (38) has been synthesised from 6-bromo-2-naphthol utilising an oxidative phenolic coupling as the key step. Their biological origin is uncertain, though a fungal source involving the acetate-malonate pathway is a possibility (91T1095).

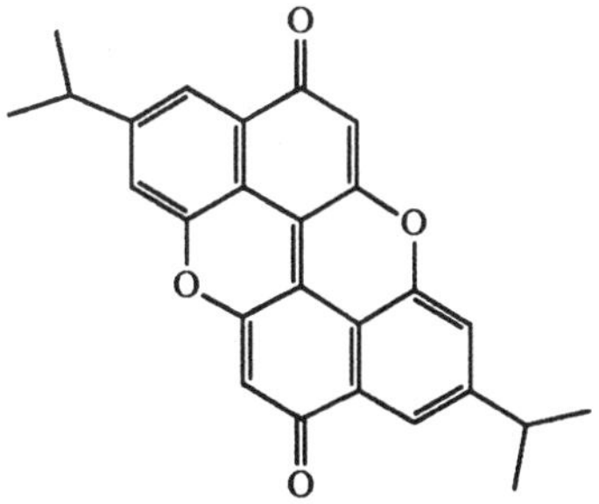

(38) 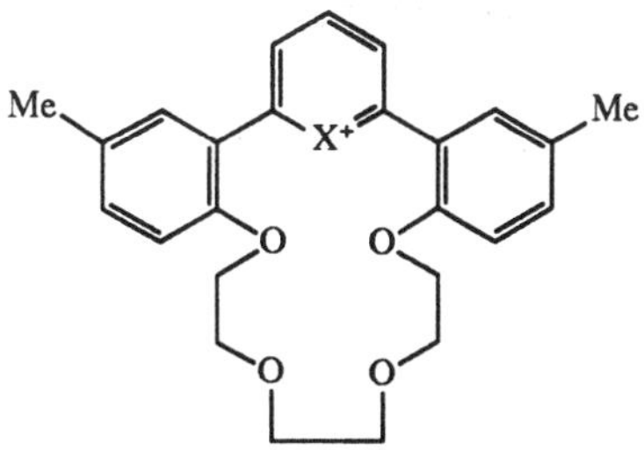 (39)

6.4.1.9 Pyrylium Salts

The synthesis, reactions and properties of benzo[c]pyrylium salts have been reviewed (90AHC158).

An additional method for changing the anion of pyrylium salts is also helpful in increasing their solubility. Treatment of the pyrylium salt with acyloxy hydroborates rapidly strips off the anion derived from a strong acid and formation of the new carboxylate salt competes favourably with reduction of the pyrylium species (91BSB665).

The ^{13}C chemical shift of C-4 in 2,6-disubstituted pyrylium salts provides a good measure of the relative electron donating ability of the substituents (91JOC126).
Corands incorporating pyrylium or thiopyrylium units (39, X = O or S) can be obtained from a common macrocyclic 1,5-pentandione precursor. The heterocyclic moiety sits in the cavity of the macro-ring (91T1977).

6.4.2 HETEROCYCLES CONTAINING ONE SULPHUR ATOM

6.4.2.1 Thiopyrans and analogues

Yet more examples of the synthesis of thiopyrans using Diels-Alder methodology have appeared. Thioaldehydes can be generated from aldehydes by reaction with bis(trimethylsilyl)sulphide in the presence of $CoCl_2.6H_2O$ and the subsequent reaction with cyclohexadiene shows high endo-selectivity. However, catalysis of the thionation by trimethylsilyl triflate allows selection of the adduct stereochemistry by varying the aldehyde: sulphurating reagent ratio (Scheme 11) (91JOC7323).

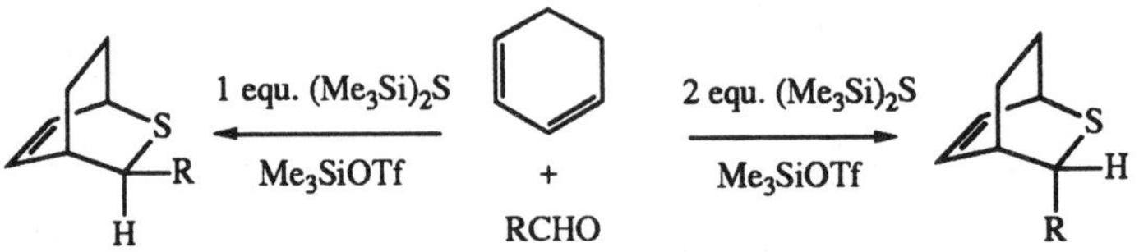
1 equ. (Me3Si)2S
Me3SiOTf
+
RCHO
2 equ. (Me3Si)2S
Me3SiOTf
S
R
H
S
H
R

Scheme 11

The reaction of thiophosgene with 2-morpholinobutadienes affords the immonium salts from which thiopyran and thiopyranone derivatives can be obtained (91SL487).

The heteroatom can also be derived from the diene component. In the first example of such, the 1-thiabutadiene, generated by thermolysis of a thiochalcone, and (-) dimenthyl fumarate reacted with 100% endo selectivity to give the two 3,4-*cis*-cycloadducts. Removal of the chiral auxiliary from the purified major adduct using LAH gave the optically pure diols (91JCS(P1)2281).

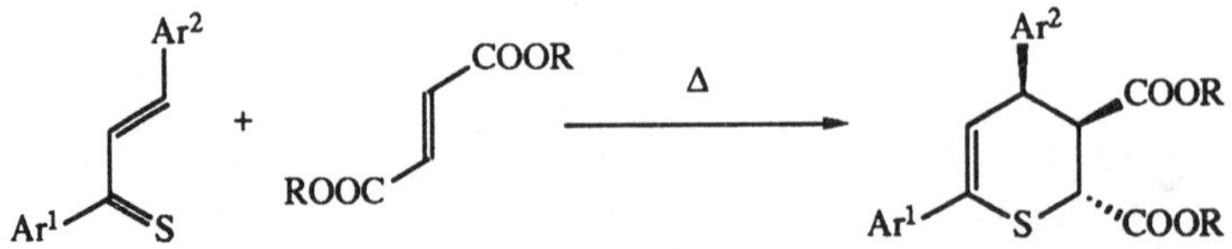
Ar2
Ar1
S
+
COOR
ROOC
Δ
Ar2
COOR
Ar1
S
COOR

The thianaphthylium salts (40) undergo [4+2]-cycloaddition reactions with butadienes to give sulphonium salts (41) but substitution at the 1-position leads to 1H-[2]thiochromenes (42) on reaction with furan. Cleavage of the C-S bond in (41) occurs on treatment with nucleophiles leading to further 1-substituted thiochromenes (91TL5571).

(41) (40) (42)

Nucleophilic substitution of the 3-nitro group in the phthalimide (43) by thiosalicylic acid provides the precursor for a range of 3,4-disubstituted thioxanthones (44). Similar displacement of a 4-chloro substituent enables the analogous 1,2- and 2,3- derivatives to be prepared, whilst functional group interconversions further extend the value of this route (91HCA1119).

PPA, Δ

(43) (44)

The ten-membered 3-ene-1,6-diyne ring system (45) undergoes a Myers-like intramolecular cyclisation on treatment with selenium dioxide. A biradical is postulated which then affords an isothiochroman-4-one (91CC694).

SeO_2, CCl_4

(45) R = H or Cl

Several condensed derivatives of thiopyran have been reported. The sulphur and selenium analogues of benzo[1,2-b][5,4-b']dipyran-2,8-dione (46) have been obtained from 4,6-dibromoisophthalaldehyde using biphasic Wittig methodology (91BSB1) and propargyl bromide reacts with 4-hydroxy-2H-[1]benzothiopyran-2-thione under phase transfer conditions to give thiopyrano[2,3-

b][1]benzothiopyran-5(4H)-one (47). However, under normal alkylation conditions, a thieno[2,3-b]chromone results (91SL595).

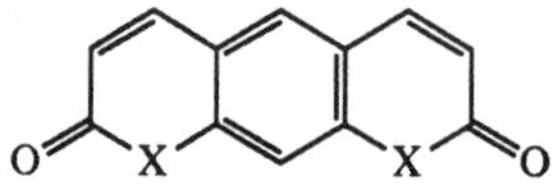

(46) X = S or Se

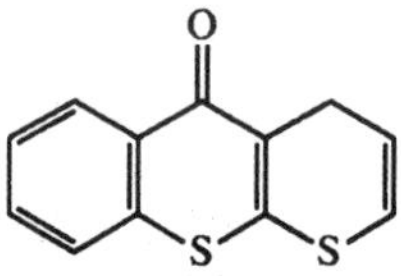

(47)

6.4.3 HETEROCYCLES CONTAINING TWO OR MORE OXYGEN ATOMS

6.4.3.1 Dioxins

The use of 2,2-dimethyl-1,3-dioxan-4,6-dione, Meldrum's acid, in organic synthesis has been reviewed (91H529).

The strong Schwesinger PN bases enable 1,3-dioxan-4-ones to be doubly alkylated diastereoselectively, whereas their lithium enolates can only be mono-alkylated with more reactive electrophiles (91CB1837). The reaction of the dioxinone lithium dienolate (48) with acetaldehyde affords the 5-substituted derivative with high diastereoselectivity. However, ketones and α,β-unsaturated and aromatic aldehydes react at the exocyclic site (91CB1845).

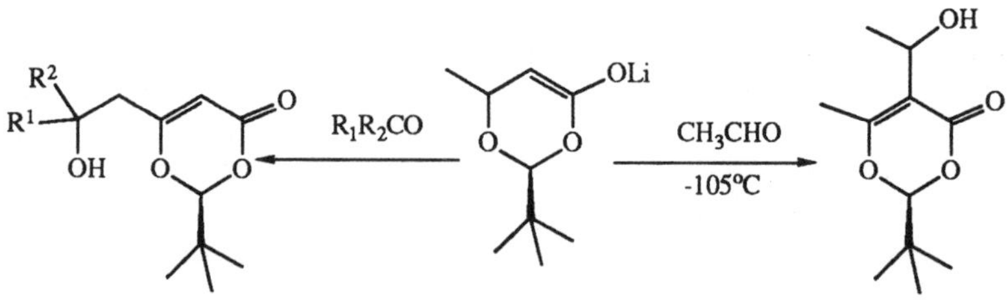

(48)

Fluorination of the 1,3-dioxinone (49) can be achieved using elemental fluorine (Scheme 12) (91CC699) and the 5-trifluoromethyl derivative is available from the 5-iodo compound by reaction with CF_3I (91CC1241).

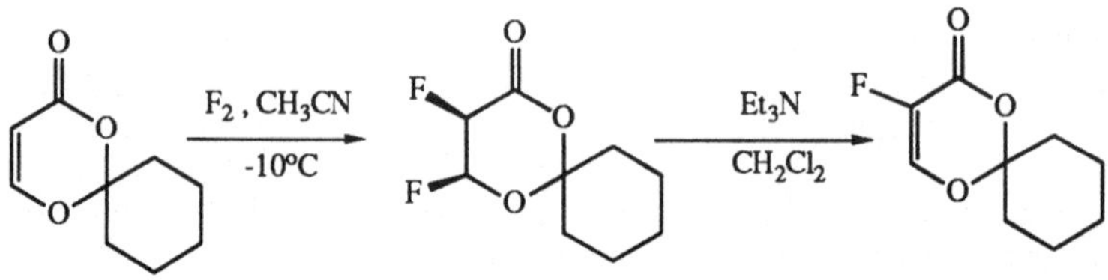

(49)

Scheme 12

3,3-Disubstituted 1,2-dioxetanes form 1,4-dioxanes as the cycloadducts with electron rich alkenes, probably via a 1,6-dipolar species (91AG(E)1365).

Mercury(II) oxide catalyses the cyclisation of prop-2-ynyloxyacetic acid and its derivatives to 6-exo-methylene-1,4-dioxan-2-ones (91JCR(S)165).

Benzo[b]naphtho[2,3-e]dioxin-6,11-quinones (50) feature an electron acceptor function attached to a donor system through two oxygen bridges, a situation offering the potential for an electrically conducting organic material (91JOC1569).

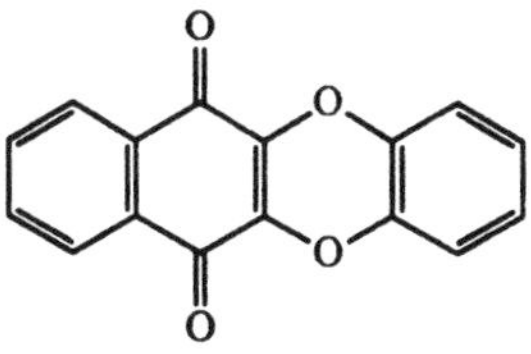

(50) 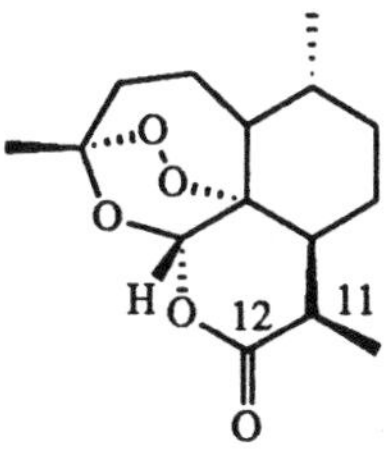 (51)

The first examples of cholestano-1,4-benzodioxanes have been synthesised for use as model compounds to enable the absolute configuration of the flavanolignan (-)-silandrin to be established from CD data (91LA633).

Addition of bromine to the double bond of 1,4-dioxins, 1,4-oxathiins and 1,4-dithiins in the presence of 1,2-diols, 1,2-dithiols or 1,2-mercaptoalcohols leads to the formation of the *cis*-isomers of various oxa- and thia-decalins. In a similar manner, bicyclic dioxins and the related sulphur analogues afford tetraheterapropellanes (91ACS410).

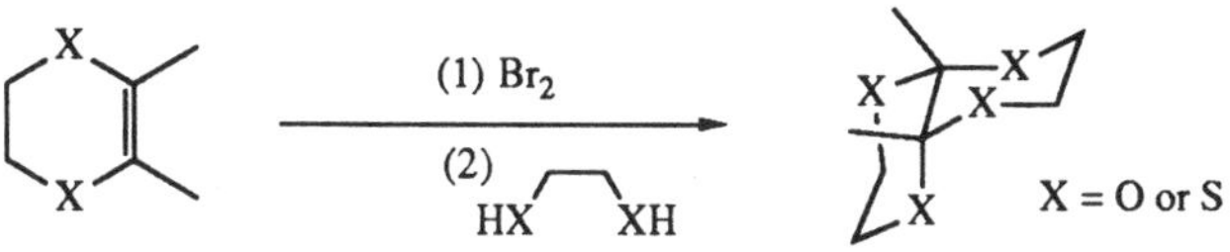

(1) Br_2 (2) HX‿XH; X = O or S

6.4.3.2 Trioxins

A three step but one-pot synthesis of 1,2,4-trioxanes from readily available materials involves an intramolecular oxymercuriation (Scheme 13) (91CC947) and the 5-keto derivative is readily synthesised by the reaction of aldehydes or ketones with trimethylsilyl α[(trimethylsilyl)peroxy]alkanoates catalysed by trimethylsilyl triflate. The trioxanones are stable to heat and do not seem to be prone to spontaneous fragmentation as previously suggested. However, they show no anti-malarial activity (91HCA1239).

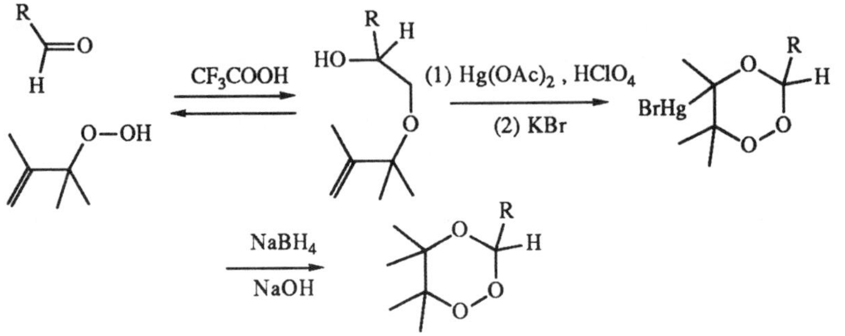

Scheme 13

Interest in the anti-malarial artemisinins (51) has been maintained and their chemistry (91H1593) and circular dichroism (91ACSI183) have been reviewed.

Reduction of the 12-keto function and dehydration of the resulting alcohol has yielded the 11,12-unsaturated analogue, from which the 11,12-epoxide has been synthesised. Facile ring opening of the 3-membered ring provided a range of 12-substituted derivatives of 11-hydroxydihydroartemisinin (91CCC1037).

6.4.4 HETEROCYCLES CONTAINING TWO OR MORE SULPHUR ATOMS

6.4.4.1 Dithiins

The thiocarbamoyl derivative (52; X = $OC(S)NMe_2$) of 2,2',6,6'-tetrahydroxybiphenyl undergoes a Newman-Kwart rearrangement on thermolysis to the carbamoylthiobiphenyl (52; X = $SC(O)NMe_2$). Alkaline hydrolysis affords the symmetrical bisdisulphide (53) (91PS51).

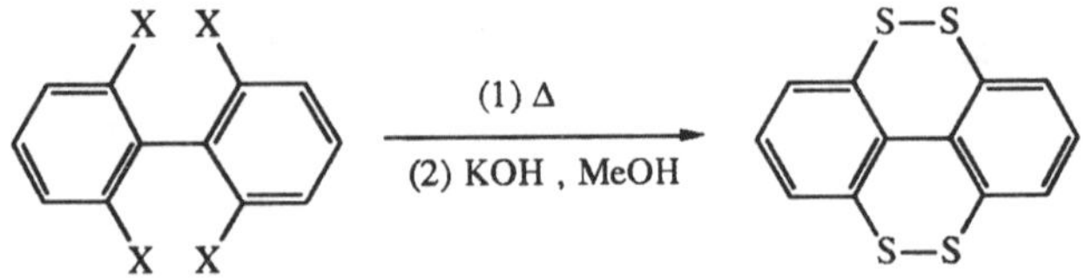

(52) (53)

The enantioselective formation of 2-substituted 1,3-dithiane 1-oxides and 1,3-dioxides involves the sequential acylation, oxidation and deacylation of 2-substituted dithianes (91SL80). The high diastereoselectivity observed in the reaction of the anion derived from *trans*-1,3-dithiane 1,3-dioxide with aromatic aldehydes is worthy of note (91TL7743).

The oxidative fluorodesulphurisation of 1,3-dithianes provides a convenient route to difluoromethylene compounds and can be effected by n-$Bu_4N^+H_2F^-$ (91SL909) or by 4-iodotoluene difluoride (91SL191).

The synthesis of 2,3-dihydro-1,4-dithiins by the rearrangement of 1,3-dithiolanes has been exploited in several directions. When the dithiolanes possess two different enolisable positions both possible dithiin derivatives are formed on reaction with bromine in anhydrous chloroform(91S223). Under the same conditions, the 1,3-dithiolane derivatives of cyclohexanones yield 1,4-benzodithiins, aromatisation being consequential on successive eliminations of HBr (91T4187). Benzo[b]pyrano[3,4-b][1,4]dithiin derivatives result in a similar manner from chroman-4-ones and naphthodithiins are formed from tetralones (91JOC6748). The acid catalysed ring expansion of 2-(1-hydroxyalkyl)-1,3-dithiolanes can lead to both the expected 2,3-dihydrodithiin and the kinetically favoured 2-alkylidene derivative (91S575).

Application of the acid catalysed rearrangement to 1,3-dithiolane sulphoxides leads predominantly to the dihydrodithiin (55). However, under neutral conditions, the *cis* and *trans* sulphoxides behave differently and undergo a sigmatropic rearrangement to the sulphenic acids (54) and (56) and hence to the dithiins (55) and (57), respectively (Scheme 14) (91T8091).

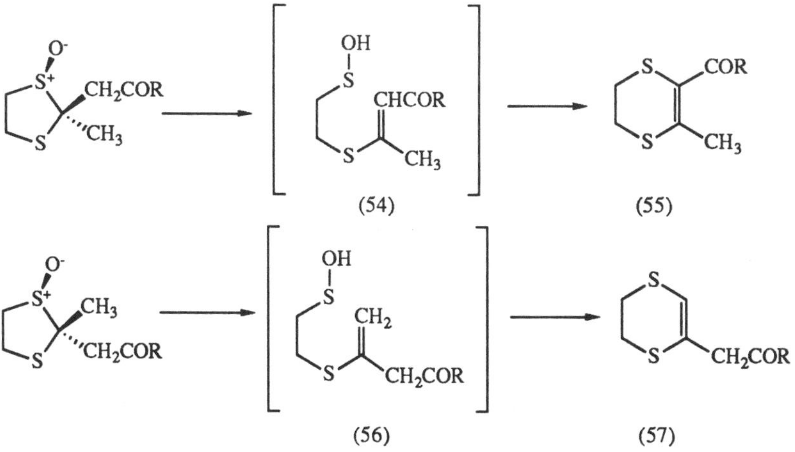

O^- S^+ CH_2COR S CH_3 → [OH S CHCOR S CH_3] → S COR S CH_3

(54) (55)

O^- S^+ CH_3 S CH_2COR → [OH S CH_2 S CH_2COR] → S S CH_2COR

(56) (57)

Scheme 14

Expansion of the 1,2-deselenolane system occurs on treatment with dimethyl diazomethanephosphonate leading to 2-dimethoxyphosphoryl-1,3-diselenanes (58). Equilibration of the diastereoisomeric *cis* and *trans* derivatives indicates Se-C-P anomeric interactions (91TL4189).

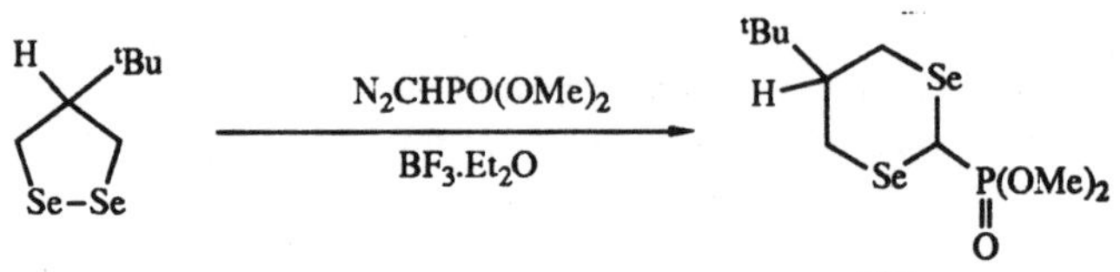
H
ᵗBu
Se−Se
N₂CHPO(OMe)₂
BF₃.Et₂O
ᵗBu
H
Se
Se
P(OMe)₂
O

(58)

6.4.4.2 Trithiins

Naturally occurring 1,2,3-trithianes have been reviewed (90SR257).

Reaction of benzopentathiepin with alkylidene phosphoranes affords good yields of 1,2,4-benzotrithiins (59) (91TL6345).

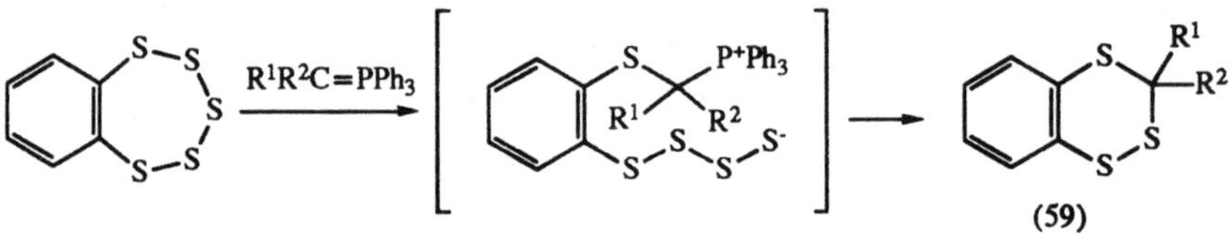
R¹R²C=PPh₃
P⁺Ph₃
R¹
R²
S⁻
R¹
R²

(59)

6.4.4.3 Tetrathiins

The synthesis, reactions and general properties of 1,2,4,5-tetrathianes have been reviewed (91SR199 and 233).

6.4.5 HETEROCYCLES CONTAINING BOTH OXYGEN AND SULPHUR IN THE SAME RING

6.4.5.1 Oxathiins

Vinyl silyl ketones act as a heterodiene on reaction with thiones giving high yields of 4H-1,3-oxathiins (60). Oxidation with mCPBA affords the S-oxides which are considerably more stable than the sulphides (91TL2971).

The α-oxosulphines derived from thiochroman-4-one 1,1-dioxides can also be trapped in Diels-Alder reactions yielding benzothiopyrano[4,3-b][1,4]oxathiins (61) (91PS129).

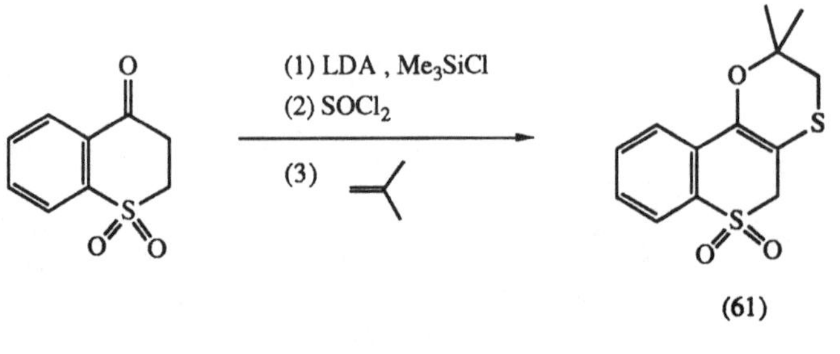
(1) LDA , Me₃SiCl
(2) SOCl₂
(3)

(61)

6.4.5.2 Dioxathiins

The hydroxysulphoxide derived from *trans*-stilbene oxide is converted into the new 1,5-dioxa-2-thiane system (62) on treatment with sulphuryl chloride. Photolysis of the *cis*-diphenyl derivative gives the isomeric sultine in which both H-3 and H-4 are axial (91CJC185).

(60) (62)

6.4.6 REFERENCES

90AHC158 — E.V. Kuznetsov, I.V. Scherbakova and A.T. Balaban, *Adv. Heterocycl. Chem.*, 1990, 50, 158.

90SR257 — L. Teuber, *Sulfur Rep.*, 1990, **9**, 257.

91ACS410 — I.R. Fjeldskaar and L. Skattebol, *Acta Chem. Scand.*, 1991, **45**, 410.

91ACSI183 — C.Y. Shen and Y. Li, *Acta Chim. Sinica*, 1991, **49**, 183.

91AG(E)17 — T. Goto and T. Kondo, *Angew. Chem. Int. Ed. Engl.*, 1991, **30**, 17.

91AG(E)1365 — W. Adam, S. Andler and M. Heil, *Angew. Chem. Int. Ed. Engl.*, 1991, **30**, 1365.

91AG(E)1658 — F. Stein, M. Deutsch, R. Lackmann, M. Noltemeyer and A. de Meijere, *Angew. Chem. Int. Ed. Engl.*, 1991, **30**, 1658.

91AG(E)1709 — H.J. Kabbe, H. Heitzer and L. Born, *Angew. Chem. Int. Ed. Engl.*, 1991, **30**, 1709.

91BSB1 — A. Jakobs, L. Christiaens and M. Renson, *Bull. Soc. Chim. Belg.*, 1991, **100**, 1.

91BSB665 — A. Dinculescu, T.S. Balaban, C. Popescu, D. Toader and A.T. Balaban, *Bull. Soc. Chim. Belg.*, 1991, **100**, 665.

91BSF189 — A. Danan and B.S. Kirkiacharian, *Bull. Soc. Chim. Fr.*, 1991, **128**, 189.

91CB881 — L.F. Tietze, U. Hartfiel, T. Hubsch, E. Voss and J. Wichmann, *Chem. Ber.*, 1991, **124**, 881.

91CB1837 — T. Pietzonka and D. Seebach, *Chem. Ber.*, 1991, **124**, 1837.

91CB1845 — D. Seebach, U. Misslitz and P. Uhlmann, *Chem. Ber.*, 1991, **124**, 1845.

91CC287 — R.G.F. Giles, I.R. Green, L.S. Knight, V.R. Lee Son, R.W. Rickards and B.S. Senanayaki, *J. Chem. Soc. Chem. Commun.*, 1991, 287.

91CC694	K. Toshima, K. Ohta, T. Ohtake and K. Tasuta, *J. Chem. Soc. Chem. Commun.*, 1991, 694.
91CC699	M. Sato, C. Kaneko, T. Iwaoka, Y. Kobayashi and T. Iida, *J. Chem. Soc. Chem. Commun.*, 1991, 699.
91CC947	A.J. Bloodworth and A. Shah, *J. Chem. Soc. Chem. Commun.*, 1991, 947.
91CC972	L. Crombie, J.L. Josephs, J. Larkin and J.B. Weston, *J. Chem. Soc. Chem. Commun.*, 1991, 972.
91CC1241	T. Iwaoka, M. Sato and C. Kaneko, *J. Chem. Soc. Chem. Commun.*, 1991, 1241.
91CCC1037	O. Petrov and I. Ognyanov, *Collect. Czech. Chem. Comm.*, 1991, **56**, 1037.
91CJC185	L. Breau, N.K. Sharma, I.R. Butler and T. Durst, *Can. J. Chem.*, 1991, **69**, 185.
91CL445	J.A.S. Cavaleiro, J. Elguero, L.M. Jimeno and A.M.S. Silva, *Chem. Letters*, 1991, 445.
91CSR211	H.G. Davies and R.H. Green, *Chem. Soc. Rev.*, 1991, 211.
91CSR271	H.G. Davies and R.H. Green, *Chem. Soc. Rev.*, 1991, 271.
91H529	B.C. Chen, *Heterocycles*, 1991, **32**, 529.
91H1593	S.S. Zaman and R.P. Sharma, *Heterocycles*, 1991, **32**, 1593.
91HCA27	D. Obrecht, *Helv. Chim. Acta*, 1991, **74**, 27.
91HCA1119	W. Fischer, *Helv. Chim. Acta*, 1991, **74**, 1119.
91HCA1239	C.W. Jefford, J. Currie, G.D. Richardson and J.-C. Rossier, *Helv. Chim. Acta*, 1991, **74**, 1239.
91JA1830	M. Hirama, T. Noda, S. Yasuda and S. Ito, *J. Amer. Chem. Soc.*, 1991, **113**, 1830.
91JA2336	S.D. Burke and J. Rancourt, *J. Amer. Chem. Soc.*, 1991, **113**, 2336.
91JCR(S)165	M. Yamamoto, T. Dewa, H. Munakata, S. Kohmoto and K. Yamada, *J. Chem. Research (S)*, 1991, 165.
91JCR(S)279	W.V. Murray and S.K. Hadden, *J. Chem. Research (S)*, 1991, 279.
91JCS(P1)89	M.E. Botha, R.G.F. Giles, C.M. Moorhoff, L.M. Engelhardt, A.H. White, A. Jardine and S.C. Yorke, *J. Chem. Soc. Perkin Trans. 1*, 1991, 89.
91JCS(P1)169	D.P. Bradshaw, D.W. Jones and J. Tideswell, *J. Chem. Soc. Perkin Trans. 1*, 1991, 169.
91JCS(P1)219	O.E.O. Hormi, C. Peltonen and R. Bergstrom, *J. Chem. Soc. Perkin Trans. 1*, 1991, 219.
91JCS(P1)639	P.I. Van Broeck, D.J. Vanderzande, E.G. Kiekens and G.J. Hoornaert, *J. Chem. Soc. Perkin Trans. 1*, 1991, 639.
91JCS(P1)667	S.V. Ley et al., *J. Chem. Soc. Perkin Trans. 1*, 1991, 667.

91JCS(P1)1683 D.A. Bleasdale and D.W. Jones *J. Chem. Soc. Perkin Trans. 1*, 1991, 1683.

91JCS(P1)2281 S. Motoki, T. Saito, T. Karakasa, H. Kato, T. Matsushita and S. Hayabashi, *J. Chem. Soc. Perkin Trans. 1*, 1991, 2281.

91JCS(P1)2363 G. Pattenden, N.A. Pegg and R.W. Kenyon, *J. Chem. Soc. Perkin Trans. 1*, 1991, 2363.

91JCS(P1)2763 D.A. Buckle, D.S. Eggleston, C.S.V. Houge-Frydrych, I.L. Pinto, S.A. Readshaw, D.G. Smith and R.A.B. Webster, *J. Chem. Soc. Perkin Trans. 1*, 1991, 2763.

91JCS(P1)3005 K. Wahala and T.A. Hase, *J. Chem. Soc. Perkin Trans. 1*, 1991, 3005.

91JCS(P2)891 D. Amic and N. Trinajstic, *J. Chem. Soc. Perkin Trans. 2*, 1991, 891.

91JCS(P2)1287 T.V. Mistry, Y. Cai, T.H. Lilley and E. Haslam, *J. Chem. Soc. Perkin Trans. 2*, 1991, 1287.

91JHC241 M. Sato, Y. Abe, K. Takayama, K. Sekiguchi, C. Kaneko, N. Inoue, T. Furuya and N. Inuka, *J. Heterocycl. Chem.*, 1991, **28**, 241.

91JHC819 S. Babu and M.J. Pozzo, *J. Heterocycl. Chem.*, 1991, **28**, 819.

91JOC126 D. Farcasiu and S. Sharma, *J. Org. Chem.*, 1991, **56**, 126.

91JOC1325 S.E. Kelly and B.C. Vanderplas, *J. Org. Chem.*, 1991, **56**, 1325.

91JOC1481 J.W. Huffmann, X. Zhang, M.-J. Wu, H.H. Joyner and W.T. Pennington, *J. Org. Chem.*, 1991, **56**, 1481.

91JOC1549 H.H. Seltzman, Y.-A. Hsieh, C.G. Pitt and P.H. Reggio, *J. Org. Chem.*, 1991, **56**, 1549.

91JOC1569 T. Czekanski, M. Hanack, J.Y. Becker, J. Bernstein, S. Bittner, L. Kaufman-Orenstein and D. Peleg, *J. Org. Chem.*, 1991, **56**, 1569.

91JOC2058 Y.-G. Lee, K. Ishimaru, H. Iwasaki, K. Ohkata and K. Akiba, *J. Org. Chem.*, 1991, **56**, 2058.

91JOC2081 J.W. Huffmann, H.H. Joyner, M.D. Lee, R.D. Jordan and W.T. Pennington, *J. Org. Chem.*, 1991, **56**, 2081.

91JOC2274 J.E. Backvall and P.G. Andersson, *J. Org. Chem.*, 1991, **56**, 2274.

91JOC3207 P. De Shong and P.J. Rybczynski, *J. Org. Chem.*, 1991, **56**, 3207.

91JOC3763 B.I. Alo, A. Kandil, P.A. Patel, M.J. Sharp, M.A. Siddiqui and V. Sneikus, *J. Org. Chem.*, 1991, **56**, 3763.

91JOC4563 W.A. Donaldson, C. Tao, D.W. Bennett and D.S. Grubisha, *J. Org. Chem.*, 1991, **56**, 4563.

91JOC4884 D. Nagarathnam and M. Cushman, *J. Org. Chem.*, 1991, **56**, 4884.

91JOC4963	S.W. McCombie, W.A. Metz, D. Nazareno, B.B. Shankar and J. Tagat, *J. Org. Chem.*, 1991, **56**, 4963.
91JOC5052	K. Ohkata, Y.-G. Lee, Y. Utsumi, K. Ishimaru and K. Akiba, *J. Org. Chem.*, 1991, **56**, 5052.
91JOC5437	C.-G. Huang, D. Shukla and P. Wan, *J. Org. Chem.*, 1991, **56**, 5437.
91JOC6271	Y. Venkateswarlu, D.J. Faulkner, J.L.R. Steiner, E. Corcoran and J. Clardy, *J. Org. Chem.*, 1991, **56**, 6271.
91JOC6748	J. Jeko, T. Timar and J.C. Jaszberenyi, *J. Org. Chem.*, 1991, **56**, 6748.
91JOC6865	C. Siegel, P.M. Gordon, D.B. Uliss, G.R. Handrick, H.C. Dalzell and R.K. Razdan, *J. Org. Chem.*, 1991, **56**, 6865.
91JOC7055	J.B. Christensen, I. Johannsen and K. Bechgaard, *J. Org. Chem.*, 1991, **56**, 7055.
91JOC7150	T. Shimo, J. Tajima, T. Suishu and K. Somekawa, *J. Org. Chem.*, 1991, **56**, 7150.
91JOC7292	W. Adam, D. Golsch, L. Hadjiarapoglou and T. Patonay, *J. Org. Chem.*, 1991, **56**, 7292.
91JOC7323	A. Capperucci, A. Degl'Innocenti, A. Ricci, A. Mordin and G. Reginato, *J. Org. Chem.*, 1991, **56**, 7323.
91LA633	S. Antus, E. Baitz-Gacs, G. Snatzke and T.S. Toth, *Liebigs Ann. Chem.*, 1991, 633.
91PS51	S. Cossu, O. de Lucchi, E. Piga and G. Valle, *Phosphorus Sulfur*, 1991, **63**, 51.
91PS129	I.W.J. Still, *Phosphorus Sulfur*, 1991, **58**, 129.
91S223	R. Caputo, C. Ferreri and G. Palumbo, *Synthesis*, 1991, 223.
91S569	G. Solladie and A. Girardin, *Synthesis*, 1991, 569.
91S575	C.A.M. Alfonso, M.T. Barros, L.S. Godinho and C.D. Maycock, *Synthesis*, 1991, 575.
91S681	A. Meou, N. Bouanah, A. Archelas, X.M. Zhang, R. Guglielmitti and R. Furstoss, *Synthesis*, 1991, 681.
91S851	C. Siegel, P.M. Gordon and R.K. Razdan, *Synthesis*, 1991, 851.
91S879	B.S. Kirkiacharian, A. Danan and P.G. Koutsourakis, *Synthesis*, 1991, 879.
91S937	H.-J. Bertram, S. Bohm and L. Born, *Synthesis*, 1991, 937.
91SC1455	V. Satyanarayana, C.P. Rao, G.L.D. Krupadanam and G. Srimannarayana, *Synth. Commun.*, 1991, 1455.
91SL80	P.C.B. Page and E.S. Namwinda, *Syn. Lett.*, 1991, 80.
91SL113	M. Matsui, K. Shibata, H. Murumatsu, H. Sawada and M. Nakayama, *Syn. Lett.*, 1991, 113.
91SL191	W.B. Motherwell and J.A. Wilkinson, *Syn. Lett.*, 1991, 191.

91SL195	A. Teniou, L. Toupet and R. Grée, *Syn. Lett.*, 1991, 195.
91SL487	J. Barluenga, F. Aznar and C. Valdes, *Syn. Lett.*, 1991, 487.
91SL553	P. Stoss and P. Merrath, *Syn. Lett.*, 1991, 553.
91SL595	K.C. Majumdar, A.T. Khan and S. Saha, *Syn. Lett.*, 1991, 595.
91SL611	J.P. Ferezou, M. Julia, R. Khourzon, A. Pancrazi and P. Robert, *Syn. Lett.*, 1991, 611.
91SL614	J.P. Ferezou, M. Julia, L.W. Liu and A. Pancrazi, *Syn. Lett.*, 1991, 614.
91SL618	J.P. Ferezou, M. Julia, L.W. Liu and A. Pancrazi, *Syn. Lett.*, 1991, 618.
91SL777	F. Meyer and A. de Meijere, *Syn. Lett.*, 1991, 777.
91SL823	I. Kadota, V. Gevorgyan, J. Yamada and Y. Yamamoto, *Syn. Lett.*, 1991, 823.
91SL895	W.A. Donaldson and C. Tao, *Syn. Lett.*, 1991, 895.
91SL909	M. Kuroboshi and T. Hiyama, *Syn. Lett.*, 1991, 909.
91SR199	W. Franck, *Sulfur Rep.*, 1991, **10**, 199.
91SR233	W. Franck, *Sulfur Rep.*, 1991, **10**, 233.
91T1095	W.G. Prowse, K.I. Arnot, J.A. Recka, R.H.Thompson and J.R. Maxwell, *Tetrahedron*, 1991, **47**, 1095.
91T1877	T. Sakamoto, Y. Kondo, A. Yasuhara and H. Yamanaka, *Tetrahedron*, 1991, **47**, 1877.
91T1977	G. Doddi, G. Ercolani and P. Mencarelli, *Tetrahedron*, 1991, **47**, 1977.
91T4187	R. Caputo, C. Ferreri, G. Palumbo and F. Russo, *Tetrahedron*, 1991, **47**, 4187.
91T4603	R. Ruzziconi, Y. Naruse and M. Schlosser, *Tetrahedron*, 1991, **47**, 4603.
91T4775	S.Y. Dike, J.R. Merchant and N.Y. Sapre, *Tetrahedron*, 1991, **47**, 4775.
91T5071	D. Nagarathnam and M. Cushman, *Tetrahedron*, 1991, **47**, 5071.
91T8091	W.S. Lee, K. Lee, K.D. Nam and Y.J. Kim, *Tetrahedron*, 1991, **47**, 8091.
91T9297	G. Petrillo, M. Novi, C. Dell'Erba and C.Tavani, *Tetrahedron*, 1991, **47**, 9297.
91TL827	T. Eszenyi, T. Timar and P. Sebok, *Tetrahedron Lett.*, 1991, 827.
91TL935	M. Terada, K. Mikami and T. Nikai, *Tetrahedron Lett.*, 1991, 935.
91TL1723	A. Stambouli, M. Chastrette and M. Soufiaoui, *Tetrahedron Lett.*, 1991, 1723.
91TL2971	B.F. Bonini S. Masiero, G. Mazzanti and P. Zani, *Tetrahedron Lett,*. 1991, 2971.

91TL3209 I.G. Collado, R. Hernandez-Galan, G.M. Massanet, F. Rodriguez-Luis and J. Salva, *Tetrahedron Lett.*, 1991, 3209.

91TL4051 J. Bloxham and C.P. Dell, *Tetrahedron Lett.*, 1991, 4051.

91TL4189 M. Mikolajczyk, M. Mikina and P. Graczyk, *Tetrahedron Lett.*, 1991, 4189.

91TL5055 N.H. Lee, A.R. Muci and E.N. Jacobsen, *Tetrahedron Lett.*, 1991, 5055.

91TL5295 G.H. Posner, T.D. Nelson, C.M. Kinter and K. Afarinkia, *Tetrahedron Lett.*, 1991, 5295.

91TL5571 H. Shimizu, S. Miyazaki, T. Kataoka and M. Hori, *Tetrahedron Lett.*, 1991, 5571.

91TL5619 O.V. Singh, R.S. Kapil, C.P. Garg and R.P. Kapoor, *Tetrahedron Lett.*, 1991, 5619.

91TL6345 R. Sato and K. Chino, *Tetrahedron Lett.*, 1991, 6345.

91TL7251 M. Tiabi and H. Zamarlik, *Tetrahedron Lett.*, 1991, 7251.

91TL7727 K.C. Santhosh and K.K. Balasubramanian, *Tetrahedron Lett.*, 1991, 7727.

91TL7739 X. Garcias, P. Ballester and J.M. Saa, *Tetrahedron Lett.*, 1991, 7739.

91TL7743 V.K. Aggarwal, R.J. Franklin and M.J. Rice, *Tetrahedron Lett.*, 1991, 7743.

91TL7759 J.D. King and P. Quayle, *Tetrahedron Lett.*, 1991, 7759.

CHAPTER 7

Seven-Membered Rings

JOHN M. KANE AND NORTON P. PEET
Marion Merrell Dow Research Institute, Cincinnati, OH, USA

INTRODUCTION

Many new and useful synthetic routes to 7-membered heterocyclic compounds appeared in the 1991 literature. In Volumes 1 and 2 we highlighted rearrangement chemistry and natural product synthesis, respectively, and last year in Volume 3 we highlighted various modes of 7-membered heterocycle construction. This year we highlight selected, novel methods of 7-membered heterocycle synthesis. In the past we have attempted to present comprehensive reviews. Since most readers now have access to automated searching methods we have decided to focus in greater detail on only a few of the synthetic methods which we felt might be of greater interest to our readers.

This chapter includes syntheses of 7-membered rings containing nitrogen, oxygen and sulfur. Coverage is by reaction type rather than ring system. As a result, the synthesis of a variety of 7-membered heterocycles may be described in the same section. We have classed the reaction types which we focus on as follows: reductive coupling of nitriles; free radical processes; 1,7-electrocyclizations; ring-expansion reactions; rearrangements; cycloaddition reactions; acyliminium ion cyclizations; ylide-mediated reactions; and metal-catalyzed processes. It is recognized that certain ring-building strategies are not unique to only one of these categories, and in some cases we have arbitrarily positioned a synthesis in the 1,7-electrocyclization section, for example, when it also would have fit in the ring-expansion or rearrangement sections.

Reaction types which were used in the construction of 7-membered heterocycles in 1991 which we have not covered, due to their established nature, include the following: Friedel-Crafts alkylations and acylations; Beckmann rearrangements; Baeyer-Villager oxidations; Schmidt rearrangements; Michael reactions; ene reactions; lactonizations and lactamizations; a variety of simple displacement reactions; iodolactonizations; reactions which form cyclic imines or amidines; reductive aminations; ketalizations, hemiketalizations, and related reactions involving other heteratom analogs; reactions involving diols, amino alcohols, diamino alkanes or phenylenediamine; intramolecular electrophilic cyclizations driven by the enamine character of substituted indoles and pyroles; formation of anhydro sugars and nucleosides; and cyclization reactions of isatoic anhydride with amino acids.

Many of the 7-membered ring heterocycles prepared in the past year were designed for biological activity. In our last review we highlighted the synthesis of one TIBO derivative, and several syntheses of related compounds appeared in 1991 <91BMC(1)531> <91JMC(34)746> <91JMC(34)2231><91JMC(34)3187><91T(47)-7451>. Unfortunately, it appears that non-nucleoside reverse transcriptase inhibitors will probably not be important therapeutic agents.

Two reviews on specific 7-membered heterocycles appeared in 1991. One review covered bicyclic diazepines <91CHC(50)1>, while the other describes azepines, azocines and azonines <91T(47)9131>.

Reductive Coupling of Nitriles

In order to expand the scope of titanium-mediated chemistry, reductive cyclizations of dinitriles with titanium tetrachloride and zinc in tetrahydrofuran were investigated <91H(32)2339>. It was found that the simple dinitriles glutaronitrile, adiponitrile and pimelonitrile afforded a series of biscycloalkylpyrazines in moderate yield. Extension of this ring forming reaction to N,N-*bis*-(2-cyanoethyl)-anilines of general structure 1 gave *bis*-azepinopyrazines 2, in yields ranging from 27-33%. Although the mechanism of this intriguing cyclization was not discussed, it probably involves the condensation of an initially formed 1-aryl-4,5-diaminoazepine with a corresponding 4,5-diiminoazepine to form the central pyrazine ring in the last step of the reaction.

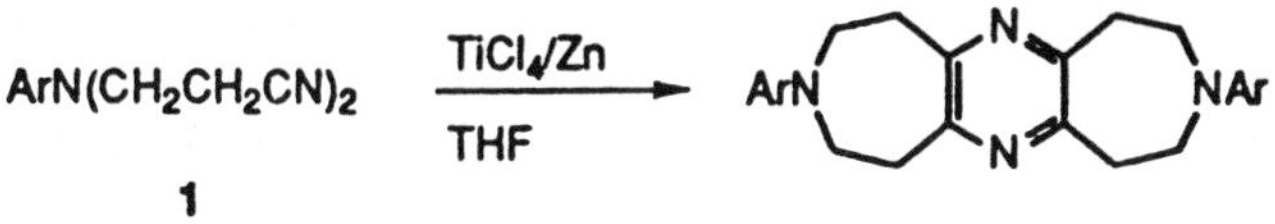

Free Radical Processes

Cyclizations in this category were dominated by processes utilizing tributyltin hydride. In a process termed "radical translocation reactions" of o-iodoanilides, Curran and co-workers <91JOC(56)4335> obtained a mixture of cyclopentane 4 and benzazepinone 5 when acetylenic o-iodoanilide 3 was treated with tributyltin hydride.

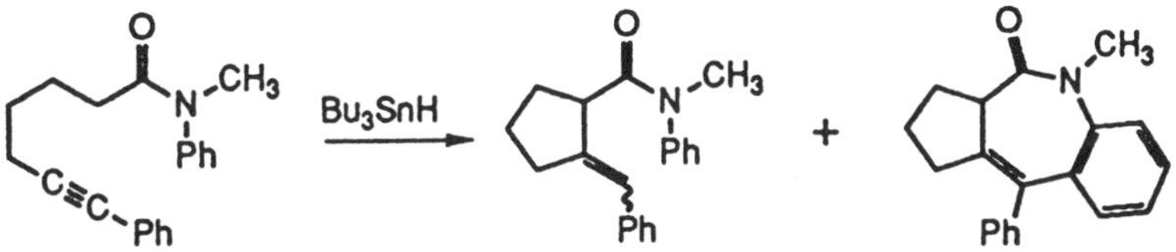

Burke and Rancourt <91JA(113)2335> obtained a mixture of products when olefinic substrate 6 was treated with tributyltin hydride and an initiator. In order of predominance, the structures produced were 6-exo anti-pyran **7**, 6-exo syn-pyran **8**, 7-endo oxepane **9**, and the simple acyclic reduced olefin **10**. This study was undertaken to assess the regio- and stereochemistry of the newly formed vicinal centers in the pyrans, which are component parts of many natural products.

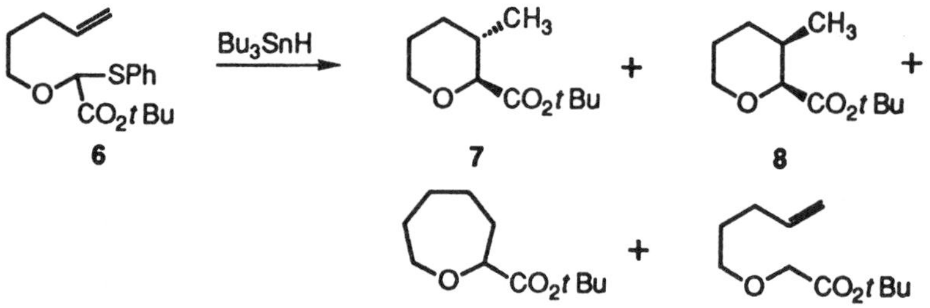

When substituted pyridines were N-alkylated with 1,5-dibromopropane, the resulting N-bromopentylpyridinium bromide could be reductively cyclized to pyridinoazepinium salts using the standard tributyltin hydride protocol <91T(47)4077>.

Clark, Jones, McCarthy and Storey showed that anilides **11**, where R=alkyl and X=Br and I, undergo reduction with tributyltin hydride to give the corresponding anilides **11**, where X=H, and benzazepinones **12**. Anilides **11** (X=H) were the major products. When R=CH_3, the yield of 7-exo cyclization product **12** was 13%; when R=$CH_2C_6H_5$, the yield of **12** was 25%.

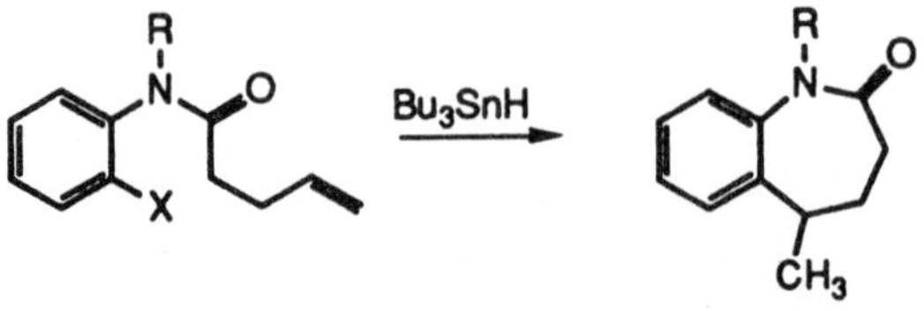

With cyclization substrates related to **11** with one less methylene group between the olefin and the carbonyl group, only trace amounts of benzotriazepines were formed.

A remarkably efficient free radical cyclization involving ring expansion was reported by Dowd and Choi <91T(47)4847>. These authors treated piperidinone **14** with tributyltin hydride and AIBN to produce azepinone **15** in 71% yield.

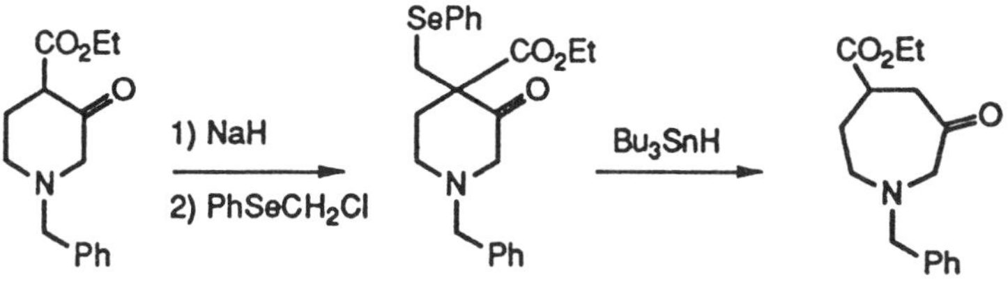

CO_2Et, O, N, Ph — 1) NaH, 2) $PhSeCH_2Cl$ → SePh, CO_2Et, O, N, Ph — Bu_3SnH → CO_2Et, O, N, Ph

13 **14** **15**

This ring expansion was based on an earlier carbocyclic ring expansion reported by the same authors. Interesting bridged systems were reported by Grigg and co-workers, using a palladium acetate-triphenylphosphine mixture for effecting the cyclization of aryl halides to proximate alkenes <91T(47)9703>. Thus, sulfonamide **16** cyclizes to 6-exo and 7-endo bridged compounds **17** and **18**. The yield for the mixture of **17**

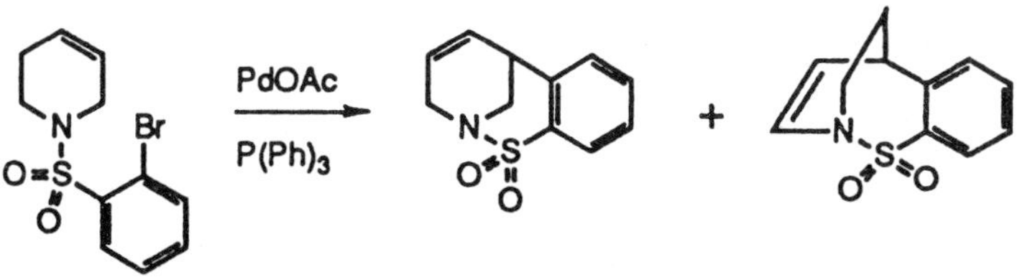

PdOAc, $P(Ph)_3$ → +

16 **17** **18**

and **18** was 89%. The 3-pyrroline analog of **16** also underwent cyclization to give the 6-exo product analogous to **17**, in 65% yield. The 3-pyrroline aryl iodide, however, did not cyclize under the same conditions.

1,7-Electrocyclizations

One class of 1,7-electrocyclization reactions involved the addition of appended trigonal centers back to aromatic nuclei. Maier and Eberbach <91-HCA(74)1095> reported the cyclization of ylide **19** to pyrido[1,2-*a*]azepine **20**, followed by oxidation to pyridoazepinone **21**.

H_2O_2, H_2O

19 **20** **21**

In a related study, cyclization of an appended nitrile ylide onto one of two aromatic rings in a competition experiment was studied <91CC658>.

In an interesting ring expansion reaction of a benzotriazole derivative, Katritzky, Fan and Greenhill <91JOC(56)1299> reported the preparation of 4-(benzotriazol-1-yl)-6*H*-benzo[*c*]tetrazolo[1,5-*e*]-[1,2,5]triazepine. A key step in the mechanism presented for this transformation is the intramolecular, electrophilic cyclization of an intermediate enamine anion onto an aryldiazonium group.

Ring Expansion Reactions

In a reaction which may serve as a model for the inactivation of plasma amine oxidase by cyclopropylamine, Sayre and co-workers <91JOC(56)1353>

treated 3,5-di-tert-butyl-1,2-benzoquinone (**22**) with 1-phenyl-cyclopropylamine to give dihydrobenzoxazepine **25**.

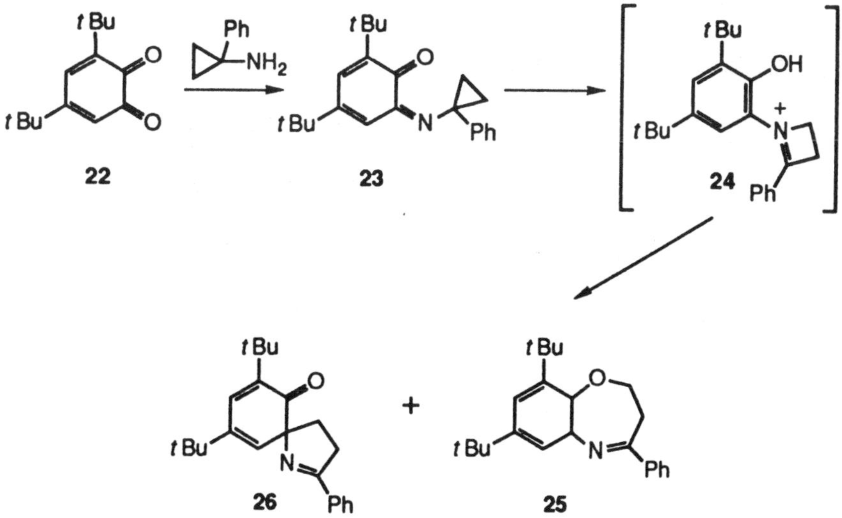

The expansion was best explained by the intermediacy of azetidinium species **24**, since the azaspirodecatrienone **26** was co-produced. It was also shown that **22** induced oxidative cleavage of a series of cyclopropylamines, even when the latter were incapable of forming initial Schiff base intermediates.

In studies related to the synthesis of vinblastine, Kuehne and co-workers <91JOC(56)513> treated chloroimine **27** with dimethyl malonate thallium enolate to afford indoloazepine **28**. Compound **27** arises

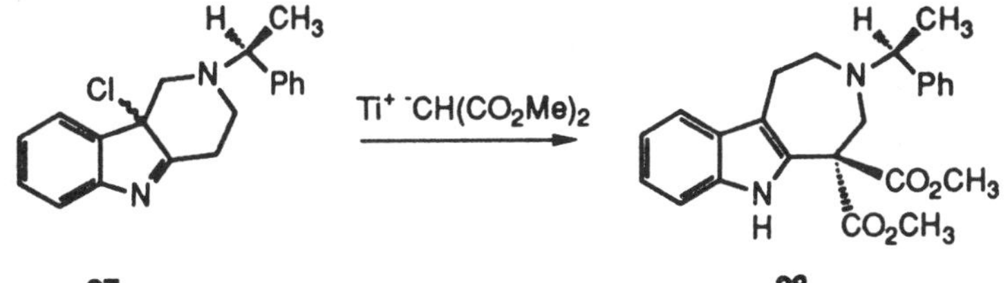

from chlorination of a γ-carboline which was made via a Fischer indole synthesis. The ring expansion of **27** to **28** is initiated by attack of the malonate on the imine, followed by regeneration of an alkylated malonate anion. This new malonate anion can attack the neighboring methylene center with subsequent breaking of the carbon-carbon bond connecting this methylene center to the indole ring and extrusion of chloride.

A transannular ring opening was used to generate 11-oxodibenzo[*b*,*c*]thiepin 30 <91JHC(28)1891> by treatment of acetoxy sulfone 29 with sodium hydroxide in methanol. In this ring contraction, hydrolysis of the acetate gave a hydroxy sulfone which collapsed to the keto sulfone by abstraction of the alcohol proton followed by alkoxide-induced breakage of the transannular carbon-carbon bond. This reaction had been previously reported by other authors, but in the present study the relative stereochemistry of 30 was assigned by comparison with an authentic sample prepared unambiguously. Compound 29 was accessed through a photoirradiation of 3-acetoxybenzo[*b*]thiophene-1,1-dioxide in the presence of cyclohexene.

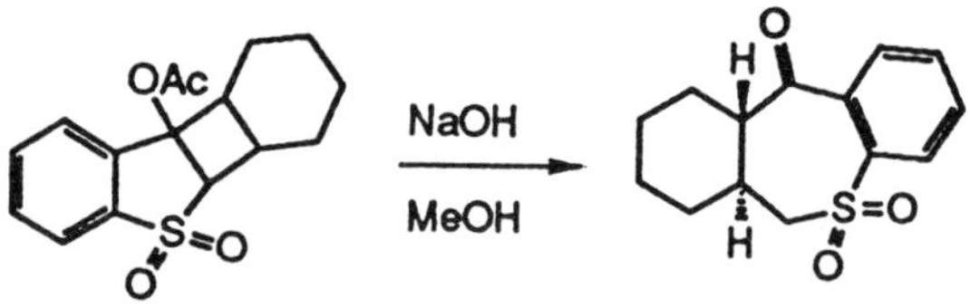

A route to 6-cyano-1,2-diazepinones from isoxazolo[3,4-*d*]pyridazin-7(6*H*)-ones was reported <91IF(46)435>. Thus, treatment of 31 with sodium hydroxide in ethanol gave a 78% yield of diazepinone 32 and a 12% yield of 4-cyano-5-(2-arylethenyl)pyrazole 33. Substituents in addition to the methyl, phenyl and fluoro groups on 31 were used. The diazepinone was always the major (61-78%) product, with low (8-12%) yields of the pyrazole, a consistent byproduct, being produced. This mechanistically interesting transformation is of practical utility for the preparation of a wide variety of substituted 1,2-diazepin-5-ones.

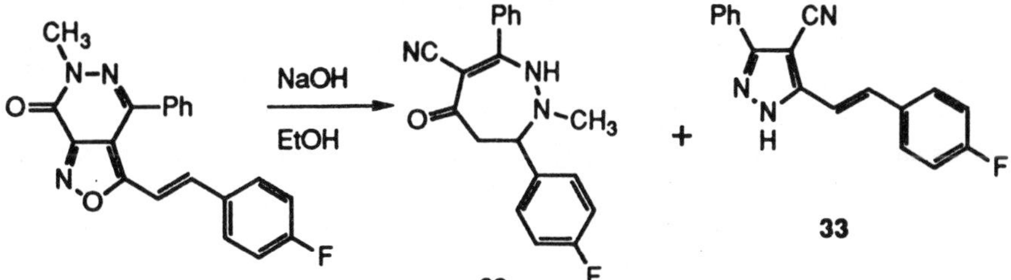

Rearrangements

Two Meisenheimer rearrangements were reported to give 7-membered rings containing nitrogen and oxygen. A series of 3-alkyl-2-aryltetrahydro-1,3-oxazine N-oxides, prepared from the corresponding 1,3-oxazines with *m*-chloroperoxybenzoic acid (MCPBA), were treated in dimethylacetamide to afford 2-alkyl-7-aryltetrahydro-1,6,2-dioxazepines in 48-75% yield <91JHC-(28)1789>. In a more complex example, Kurihara *et al.* <91CPB(39)811> described the rearrangement of *cis*-azetidine 34 to oxazepine 35 upon treatment with MCPBA, in 80% yield. The structure of 35 was unambiguously confirmed by x-ray analysis. It appears that N-oxidation of 34 occurs from the α-face, followed by [2,3]-Meisenheimer rearrangement. These authors report additional Meisenheimer rearrangements on derivatives of 34 in a later paper <91CL1781>.

Peet and LeTourneau <91H(32)41> reported a complex rearrangement of tetrahydrobenzo[1,2-*c*:3,4-*c*]-dipyrazole 36 under methylating conditions (sodium hydride, methyl iodide and dimethylformamide). In addition to the expected methylated products 37 and 38,

fused diazepinone 39 was produced. A mechanism was postulated for the rearrangement of 36 to 39, which involved the expansion of a pyrazole ring, two methylations and a condensation involving dimethylformamide.

The use of enol variants in the Cope rearrangement of *cis*-2,3-divinyl epoxides, which produced oxepins in 78-94% yield, was reported by Chou and White <91TL(32)157>. In a subsequent paper these authors <91TL(32)7637> reported the thermolysis of *trans*-2,3-epoxides to give dihydrofurans, mainly, but accompanied also by oxepins. The latter products are suggested to form by isomerization of the *trans*-epoxides to *cis*-epoxides, and this partial conversion proceeds through a carbonyl ylide in stereospecific fashion.

A polyphosphoric acid (PPA) rearrangement of diazepine 40 gave a mixture of rearrangement products <91JCS(P1)2139>. The expected product was 5-benzoyl-2,3,4,5-tetrahydro-1*H*-benz[*b*]azepine (41), the result of a [3,3]sigmatropic rearrangement. Co-produced were 2-benzoyl-1-phenylpyrrolidine (42) and 2,5-*bis*-(3-anilinopropyl)-3,6-diphenylpyrazine (43). Structures 41 and 43 were deduced on the basis of spectral data, while pyrrolidine 42 was independently synthesized.

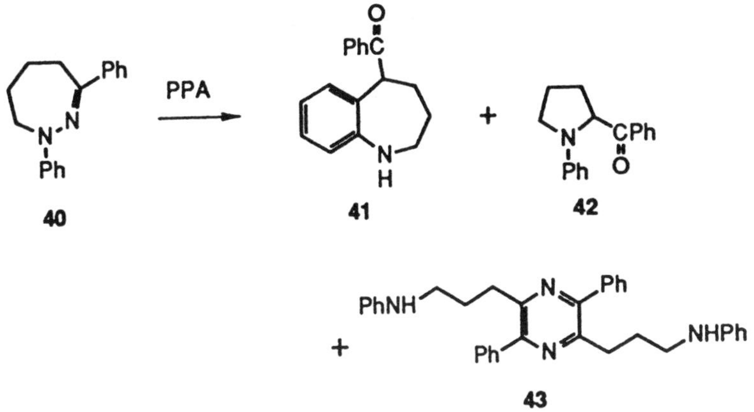

Treatment of oxazole 44 with rhodium (II) acetate followed by dimethyl acetylene dicarboxylate (DMAD) gave oxepino[2,3-*d*]isoxazole 45 <91TL(32)1161>. This rearrangement and addition process is thought to proceed through adduct 46, which forms after nitrogen extrusion. Collapse of 46 to zwitterion 47 then leads to epoxide 48, which undergoes Cope rearrangement to the fused oxepin 45.

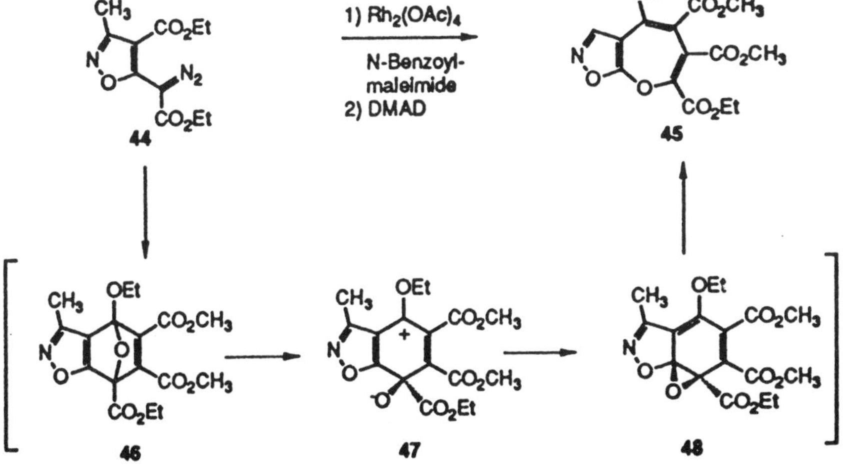

Cycloaddition Reactions

The Diels-Alder reaction is arguably the most powerful ring-forming reaction available to the synthetic chemist. Over the years it has been used to prepare a variety of 6-membered carbocyclic and heterocyclic products and it has figured prominently in the total synthesis of complex natural products. Electronically identical 1,3-dipolar cycloaddition reactions have been extensively employed in the synthesis of 5-membered heterocycles. In recent years, intramolecular versions of these reactions have increased in popularity. These intramolecular reactions, in addition to forming 5- and 6-membered rings, offer the possibility of 7-membered ring formation as the length of the tether connecting the diene (1,3-dipole) and the dienophile (dipolarophile) is manipulated.

An elegant application of a Diels-Alder strategy leading to the formation of a 7-membered heterocycle was reported by Jacobi and co-workers during their total synthesis of (-)-norsecurinine <91JA(113)-5384>. Thus, conjugate addition of pyrrolidine 49 to the α,β-unsaturated ketone 50 afforded intermediate 51 which underwent an intramolecular Diels-Alder reaction upon warming in mesitylene. The primary cycloadduct 52 spontaneously eliminated acetonitrile affording pyrrolo[1,2-*a*]azepine 53 which was isolated in approximately 50% yield. Several subsequent steps transformed azepine 53 to (-)-norsecurinine (54).

The most commonly encountered intramolecular 1,3-dipolar cycloaddition reactions involve the intermediacy of either nitrile oxides or nitrones. Therefore, it is not surprising that both of these dipoles have been used to prepare 7-membered heterocycles. Hassner and co-workers have reported several intramolecular nitrile oxide cycloadditions. For example, dehydration of nitroindole 55 under standard conditions gave nitrile oxide 56 which underwent an intramolecular 1,3-dipolar cycloaddition forming tetracyclic azepine 57 <91JOC(56)896>. Under these reaction conditions, azepine 57 partially decomposed to urethane 58 by rearomatization and trapping of the resultant oxime by the dehydrating agent. In a separate study, Hassner reported intramolecular nitrile oxide cycloadditions in which the 1,3-dipole and the olefinic dipolarophile were tethered by an ether-containing linkage. When this tether was the appropriate length, an isoxazolo[3,4-c]oxepine was formed in moderate yield <91CB(124)1181>.

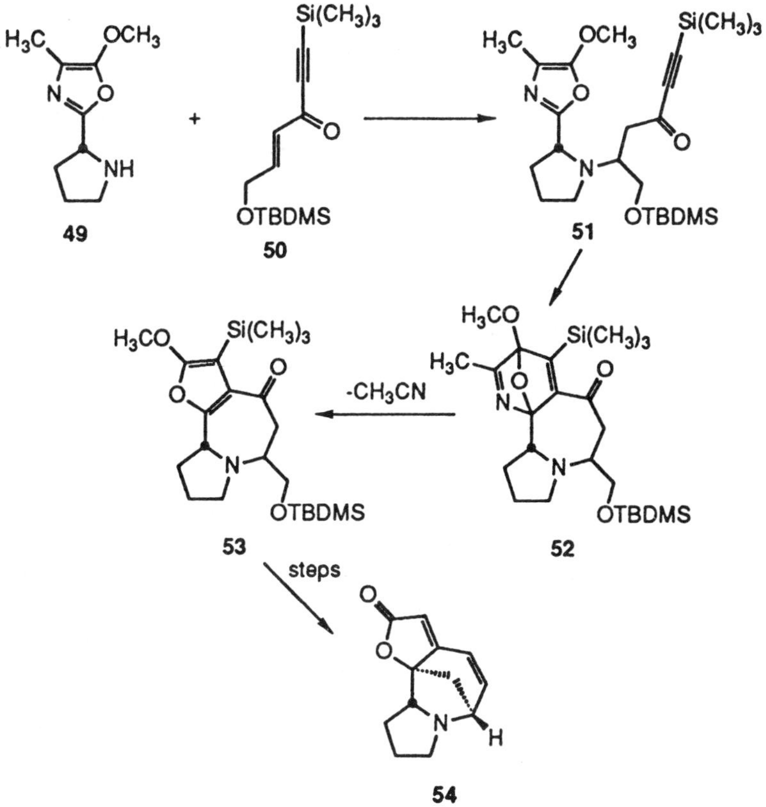

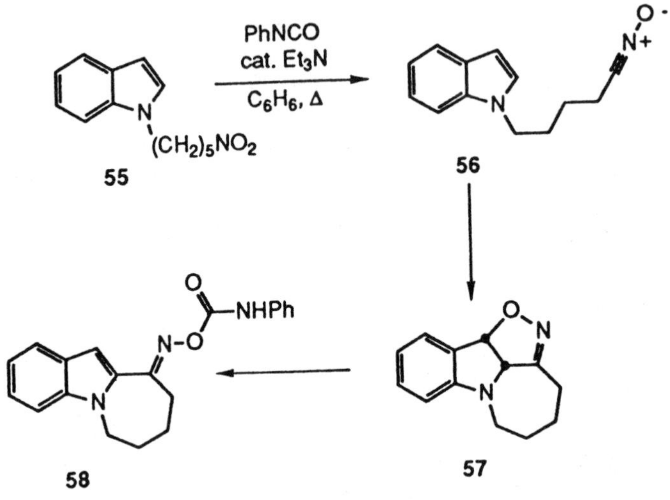

In most instances, the intramolecular cycloaddition of nitrile oxides proceeds with high regioselectivity such that the oxygen atom of the nitrile oxide reacts with the distal end of the attached dipolarophile. Such high regioselectivity is not always observed in the intramolecular cycloadditions of nitrones. For example, Grigg and co-workers have recently generated nitrones by the conjugate addition of oximes **59** to divinylsulfone (**60**). These nitrones underwent 1,3-dipolar cycloaddition via conformations **61** and **62** to yield both bridged oxathiazepines **63** and bridged oxathiazines **64** <91T(47)8297>. Extending this

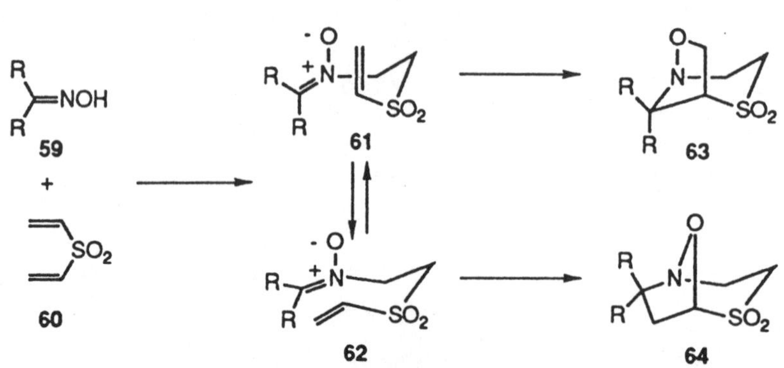

methodology, Grigg has also prepared bridged oxazepines <91T(47)8297> and bridged oxadiazepines <91T-(47)4495> as well as bridged and nonbridged oxepine derivatives <91T(47)4495>. In some closely related intramolecular nitrone cycloadditions, Holmes and co-workers have prepared several bridged oxazepine derivatives which have been subsequently utilized in the synthesis of indolizidine alkaloids <91JCS(P1)175> <91JOC(56)1393>. A rather unusual intramolecular cycloaddition was proposed by Joule and co-workers to explain the formation of a 3,2-benzoxazepine derivative from the nitration of 2-methoxy-1-nitronaphthalene <91CC559>.

Acyliminium Ion Cyclizations

Acyliminium ion-initiated cyclizations are becoming an increasingly popular method for the preparation of nitrogen-containing heterocycles. The pioneering work of Speckamp has clearly demonstrated the versatility of these cyclizations and their use in the synthesis of more complex natural products has been documented. In general, acyliminium ion cyclizations have been used to prepare 5- and 6-membered heterocycles, although the formation of some bridged 7-membered heterocycles has also been reported. The synthesis of nonbridged 7-membered heterocycles by this method has received less attention.

In 1991 two groups described acyliminium ion-mediated syntheses of nonbridged azepine derivatives. In a continuation of their previous studies, Flynn and co-workers reported the syntheses of two benzazepines as "anti"-phe-gly dipeptide mimetics <91-BMC(1)309>. Thus, treatment of oxazolidinone 65 with a variety of acids produced acyliminium ion 66 which was intercepted by the tethered aromatic ring yielding 2-benzazepine 67 in good yields. Unfortunately, these conditions resulted in the racemization of the phenylalanine-derived chiral center. Alternatively, tri-

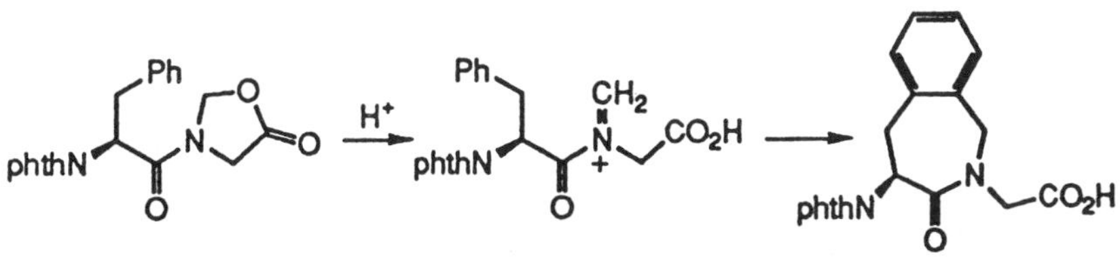

trifluoromethanesulfonic acid-induced cyclization of α-methoxyglycine derivative **68** proceeded via acyliminium ion **69** to yield 2-benzazepine **70** in 82% yield as a separable 11:1 mixture of diastereomers.

68 **69** **70**

X = H, Y = CO_2CH_3 7%
X = CO_2CH_3, X = H 75%

Acyliminium ions were also used to prepare ring-fused azepine derivatives. Thus, treatment of 2-methoxypyrrolidine derivative **71** with $TiCl_4$ resulted in the formation of acyliminium ion **72** which cyclized yielding pyrrolo[1,2-a]azepine **73** in moderate to good yield <91T(47)583>. The starting 2-methoxypyrrolidines were prepared by anodic oxidation of the corresponding acylated pyrrolidines.

$TiCl_4$

71 **72** **73**

Ylide-Mediated Reactions

In 1991 several 7-membered heterocyclic ring systems were prepared by ylide-mediated processes. In general, these processes involved either ylide-mediated cyclizations or ylide-mediated rearrangements. Among the ylide-mediated cyclizations, those involving iminophosphoranes have been extensively studied. The Molina group has been especially active in this area, employing these cyclizations in the syntheses of a variety of heterocyclic systems. In general, the synthesis of heterocycles by this method involves the formation of an imine bond in the ring-forming step (aza-Wittig reaction). For example, Nitta and co-workers have prepared several 5-azaazulene derivatives **74** by the reactions of substituted (vinylimino)phosphoranes **75** and 2-formyl-6-dimethylaminofulvene (**76**) <91TL(32)6727>. A slightly different

75 **76** **74**

mode of cyclization was observed by Molina and coworkers for the reaction of bis(iminophosphorane) **77** and aromatic acid chlorides. In this instance, the initial reaction presumably formed imidoyl chloride **78** which cyclized by attack of the iminophosphorane nitrogen yielding **79**. Cleavage of the phosphorous-nitrogen bond then gave the 1,3-benzodiazepines **80** which were isolated in 55-65% yields <91TL(32)4401>.

77 **78** **80** **79**

Ylide-mediated rearrangements have also been used to prepare 7-membered heterocycles. For example, treatment of a dichloromethane solution of N-aminophthalimide (**81**) and cepham **82** with lead tetraacetate presumably generated sulfilimine **83** via nitrene addition to the sulfur atom of the cepham. Subsequent [2,3]sigmatropic rearrangement afforded the ring-fused thiadiazepines **84** which were obtained in 28-73% yields <91H(32)2193>. In a somewhat similar fashion, sulfur ylides have been generated by rhodium carbenoid addition to thioethers. For example, treatment of allyl ether **85** with rhodium (II) acetate afforded bridged oxepane **86** in 34% yield. The reaction presumably involved the addition of a rhodium carbenoid to the thioether, generating sulfonium ylide **87**, which then underwent a [2,3]sigmatropic rearrangement yielding **86** <91TL(32)6159>.

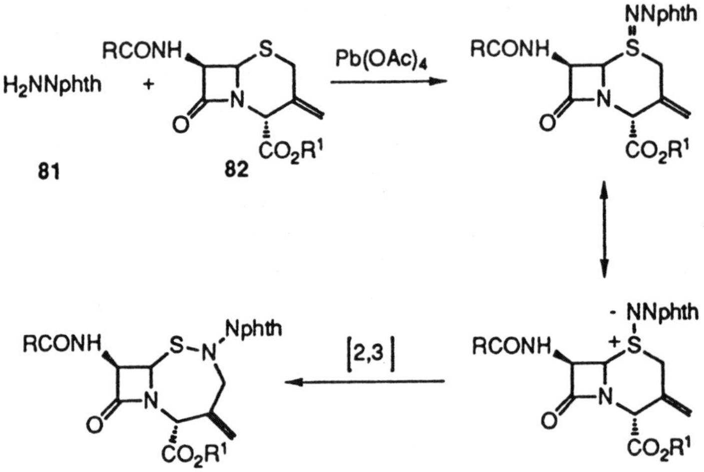

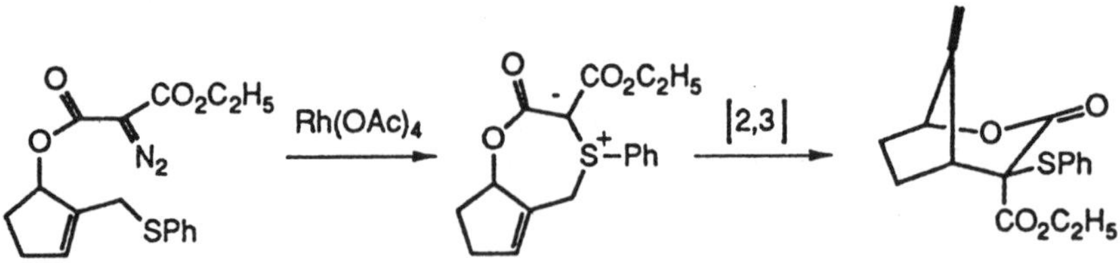

Other sulfur ylides have been implicated in the syntheses of 7-membered heterocycles. Thus, treatment of 9-methyl-10-aza-9-thiaphenanthrene (**88**) with LDA generated the red-colored carbanion **89** which was successfully alkylated with either methyl or ethyl iodide. Reaction of this carbanion with trimethylsilyl chloride, however, did not yield the expected trimethylsilyl derivative **90** but rather the ring expanded thiazepine **91** which was isolated in 27% yield. In this case, **90** presumably underwent a 1,3-proton shift yielding **92** which went on to product by a ring opening/ring closure sequence <91TL(32)4359> <91JCS(P1)-1733>.

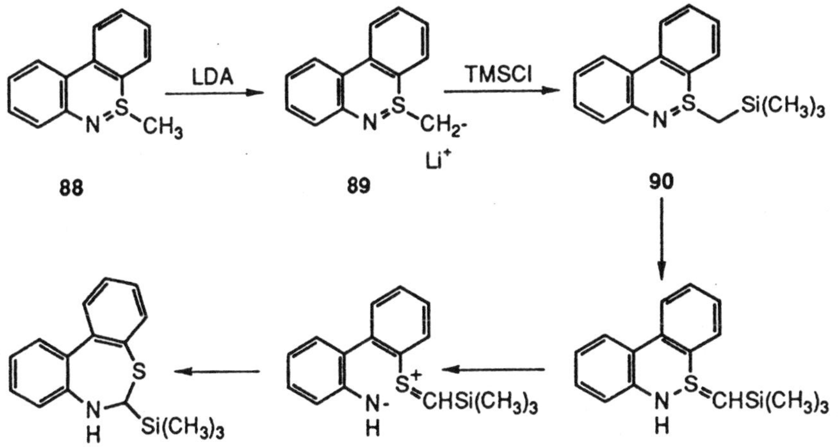

Metal-Catalyzed Processes

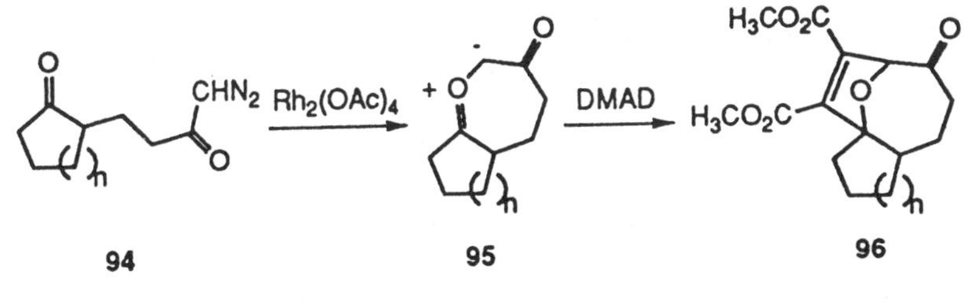

The catalysis of organic reactions by different metals has dramatically increased the synthetic chemist's effectiveness, allowing him to perform synthetic transformations which would have been previously unthinkable. Of the various metal-catalyzed processes, the chemistry of rhodium carbenoids has been effectively employed in the synthesis of 7-membered heterocycles. The formation of 7-membered heterocycles by this method involved either carbenoid addition to a heteroatom or carbenoid insertion into a heteroatom-hydrogen bond. As an illustration of the former process, Padwa and co-workers reported the synthesis of 7-membered oxygen heterocycles by rhodium catalyzed carbonyl ylide formation. Thus, treatment of diazoketone 94 with rhodium (II) acetate generated carbonyl ylide 95 which was trapped by dimethyl acetylenedicarboxylate (DMAD) yielding bridged oxepane 96 <91JOC(56)3271>. We distinguished this reaction from the reactions described in the section dealing with ylide-mediated reactions on the basis that, in this instance, the carbonyl ylide did not mediate the 7-membered ring-forming step but was rather a consequence of the ring-forming step. In some closely related experiments, Doyle and Padwa reported the synthesis of oxazepine derivatives by this type of reaction. In this case, however, the intermediate carbonyl ylide could not be trapped by dipolarophiles and instead underwent an internal proton transfer to yield the observed oxazepines <91JOC(56)820>.

Carbenoid insertions into heteroatom-hydrogen bonds have also been described as an effective method of preparing 7-membered heterocycles. For example, Moody and co-workers have found that rhodium (II) acetate-catalyzed decomposition of diazoketones **97** resulted in the formation of oxepanes **98** in 48-73% yield <91JCS(P1)1>. The synthetic potential of this route was further demonstrated by the synthesis of the *cis*-2,7-disubstituted oxepane skeleton of isolaurepinnacin <91JCS(P1)9>.

$Rh_2(OAc)_4$

97 **98**

The intramolecular, Lewis acid-catalyzed addition of allylic metal derivatives to aldehydes is a powerful annulation method. During the past year, it has been applied to the synthesis of oxepane derivatives. For example, the low temperature reaction of allylstannane **99** and boron trifluoride etherate afforded bicyclic oxepane **100** <91TL(32)4505>. Application of this methodology to the synthesis of a tetracyclic portion of the polyether brevetoxin B was also described <91TL(32)7069>. Using the corresponding acetals as the electrophilic component in these reactions has also been reported to yield oxepane derivatives <91SL823>.

No section dealing with metal-catalyzed processes would be complete without a reference to a palladium-catalyzed reaction. In the present context, Kozikowski and Ma have reported that the palladium-catalyzed cyclization of bromophenylindole derivative **101** afforded ring-fused azepine **102** in 81% yield <91-

$BF_3 \cdot OEt_2$, CH_2Cl_2, -78°

99 → **100**

TL(32)3317>. The formation of azepine derivatives has also been described by the palladium-catalyzed cyclization of allenic amines and the silver-catalyzed cyclization of allenic oximes <91JCS(P1)659>. The latter reaction generated a 7-membered cyclic nitrone which was trapped by the addition of a dipolarophile.

$Pd(PPh_3)_4$, KOAc, DMA, 160°

101 → **102**

VOLUME REFERENCES

91BMC(1)309	G.A. Flynn, T.P. Burkholder, E.W. Huber and P. Bey; Bioorg. Med. Chem. Lett., 1991, 1, 309.
91BMC(1)531	C.Y. Ho and M.J. Kukla; Bioorg. Med. Chem. Lett., 1991, 1, 531.
91CB(124)1181	A. Hassner and W. Dehaen; Chem. Ber., 1991, 124, 1181.
91CC559	P. Balczewski, R.L. Beddoes and J.A. Joule; J. Chem. Soc., Chem. Commun., 1991, 559.
91CC658	K.E. Cullen and J.T. Sharp; J. Chem. Soc., Chem. Commun., 1991, 658.
91CHC(50)1	R.I. Fryer; Ed., The Chemistry of Heterocyclic Compounds, Vol. 50, Bicyclic Diazepines, Wiley Interscience, New York, NY, 1991.
91CL1781	T. Kurihara, K. Ohuchi, M. Kawamoto, S. Harusawa and R. Yoneda; Chem. Lett., 1991, 1781.
91CPB(39)811	T. Kurihara, M. Doi, K. Hamaura, H. Ohishi, S. Harusawa and R. Yoneda; Chem. Pharm. Bull., 1991, 39, 811.
91H(32)41	N.P. Peet and M.E. LeTourneau; Heterocycles, 1991, 32, 41.
91H(32)2193	N.J. Snyder, J.W. Paschal, T.K. Elzey and D.O. Spry; Heterocycles, 1991, 32, 2193.
91H(32)2339	J.-X. Chen, J.-P. Jiang, W.-X. Chen and T.-Y. Kao; Heterocycles, 1991, 32, 2339.

91HCA(74)1095	W. Maier and W. Eberbach; Helv. Chim. Acta, 1991, 74, 1095.
91IF(46)435	V.D. Piaz, G. Ciciani and M.P. Giovannoni; Il Farmaco, 1991, 46, 435.
91JA(113)2335	S.D. Burke and J. Rancourt; J. Am. Chem. Soc., 1991, 113, 2335.
91JA(113)5384	P.A. Jacobi, C.A. Blum, R.W. DeSimone and U.E.S. Udodong; J. Am. Chem. Soc., 1991, 113, 5384.
91JCS(P1)1	M.L. Davies, C.J. Moody and R.J. Taylor; J. Chem. Soc., Perkin Trans. 1, 1991, 1.
91JCS(P1)9	M.L. Davies and C.J. Moody; J. Chem. Soc., Perkin Trans. 1, 1991, 9.
91JCS(P1)175	I. Collins, M.E. Fox, A.B. Holmes, S.F. Williams, R. Baker, I.J. Forbes and M. Thompson; J. Chem. Soc., Perkin Trans. I, 1990, 175.
91JCS(P1)659	R. Shaw, D. Lathbury, M. Anderson and T. Gallagher; J. Chem. Soc., Perkin Trans. 1, 1991, 659.
91JCS(P1)1733	H. Shimizu, K. Ikedo, K. Hamada, M. Ozawa, H. Matsumoto, K. Kamata, H. Nakamura, M. Ji, T. Kataoka and M. Hori; J. Chem. Soc., Perkin Trans. I, 1991, 1733.
91JCS(P1)2139	T. Benincori, E. Brenna and F. Sannicolo; J. Chem. Soc., Perkin Trans. 1, 1991, 2139.
91JHC(28)1789	M.L. Lann, S. Saba and M. Shoja; J. Heterocyclic Chem., 1991, 28, 1789.
91JHC(28)1891	M. Kurokawa, H. Uno, A. Itogawa, F. Sato, S. Naruto and J. Matsumoto; J. Heterocyclic Chem., 1991, 28, 1891.

91JMC(34)746	M.J. Kukla, H.J. Breslin, R. Pauwels, C.L. Fedde, M. Miranda, M.K. Scott, R.G. Sherrill, A. Raeymaekers, J. Van Gelder, K. Andries, M.A.C. Janssen, E. De Clerq and P.A.J. Janssen; J. Med. Chem., 1991, 34, 746.
91JMC(34)2231	K.D. Hargrave, J.R. Proudfoot, K.G. Grozinger, E. Cullen, S.R. Kapadia, U.R. Patel, V.U. Fuchs, S.C. Mauldin, J. Vitous, M.L. Behnke, J.M. Klunder, K. Pal, J.W. Skiles, D.W. McNeil, J.M. Rose, G.C. Chow, M.T. Skoog, J.C. Wu, G. Schmidt, W.W. Engel, W.G. Eberlein, T.D. Saboe, S.J. Campbell, A.S. Rosenthal and J. Adams; J. Med. Chem., 1991, 34, 2231.
91JMC(34)3187	M.J. Kukla, H.J. Breslin, C.J. Diamond, P.P. Grous, C.Y. Ho, M. Miranda, J.D. Rogers, R.G. Sherrill, E. De Clerq, R. Pauwels, K. Andries, L.J. Moens, M.A.C. Janssen, P.A.J. Janssen; J. Med. Chem., 1991, 34, 3187.
91JOC(56)513	M.C. Kuehne, P.A. Matson and W.G. Barnmann; J. Org. Chem., 1991, 56, 513.
91JOC(56)820	M.P. Doyle, R.J. Pieters, J. Taunton, H.Q. Pho, A. Padwa, D.L. Hertzog and L. Precedo; J. Org. Chem., 1991, 56, 820.
91JOC(56)896	W. Dehaen and A. Hassner; J. Org. Chem., 1991, 56, 896.
91JOC(56)1299	A.R. Katritzky, W.-Q. Fan and J.V. Greenhill; J. Org. Chem., 1991, 56, 1299.
91JOC(56)1353	L.M. Sayre, M.P. Singh, P.B. Kokil and F. Wang; J. Org. Chem., 1991, 56, 1353.

91JOC(56)1393	A.B. Holmes, A.L. Smith, S.F. Williams, L.R. Huges, Z. Lidert and C. Swithenbank; J. Org. Chem., 1991, 56, 1393.
91JOC(56)3271	A. Padwa, R.L. Chinn, S.F. Hornbuckle and Z.J. Zhang; J. Org. Chem., 1991, 56, 3271.
91JOC(56)4335	D.P. Curran, A.C. Abraham and H. Liu; J. Org. Chem., 1991, 56, 4335.
91SL823	I. Kadota, V. Gevorgyan, J. Yamada and Y. Yamamoto; Synlett., 1991, 823.
91T(47)583	K.D. Moeller, S.L. Rothfus and P.L. Wong; Tetrahedron, 1991, 47, 583.
91T(47)4077	J.A. Murphy and M.S. Sherburn; Tetrahedron, 1991, 47, 4077.
91T(47)4495	P. Armstrong, R. Grigg, F. Heaney, S. Surendrakumar and W.J. Warnock; Tetrahedron, 1991, 47, 4495.
91T(47)4847	P. Dowd and S.-C. Choi; Tetrahedron, 1991, 47, 4847.
91T(47)7459	A. Marcos, C. Pedregal and C. Avendano; Tetrahedron, 1991, 47, 7451.
91T(47)8297	R. Grigg, M.J. Dorrity, F. Heaney, J.F. Malone, S. Rajviroongit, V. Sridharan and S. Surendrakumar; Tetrahedron, 1991, 47, 8297.
91T(47)9131	P.A. Evans and A.B. Holmes; Tetrahedron, 1991, 47, 9131.
91T(47)9703	R. Grigg, V. Santhakumar, V. Sridharan, P. Stevenson, A. Teasdale, M. Thornton-Pett and T. Worakun; Tetrahedron, 1991, 47, 9703.

91TL(32)157	W.-N. Chou and J.B. White; Tetrahedron Lett., 1991, 32, 157.
91TL(32)1161	L. Aszmann, T. Debaerdemaeker and W. Friedrichsen; Tetrahedron Lett., 1991, 32, 1161.
91TL(32)2829	A.J. Clark, K. Jones, C. McCarthy and J.M.D. Storey; Tetrahedron Lett., 1991, 32, 2829.
91TL(32)3317	A.P. Kozikowski and D. Ma; Tetrahedron Lett., 1991, 32, 3317.
91TL(32)4359	H. Shimizu, K. Hamada, M. Ozawa, T. Kataoka and M. Hori; Tetrahedron Lett., 1991, 32, 4359.
91TL(32)4401	P. Molina, A. Arques, A. Alías and M.V. Vinader; Tetrahedron Lett., 1991, 32, 4401.
91TL(32)4505	T. Suzuki, O. Sato, M. Hirama, Y. Yamamoto, M. Murata, T. Yasumoto and N. Harada; Tetrahedron Lett., 1991, 32, 4505.
91TL(32)6159	F. Kido, Y. Kawada, M. Kato and A. Yoshikoshi; Tetrahedron Lett., 1991, 32, 6159.
91TL(32)6727	M. Nitta, Y. Iino and S. Mori; Tetrahedron Lett., 1991, 32, 6727.
91TL(32)7069	Y. Yamamoto, J. Yamada and I. Kadota; Tetrahedron Lett., 1991, 32, 7069.
91TL(32)7637	W.-N. Chou and J.B. White; Tetrahedron Lett., 1991, 32, 7637.

CHAPTER 8

Eight-Membered and Larger Rings

GEORGE R. NEWKOME
University of South Florida, Tampa, FL, USA

8.1 INTRODUCTION

In 1991, the incorporation of heteroatoms other than oxygen continued to be the trend in the creation of new heteromacrocyclic systems. With the emphases in supramolecular chemistry <91MI01> <91MI265><91MI3><91MI01> and molecular recognition <90MI524><91MI02> <91AG(E)1417>, chemists continue to design molecules which probe the mesoscopic dimensions. Numerous reviews and monographs have appeared dealing with diverse topics such as: crown ethers and cryptands <91005><91MI193>, ion extraction and transport by proton-ionizable crown ethers <91MI1>, structure and selectivity of crown ether complexes <90MI64>, functionalization of crown ethers <91MI288>, amine discrimination coloration with azophenol-dyed crowns <90MI29>, metallocrown ethers <90MI59>, interlocked and knotted rings (catanenes and rotaxanes) <91MI03><91MI802><91MI195a>, supramolecular photochemistry <91MI04>, cations with crown ethers and related macrocycles <91MI179>, organized polyaza cavity-shaped molecules <91T6851>, conductometric behavior of cation-macrocycle complexes in solution <91MI133>, crystallography of cation complexes of lariat ethers <91MI311>, transition and post-transition metal ions with mixed donor macrocycles <91MI599>, cation binding by modified ionophores <91MI497>, thermodynamic and kinetic data for macrocycle interactions with cations and anions <91CR1721>, stability constants of macrocyclic alkali metal cation complexes <91T2903>, porphyrins <91AJC1163>, sapphyrins (expanded porphyrins) <91MI127a>, complexes of nitrogen-, and sulfur-containing macroheterocycles as enzyme models <91MI2497>.

Numerous reviews <91MI445><91MI714> and exciting papers by Stoddart et al. have described the concepts of self assembly in organic synthesis. These synthetic strategies have given rise to mechanically interlocking molecular components, which give rise to catenanes and rotaxanes. They put forth the argument <91MI445> that "there are inherently simple ways of making apparently complex unnatural products from appropriate substrates without the need for reagent control or catalysis".

Because of spacial limitations, only macrocycles possessing heteroatoms and subheterocyclic rings are reviewed; lactams, lactones, and cyclic imines are excluded.

8.2 CARBON-OXYGEN RINGS

Chiral *O*-macrocycles have been constructed from varied starting materials, which incorporate different degrees of flexibility and molecular shape affording surfaces capable of enantiomeric differentiation and/or stereoselective catalysis. Activated 2',3'-*seco*nucleosides <91TL1821>, naproxen <91TL6277>, 4-phenyl-1,2,3,4-tetrahydroisoquinoline <91JOC3362>, carbohydrate orthoester derivatives <91JCS(P2)905>, tartaric acid derivatives <91CJC12><91S420>, cyclohexane-1,2-diols, prepared by enantioselective hydrolysis of their acetates using pig liver esterase <91JCS(P1)957>, selectively protected chiral threitol derivatives <91CPB530>, chiral tetraol derivatives <91LA1091>, and the disaccharide α,α-trehalose <91JOC3614> were used to instill the desired chiral spacer into the hetero-macrocycle. Cholesteric liquid crystals possessing steroidal crown ethers change their helical pitch upon addition of chiral ammonium ions and, in certain cases, result in visual color changes <91CC339>. An achiral parent podand or coronand was easily converted after complexation with boron into the corresponding chiral bicyclic coronand with a C_2-symmetry; a positive cooperativity for the bonding of different metal ions in a single host was demonstrated <91AG(E)1496>.

New synthetic catalysts with balanced conformational flexibility and preorganization of substituent have led to the design of a hemiacetal catalyst **1**, which is about 10 times more efficient than 2-pyridinone for tetramethylglucose mutarotation in benzene <91JOC3201>.

The thermal induced dehydration of 2,6-dihydroxymethyl-4-*tert*-butylphenol afforded a low yielded the oxacalix[3,4]arenes **2** <91TL1879>, whereas the related macrocyclic polyethereal trithiol **3** was synthesized (8% overall yield) in five steps from resorcinol <91IC2411>. Subsequent treatment of **3** with $[Fe_4S_4(SEt)_4]^{-2}$ afforded the black crystalline $[Fe_4S_4(c\text{-}LS_3)(SEt)]^{-2}$ cluster in nearly quantitative yield.

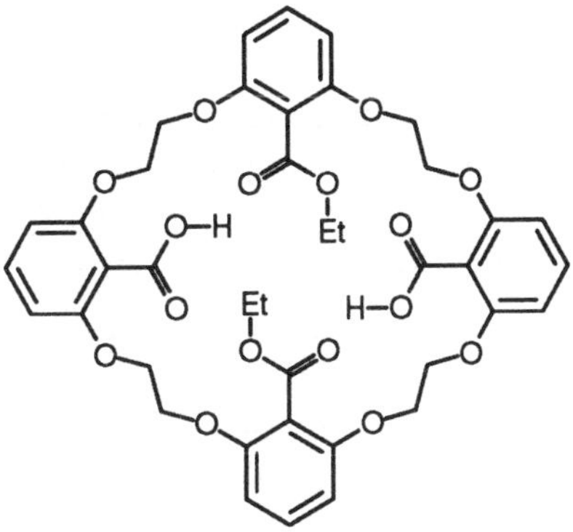

1

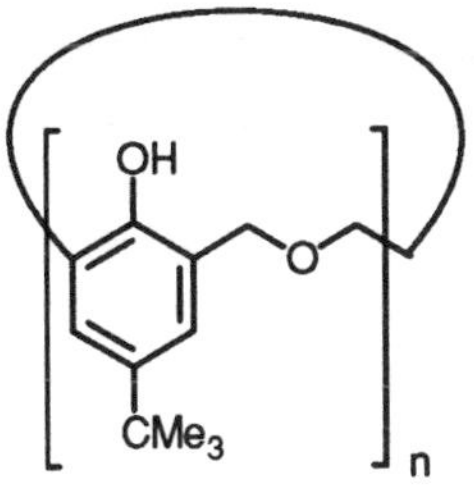

2

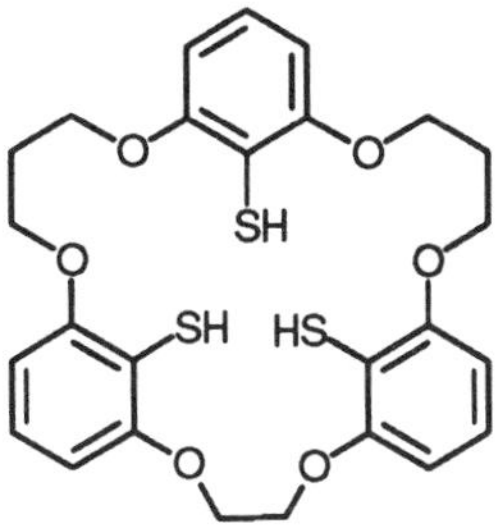

3

Halogen-metal exchange of **4** (n = 1,2) with organomagnesiums and diphenylcalcium afforded the corresponding aryl-magnesio or -calcio products <91OM3826>. A novel *heterotope* host molecule intercalated two different guests to generate complex **5** <91AG(E)1472><91AG(E)1474>.

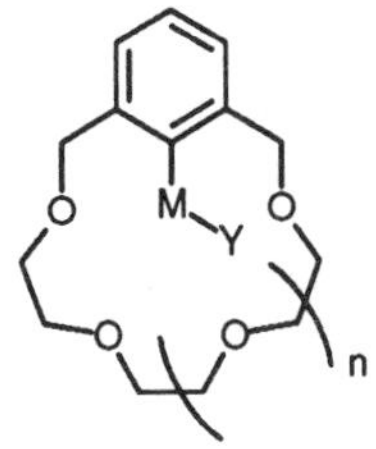

4 (n = 1; M-Y = MgAr)
(n = 2; M-Y = CaPh)
(n = 2; M-Y = $SnMe_3$)

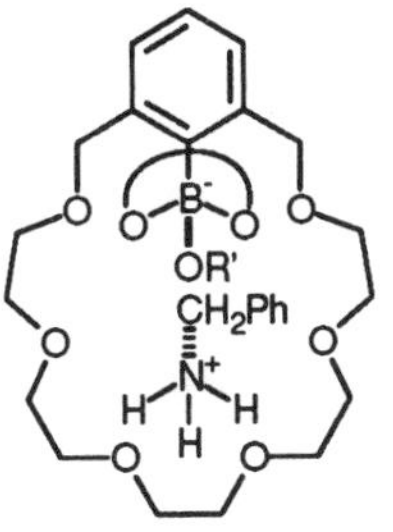

5

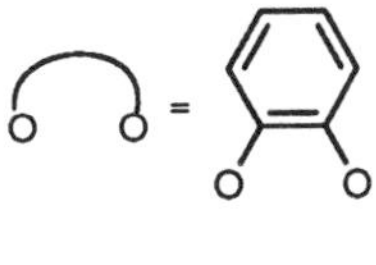

8.3 CARBON-NITROGEN RINGS

Azamacrocycles afford facile entry into lariat ethers and three-dimensional frameworks capable of encapsulating guest molecules and cations. Multiarmed macrocyclic polyamines with functionalized sidearms afforded diverse secondary donor sites <91JOC7102><91JCS(P2)87><91MI10645><91RTC124><91IC1265>. When the sidearms are tied together, more rigid cage-type cubical molecules, e.g. "kyuphane" <91JA8229>, are generated <91JA1323><91GCA29>. A novel cage-type cyclophane **6** incorporates two rigid macrocyclic rings and four bridging chiral binding sites; molecular inclusion of hydrophobic guests in aqueous media was demonstrated <91CC950>. Selective synthesis of *mono-N*-substituted tetraazamacrocycles can be easily achieved using boron triprotection <91TL639>.

New macrobicyclic ligands incorporating 2,2'-bipyridine <91HCA1163> and 3,3'-biisoquinoline as well as their related *N,N'*-dioxides <91HCA572> have been prepared and converted to the corresponding cryptates. Electroreductive crystallization of the sodium cryptate [Na^+ **7**]Br^- of the macrobicyclic *tris*(bipyridine) ligand **7** afforded an expanded alkali-metal ion atom or radical contact ion pair [Na^+ **7**]e^-; the term *sodio-cryptatium* denotes this first member of the cryptatium family <91AG(E)838>. A Wittig-type macrocyclization of a bipyridinedicarbaldehyde and an aryl *bis*-(phosphonium bromide) gave a new bipyridinophane <91CB2181>. The

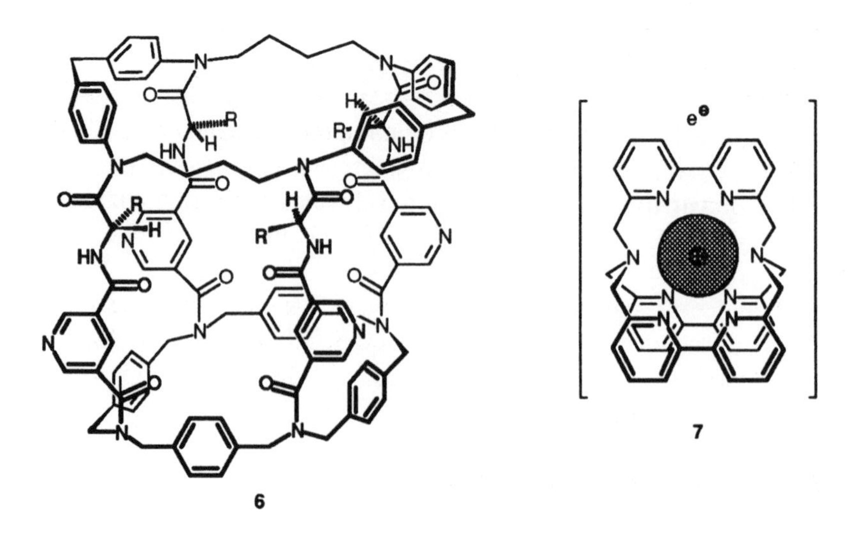

hexaimine, prepared from 'tren' and diformyl-2,2'-bipyridine, was reduced to the larger related macrobicyclic polyamine, which possesses a cage with 14 *N*-donor centers <91AG(E)1331>.

Although cofacial dimeric porphyrins have been known for over a decade, the recent synthesis of quadruple two- and three-atom azabridged cofacial tetraphenylporphyrins has been communicated <91JOC3470> and described in detail <91JA4208>. The *meso*-octaethylporphyrinogen tetraanion provides, owing to its conformational flexibility, σ and π binding pyrrolyl anions to electron-poor transition metals <91CC790>.

The first members of imidazole <91TL3333> and pyrazole <91NJC677> containing macrocycles have been prepared and characterized.

8.4 CARBON-SULFUR RINGS

In view of the interest as potential ligands for transition metals, the availability of polysulfur macrocycles has been a serious limitation to exploration of this area. The development of a automated apparatus for ring closure was shown <91MI474> to be a simple and reliable route to *S*-macrocycles. The synthesis and complexation with diverse transition metal ions <91IC3103><91CC475><91JA8663><91CC1119><91ZN209> of thiocrown ethers have afforded new insights into the different coordination modes available using this softer element.

Probably the most utilitarian use of *S*-macrocycles is their facile transformation to the corresponding *S*-free macrocycle. The extrusion of the sulfur moiety has been utilized by numerous researchers, but Vögtle et al. has consistently devised the most unusual phanes <91CB915>.

The preparation of diyne *S*-macrocycles and their aromatization by treatment with $CpCo(CO)_2$ <91CB357> or DBU in CCl_4 <91TL391> have been reported.

The microcrystalline cyclic hepta- and octa-sulfanes have been easily prepared in good yields from sulfenylchlorides and $(C_5H_5)_2TiS_5$; very large organic polysulfanes with more than 10 sulfur atoms in the ring can be synthesized via this procedure <91MI127b>.

8.5 CARBON-MERCURY RINGS

The icosahedral carborane 1,2-$C_2B_{10}H_{12}$ was readily lithiated at carbon with BuLi and subsequently reacted with $HgCl_2$ to give a cyclic tetramer **8**, which is a new class of rigid electrophilic macrocycles with an inverse charge distribution compared to 12-crown-4 <91AG(E)1507>.

8.6 CARBON-TIN RINGS

Force field parameters for the use of MM2 modeling programs for compounds containing tetrahedral tin have been developed <91O1732>; several symmetrical macrobicycles containing tin atoms at the bridgehead positions have been studied <91O1741>.

8.7 CARBON-ARSONIC RINGS

Cyclic *bis*arsonium salts were prepared and subsequently transformed to the tetraphenylborate salts <91ZOK1137>.

8.8 CARBON-OXYGEN-NITROGEN RINGS

Sauvage et al. have devised elegant synthetic pathways to interlocking macrocyclic rings, e.g. catenanes <19MI195a>. Such studies have lead to attachment of up to six peripheral 27-membered rings on a large (up to 132 atoms) macrocycle <91JA4023> and luminescence behavior of supramolecular mono- and bi-nuclear complexes <91JA4033>. Using similar phenanthroline building blocks, an interesting series of (poly)rotaxanes <91MI569> and bimolecular cyclophanes <91CB643> were created. The incorporation of other subheterocyclic moieties in a crown motif have included triazolo and pyridinono <91JHC773>, pyridino <91JOC3330> <91JOC1236>, pyrimidino <91JOC4653>, and pyrylium <91T1977> subunits.

Inclusion of a nitrogen atom in the macrocycle has afforded a site of attachment of a pendent group or for cyclization to generate macrobicycles and cryptands <91JOC3723>. The *N*-attachment of spirobenzopyran derivatives <91CC147> and azophenols <91JOC2575> gave rise to interesting chromogenic species. Photo-cleavable *C*-substituents attached to a cryptand framework retained the selective binding features and allowed the photolytically induced release of alkali ions in an aqueous solution <91HCA671>.

Alternate synthesis of cyclic (poly)aza crown ethers has been reported <91ACS621>. Diazacrown ethers were readily prepared by cyclization of 2-ethyl glycidyl ethers with *N,N'*-substituted ethylenediamines <91ZOK1366>. Tetraamminocopper(II) complexes with paraformaldehyde afforded the stable (over appreciable periods of time) oxatetraazacyclo-alkane copper(II) complexes <91AJC1227>. Azacycloundec-6-ene was prepared and subsequently used in the total synthesis of menzamine C <91T8067>.

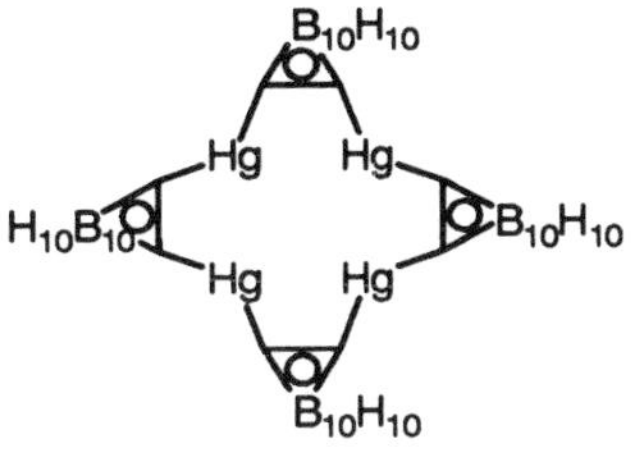

8

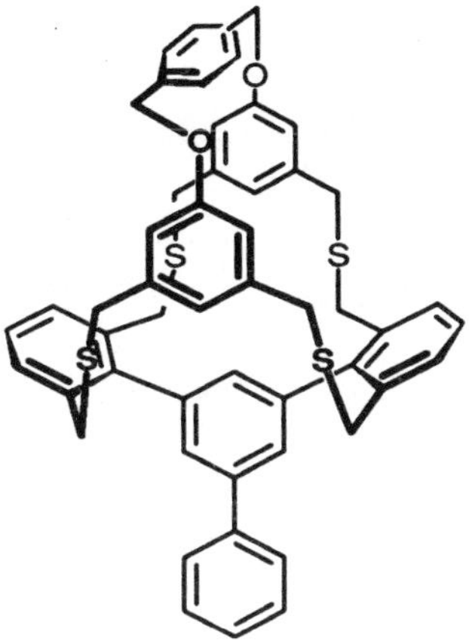

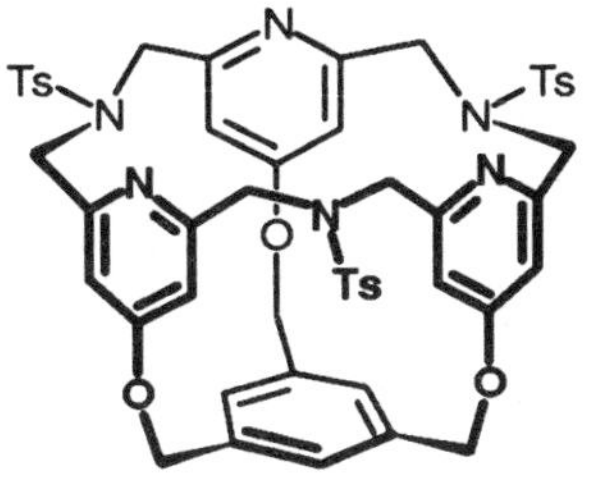

9

10

A novel molecular basket **9** was synthesized from three 4-donor-substituted pyridine units by cyclization with sulfonamides <91CB2323>.

8.9 CARBON-OXYGEN-SULFUR RINGS

The successful syntheses of *vaulted cappedophane* **10** were accomplished by capping the rigid walls of a cuppedophane precursor, thus affording insight to specifically designed microenvironments <91JA5630>. The 1,2,3-triply bridged cyclophanes were prepared by initially introducing the central oxy or dioxy bridge, followed by intramolecular coupling of the two thia bridges <91JOC>.

8.10 CARBON-NITROGEN-SULFUR RINGS

Macrocyclization of two appropriately tetrasubstituted azaphanes afforded 'a short molecular tube' **11**, which possesses a cavity large enough to accommodate a benzene molecule <91AG(E)575>. A new [14]aneN_2S_2 ligand was prepared by the B_2H_6 reduction of the corresponding diamide generated from dimethylmalonyl dichloride and 1,9-diamino-3,7-dithianonane <91JA4857>.

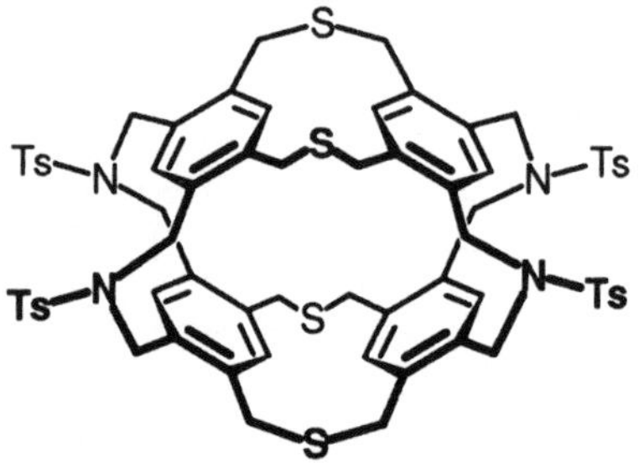

11 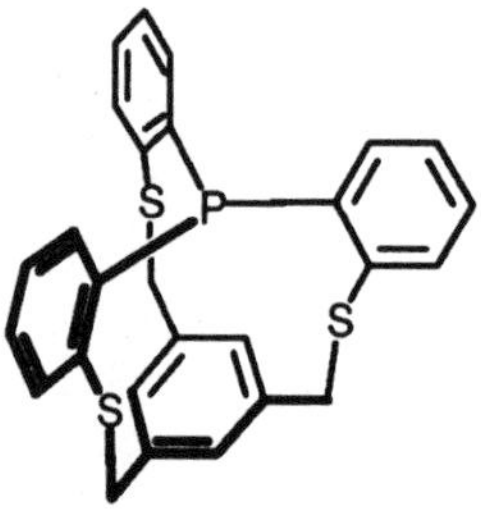 **12**

8.11 CARBON-SULFUR-PHOSPHORUS RINGS

The *P,S*-macrocycle **12** was prepared by base-promoted condensation of *tris*(2-mercaptophenyl)phosphine and *tris*(bromomethyl)benzene; the related *Si,S*-cyclophane was similarly prepared <91JA2672>. Protonation of **12** was difficult and *S*-oxidation predominated confirming the molecular strain and directivity of the phosphine moiety.

8.12 CARBON-SELENIUM-OXYGEN RINGS

The 'first' syntheses of selenium-containing benzocrown ethers have been reported <91MI247><91MI195>, thus offering a higher affinity for certain "softer" cations.

8.13 CARBON-SELENIUM-NITROGEN RINGS

Utilization of the 'selena-procedure' gave rise to substituted 2,11-diselena[3.3](2,6)pyridinophane, which underwent deselenation by irradiation in dry $P(OMe)_3$ <91CC860>. As therein noted, "...the seleno-method to functionalized heteroaromatic systems opens a wide spectrum of C-C linkages leading to new strained molecules."

8.14 REFERENCES

90MI01 "Bioorganic Chemistry Frontiers", H. Dugas, ed., Springer-Verlag: Berlin, 1990, Volume 1.

90MI29 S. Misumi & T. Kaneda; *Mem. Inst. Sci. Ind. Res., Osaka Univ.*, 1990, **47**, 29.

90MI59 M. S. Lah & V. L. Pecoraro; *Comments Inorg. Chem.*, 1990, **11**, 59.

90MI64 E. Weber; *Wiss. Ber.-Zentralinst. Festkoerperphys. Werstofforsch.*, 1990, **44**, 64.

90MI524 D. N. Reinhoudt; *Chem. Mag.*, 1990 (Oct), 524-5, 527.

91ACS621 L. Börjesson & C. J. Welch; *Acta Chem. Scand.*, 1991, **45**, 621.

91AG(E)575 F. Vögtle, A. Schröder, & D. Karbach; *Angew. Chem., Int. Ed. Engl.* 1991, **30**, 575.

91AG(E)838 L. Echegoyen, A. DeCian, J. Fischer, & J.-M. Lehn; *Angew. Chem., Int. Ed. Engl.*, 1991, **30**, 838.

91AG(E)1331 J. de Mendoza, et al.; *Angew. Chem., Int. Ed. Engl.*, 1991, **30**, 1331.

91AG(E)1417 H. J. Schneider; *Angew. Chem., Int. Ed. Engl.*, 1991, **30**, 1417.

91AG(E)1472 M. T. Reetz, C. Niemeyer, & K. Harms; *Angew. Chem., Int. Ed. Engl.*, 1991, **30**, 1472.

91AG(E)1474 M. T. Reetz, C. Niemeyer, & K. Harms; *Angew. Chem., Int. Ed. Engl.*, 1991, **30**, 1474.

91AG(E)1496 Y. Kobuke, Y. Sumida, M. Hayashi, & H. Ogoshi; *Angew. Chem., Int. Ed. Engl.*, 1991, **30**, 1496.

91AG(E)1507 X. Yang, C. C. Knobler, & F. Hawthorne; *Angew. Chem., Int. Ed. Engl.*, 1991, **30**, 1507.

91AJC1163 P. S. Clezy; *Aust. J. Chem.*, 1991, **44**, 1163.

91AJC1227 G. A. Lawrance, et al.; *Aust. J. Chem.*, 1991, **44**, 1227.

91CB357 R. Gleiter, S. Rittinger, & H. Langer; *Chem. Ber.*, 1991, **124**, 357.

91CB643 U. Lüning & M. Müller; *Chem. Ber.*, 1991, **123**, 643.

91CB915 J. Dohm, M. Nieger, K. Rissanen, & F.Vögtle; *Chem. Ber.*, 1991, **124**, 915.

91CB2181 F. Vögtle, et al.; *Chem. Ber.*, 1991, **123**, 2181.

91CB2323 J. Breitenbach, K. Rissanen, U. U. Wolf, & F. Vögtle; *Chem. Ber.*, 1991, **124**, 2323.

91CC147 K. Kimura, T. Yamashita, & M. Yokoyama; *J. Chem. Soc., Chem. Commun.*, 1991, 147.

91CC339 T. Nishi, A. Ikeda, T. Matsuda, & S. Shinkai; *J. Chem. Soc., Chem. Commun.*, 1991, 339.

91CC475 R. J. Smith, et al.; *J. Chem. Soc., Chem. Commun.*, 1991, 475.

91CC790 D. Jacoby, C. Floriani, A. Chiesi-Villa, & C. Rizzoli; *J. Chem. Soc., Chem. Commun.*, 1991, 790.

91CC860 F. Vögtle, J. Breitenbach, & M. Nieger; *J. Chem. Soc., Chem. Commun.*, 1991, 860.

91CC950 Y. Murakami, T. Ohno, O. Hayashida, & Y. Hisaeda; *J. Chem. Soc., Chem. Commun.*, 1991, 950.

91CC1119	S. J. Loeb & G. K. H. Shimizu; *J. Chem. Soc., Chem. Commun.*, 1991, 1119.
91CJC12	F. R. Fronczek, et al.; *Can. J. Chem.*, 1991, **69**, 12.
91CPB530	T. Yasukata, S. Sasaki, & K. Koga; *Chem. Pharm. Bull.*, 1991, **39**, 530.
91CR1721	R. M. Izatt, K. Pawlak, J. S. Bradshaw, & R. L. Bruening; *Chem. Rev.*, 1991, **91**, 1721.
91GCA29	M. Micheloni, N. Nardi, & B. Valtancoli; *Gazz. Chim. Ital.*, 1991, **121**, 29.
91HCA572	J.-M. Lehn & C. O. Roth; *Helv. Chim. Acta*, 1991, **74**, 572.
91HCA671	R. Warmuth, et al.; *Helv. Chim. Acta*, 1991, **74**, 671.
91HCA1163	I. Bkouche-Waksman, et al.; *Helv. Chim. Acta*, 1991, **74**, 1163.
91IC1265	D. D. Dischino, et al.; *Inorg. Chem.*, 1991, **30**, 1265.
91IC2411	M. A. Whitener, G. Peng, & R. H. Holm; *Inorg. Chem.*, 1991, **30**, 2411.
91IC3103	B. d. Groot & S. J. Loeb; *Inorg. Chem.*, 1991, **30**, 3103.
91JA1323	H. Takemura, T. Shinmyozu, & T. Inazu; *J. Am. Chem. Soc.*, 1991, **113**, 1323.
91JA2672	R. P. L'Esperance, A. P. West, Jr., D. V. Engen, & R. A. Pascal, Jr.; *J. Am. Chem. Soc.*, 1991, **113**, 2672.
91JA4023	F. Bitsch, et al.; *J. Am. Chem. Soc.*, 1991, **113**, 4023.
91JA4033	N. Armaroli, et al.; *J. Am. Chem. Soc.*, 1991, **113**, 4033.
91JA4208	B. C. Bookser & T. C. Bruice; *J. Am. Chem. Soc.*, 1991, **113**, 4208.
91JA4857	E. Kimura, et al.; *J. Am. Chem. Soc.*, 1991, **113**, 4857.
91JA8229	Y. Murakami, et al.; *J. Am. Chem. Soc.*, 1991, **113**, 8229.
91JA8663	J. M. Desper, S. H. Gellman, R. E. Wolf, Jr., & S. R. Cooper; *J. Am. Chem. Soc.*, 1991, **113**, 8663.
91JCS(P1)957	K. Naemura, H. Miyabe, & Y. Shingai; *J. Chem. Soc. Perkin 1*, 1991, 957.
91JCS(P2)87	C. J. Broan, et al.; *J. Chem. Soc. Perkin Trans 2*, 1991, 87.
91JCS(P2)905	C. Vicent, et al.; *J. Chem. Soc. Perkin 2*, 1991, 905.
91JHC773	J. S. Bradshaw, K. E. Krakowiak, P. Huszthy, & R. M. Izatt; *J. Heterocycl. Chem.*, 1991, **28**, 773.
91JOC264	Y.-H. Lai & C.-W. Tan; *J. Org. Chem.*, 1991, **56**, 264.
91JOC1236	E. Weber, H.-J. Köhler, & H. Reuter; *J. Org. Chem.*, 1991, **56**, 1236.
91JOC2575	E. Chapoteau, et al.; *J. Org. Chem.*, 1991, **56**, 2575.
91JOC3330	P. Huszthy, et al.; *J. Org. Chem.*, 1991, **56**, 3330.
91JOC3201	C. Gennari, et al.; *J. Org. Chem.*, 1991, **56**, 3201.
91JOC3362	T. M. Georgiadis, M. M. Georgiadis, & F. Diederich; *J. Org. Chem.*, 1991, **56**, 3362.
91JOC3470	R. Karaman & T. C. Bruice; *J. Org. Chem.*, 1991, **56**, 3470.
91JOC3614	C. Vicent, et al.; *J. Org. Chem.*, 1991, **56**, 3614.
91JOC3723	K. E. Krakowiak & J. S. Bradshaw; *J. Org. Chem.*, 1991, **56**, 3723.
91JOC4653	A. F. Cichy, et al.; *J. Org. Chem.*, 1991, **56**, 4653.
91JOC5630	T. K. Vinod & H. Hart; *J. Org. Chem.*, 1991, **56**, 5630.
91JOC7102	H. Tsukube, H. Adachi, & S. Morosawa; *J. Org. Chem.*, 1991, **56**, 7102.
91LA1091	E. V. Dehmlow & V. Knufinke; *Liebigs Ann. Chem.*, 1991, 1091.

91MI01	"Advances in Supramolecular Chemistry", G. W. Gokel, ed., JAI Press, Greenwich, CT, Vol. 1, 1991.
91MI02	A. D. Hamilton; in "Bioorganic Chemistry Frontiers", H. Dugas, ed., Springer-Verlag, Berlin, 1991, Volume 2.
91MI03	J.-P. Sauvage & C. Dietrich-Buchecker; in "Bioorganic Chemstry Frontiers", H. Dugas, ed., Springer-Verlag, Berlin, 1991, Volume 2.
91MI04	V. Balzani & F. Scandola; "Supramolecular Photochemistry", Ellis Harwood: Chichester, UK, 1991.
91MI05	G. W. Gokel; "Crown Ethers and Cryptands", Royal Society of Chemistry, Cambridge, UK, 1991.
91MI1	P. R. Brown, & R. A. Bartsch; *Top. Inclusion Sci.*, 1991, **2**, 1.
91MI3	M. Kozbial; *Przem. Chem.*, 1991, **70**, 3.
91MI127a	J. L. Sessler, M. J. Cyr, & A. K. Burrell; *Synlett*, 1991, 127.
91MI127b	R. Steudel & M. Kustos; *Phos. Sulfur & Silicon*, 1991, **62**, 127.
91MI133	Y. Takeda; in "Cation Binding by Macrocycles", Y. Inoue & G. W. Gokel, eds., Marcel Dekker, New York, 1991, p 133.
91MI179	E. M. Eyring & S. Petrucci; in "Cation Binding by Macrocycles", Y. Inoue & G. W. Gokel, eds., Marcel Dekker, New York, 1991, p 179.
91MI193	R. Ostaszewski & J. Jurczak; *Wiad. Chem.*, 1991, **44**, 193.
91MI195a	C. O. Dietrich-Buchecker & J.-P. Sauvage; *Bioorg. Chem. Front.*, 1991, **2**, 195.
91MI195b	H. S. Xu, W. P. Li, & X. F. Liu; *Chinese Chem. Lett.*, 1991, **2**, 195.
91MI247	A. Mazouz, J. Bodiguel, P. Meunier, & B. Gautheron; *Phos. Sulfur, & Silicon*, 1991, **61**, 247.
91MI265	K. Saigo; *Kagaku Kogyo*, 1991, **42**, 265.
91MI288	M. Hiraoka; *Kagaku Kogyo*, 1991, **42**, 288.
91MI311	F. R. Fronczek & R. D. Gandour; in "Cation Binding by Macrocycles", Y. Inoue & G. W. Gokel, eds., Marcel Dekker, NY, 1991, p 311.
91MI445	D. Philp & J. F. Stoddart; *Synlett*, 1991, 445.
91MI474	M. C. Durrant & R. L. Richards; *Chem. Ind.*, 1991, 474.
91MI497	H. Tsukube; in "Cation Binding by Macrocycles", Y. Inoue & G. W. Gokel, eds., Marcel Dekker, NY, 1991, p 497.
91MI569	C. Wu, P. R. Lecavalier, Y. X. Shen, & H. W. Gibson; *Materials*, 1991, **3**, 569.
91MI599	L. F. Lindoy; in "Cation Binding by Macrocycles", Y. Inoue & G. W. Gokel, eds., Marcel Dekker, New York, 1991, p 599.
91MI714	F. Stoddart; *Chem. Brit.*, 1991, 714.
91MI802	M. Fujita; *Kagaku (Kyoto)*, 1991, **46**, 802.
91MI2497	M. G. Voronkov & V. I. Knutov; *Usp. Khim.*, 1991, **12**, 2497.
91MI21083	J. F. Carvalho, S. P. Crofts, & S. M. Rocklage; *PCT Int. Appl. WO 91 10,645*; *Chem. Abstr.*, 1992, **116**, 21083h.
91NJC677	G. Tarrago, S. E. Kadiri, C. Marzin, & C. Coquelet; *New J. Chem.*, 1991, **15**, 677.
91O1732	J. H. Horner & M. Newcomb; *Organometallics*, 1991, **10**, 1732.
91O1741	J. H. Horner, et al.; *Organometallics*, 1991, **10**, 1741.
91O3826	P. R. Markies, et al.; *Organometallics*, 1991, **10**, 3826.

91RTC124	D. W. Swinkels, J. P. M. v. Duynhoven, C. W. Hilbers, & G. I. Tesser; *Recl. Trav. Chim. Pays-Bas*, 1991, **110**, 124.
91S420	H. Dugas & J. Vaugeois; *Synthesis*, 1991, 420.
91T1977	G. Doddi, G. Ercolani, & P. Mencarelli; *Tetrahedron*, 1991, **47**, 1977.
91T2903	B.-L. Poh; *Tetrahedron*, 1991, **47**, 2903.
91T6851	R. P. Thummel; *Tetrahedron*, 1991, **47**, 6851.
91T8067	Y. Torisawa, et al.; *Tetrahedron*, 1991, **47**, 8067.
91TL391	K. Toshima, K. Ohta, T. Ohtake, & K. Tatsuta; *Tetrahedron Lett.*, 1991, **32**, 391.
91TL639	H. Bernard, et al.; *Tetrahedron Lett.*, 1991, **32**, 639.
91TL1821	V. Škarić, V. Čaplar, D. Škarić, & M. Žinić; *Tetrahedron Lett.*, 1991, **32**, 1821.
91TL1879	P. Zerr, M. Mussrabi, & J. Vicens; *Tetrahedron Lett.*, 1991, **32**, 1879.
91TL3333	J. P. Street; *Tetrahedron Lett.*, 1991, **32**, 3333.
91TL6277	P. P. Castro & F. Diederich; *Tetrahedron Lett.*, 1991, **32**, 6277.
91ZN209	D. Sellmann, P. Lechner, M. Moll, & F. Knoch; *Z. Naturforsch.*, 1991, **41b**, 209.
91ZOK1137	L. E. Rzhechitskaya, V. S. Gamayurova, R. Z. Musin, & F. G. Khalitov; *Zh. Obshch. Khim.*, **61**, 1137.
91ZOK1366	A. I. Ismiev, V. M. Farzaliev, & M. A. Allakhverdiev; *Zhur. Org. Khim.*, 1991, **27**, 1366.

SUBJECT INDEX

CORRIGENDUM

This page was inadvertently omitted from the end of Chapter 5.6 in PHC Volume 3.

90MI111	A. Saparov, A. Taganlyev, T. Kh. Khodzhalyev, D. K. Kurbanov, and Yu. K. Khekimov, *Izv. Akad. Nauk Turkm. SSR, Ser. Fiz.-Tekh. Khim. Geol. Nauk*, 1990, 111 [*Chem. Abstr.*, 1991, **114**, 6335].
90MI314	A. Mosandl and U. Hagenauer-Hener, *Z. Lebensm.-Unters. Forsch.*, 1990, **190**, 314 [*Chem. Abstr.*, 1990, **113**, 97488].
90MI383	S. Fujisaki, K. Hanada, A. Nishida, and S. Kajigaeshi, *Kogakubu Kenkyu Hokoku (Yamaguchi Daigaku)*, 1990, **40**, 383 and 391 [*Chem. Abstr.*, 1991, **114**, 23826 and 23827].
90MI1607	M. N. Nazarov, D. E. Lozin, L. G. Kulak, S. S. Zlotskii, and D. L. Rakhmankulov, *Zh. Prikl. Khim. (Leningrad)*, 1990, **63**, 1607 [*Chem. Abstr.*, 1991, **114**, 23831].
90S271	E. Schaumann, S. Winter-Extra, K. Kummert, and S. Scheiblich, *Synthesis*, 1990, 271.
90S599	G. Kneer, J. Mattay, G. Raabe, C. Krüger, and J. Lauterwein, *Synthesis*, 1990, 599.
90SC1	R. P. Houghton and J. E. Dunlop, *Synth. Commun.*, 1990, **20**, 1.
90SC153	R. Miranda, H. Cervantes, and P. Joseph-Nathan, *Synth. Commun.*, 1990, **20**, 153.
90SC1175	R. P. Hsung, *Synth. Commun.*, 1990, **20**, 1175.
90SL209	B. A. Trofimov, Yu. M. Skvortsov, A. G. Mal'kina, and A. I. Grista, *Sulfur Lett.*, 1990, **11**, 209.
90T433	F. Bertho, A. Robert, P. Batail, and P. Robin, *Tetrahedron*, 1990, **46**, 433.
90T1553	E. Fanghänel, N. Beye, and A. M. Richter, *Tetrahedron*, 1990, **46**, 1553.
90T1783	J. R. Moran, I. Tapia, and V. Alcazar, *Tetrahedron*, 1990, **46**, 1783.
90T4573	G. H. Posner and T. D. Nelson, *Tetrahedron*, 1990, **46**, 4573.
90TL449	D. A. Jaeger, Y. M. Sayed, and A. K. Dutta, *Tetrahedron Lett.*, 1990, **31**, 449.
90TL623	M. E. Jung and W. Lew, *Tetrahedron Lett.*, 1990, **31**, 623.
90TL1007	Y. Gimbert, A. Moradpour, and S. Bittner, *Tetrahedron Lett.*, 1990, **31**, 1007.
90TL2121	C. M. Yeung and L. L. Klein, *Tetrahedron Lett.*, 1990, **31**, 2121.
90TL2135	A. Greiner and J. Y. Ortholand, *Tetrahedron Lett.*, 1990, **31**, 2135.
90TL3763	D. J. Ramon and M. Yus, *Tetrahedron Lett.*, 1990, **31**, 3763.
90USP4900830	A. Fisher and I. Karton, *U. S. Pat.*, 4 900 830 (1990) [*Chem. Abstr.*, 1990, **113**, 59155].
90ZN(B)1216	G. C. Papavassiliou, D. J. Lagouvardos, V. C. Kakoussis, and G. A. Mousdis, *Z. Naturforsch. B: Chem. Sci.*, 1990, **45**, 1216.
90ZOR138	L. P. Dement'eva and R. R. Kostikov, *Zh. Org. Khim.*, 1990, **26**, 138.
90ZOR281	O. B. Bondarenko, L. G. Saginova, T. I. Voevodskaya, A. V. Buevich, D. S. Yufit, Yu. T. Struchkov, and Yu. S. Shabarov, *Zh. Org. Khim.*, 1990, **26**, 281.
90ZOR377	E. S. Kozlov, A. A. Yurchenko, L. A. Nechitailo, and N. V. Ignat'ev, *Zh. Org. Khim.*, 1990, **26**, 377.
90ZOR553	O. B. Bondarenko, L. G. Saginova, T. I. Voevodskaya, and Yu. S. Shabarov, *Zh. Org. Khim.*, 1990, **26**, 553.
90ZOR652	M. V. Gorelik, V. Ya. Shteiman, V. A. Tidatyan, V. A. Tafeenko, and T. A. Mikhailova, *Zh. Org. Khim.*, 1990, **26**, 652.